P9-DGF-803

PERFORMANCE ANALYSIS OF DIGITAL TRANSMISSION SYSTEMS

ELECTRICAL ENGINEERING, COMMUNICATIONS, AND SIGNAL PROCESSING

Raymond L. Pickholtz, Series Editor

Anton Meijer and Paul Peeters
Computer Network Architectures

Marvin K. Simon, Jim K. Omura, Robert A. Scholtz, and Barry K. Levitt
Spread Spectrum Communications, Volume I

Marvin K. Simon, Jim K. Omura, Robert A. Scholtz, and Barry K. Levitt
Spread Spectrum Communications, Volume II

Marvin K. Simon, Jim K. Omura, Robert A. Scholtz, and Barry K. Levitt
Spread Spectrum Communications, Volume III

William W Wu
Elements of Digital Satellite Communication: System Alternatives, Analyses and Optimization, Volume I

William W Wu
Elements of Digital Satellite Communication: Channel Coding and Integrated Services Digital Satellite Networks, Volume II

Yechiam Yemini
Current Advances in Distributed Computing and Communications

Michael J. Miller and Syed V. Ahamed
Digital Transmission Systems and Networks, Volume I: Principles

Michael J. Miller and Syed V. Ahamed
Digital Transmission Systems and Networks, Volume II: Applications

Osamu Shimbo
Transmission Analysis in Communication Systems, Volume I

Osamu Shimbo
Transmission Analysis in Communication Systems, Volume II

David L. Nicholson
Spread Spectrum Signal Design: LPE and AJ Systems

Andrew Bateman and Warren Yates
Digital Signal Processing Design

Lavon B. Page
Probability for Engineering with Applications to Reliability

Dmitri Kazakos and P. Papantoni-Kazakos
Detection and Estimation

Allan R. Hambley
An Introduction to Communication Systems

Pramode K. Verma
Performance Estimation of Computer Communication Networks: A Structured Approach

Arthur F. Standing
Measurement Techniques for In-Orbit Testing of Satellites

Fred J. Ricci and Daniel Schutzer
U.S. Military Communications: AC^3I Force Multiplier

William Turin
Performance Analysis of Digital Transmission Systems

Miron Abramovici, Melvin A. Breuer, and Arthur D. Friedman
Digital Systems Testing and Testable Design

OTHER WORKS OF INTEREST

Raymond L. Pickholtz, Editor
Local Area and Multiple Access Networks

Walter Sapronov, Editor
Telecommunications and the Law: An Anthology

PERFORMANCE ANALYSIS OF DIGITAL TRANSMISSION SYSTEMS

William Turin,

AT&T Bell Laboratories

Computer Science Press

Library of Congress Cataloging-in-Publication Data

Turin, William.
Performance analysis of digital transmission systems/William Turin.
p. cm.
Includes bibliographical references.
ISBN 0-7167-8212-X
1. Digital communications–Evaluation. I. Title.
TK5103.7.T868 1990
621.382–dc20 89-25299
CIP

Printed in the United States of America

Computer Science Press

An imprint of W. H. Freeman and Company
41 Madison Avenue, New York, NY 10010
20 Beaumont Street, Oxford OX1 2NQ, England

1 2 3 4 5 6 7 8 9 0 RRD 9 9 8 7 6 5 4 3 2 1 0

CONTENTS

PREFACE

This book is intended for those who design digital communication systems. A computer network designer is interested in selecting communication channels, error protection schemes, and link control protocols. To do this efficiently, one needs a mathematical model that accurately predicts system behavior.

Two basic problems arise in mathematical modeling: the problem of identifying a system and the problem of applying a model to the system analysis. The system identification consists of selecting a certain class of mathematical objects that describe fundamental properties of the system behavior. We use a class of stochastic sequential machines with hidden states that are governed by a Markov chain to model communication systems. According to this model, a system's output symbols depend not only on its input symbols but also on the machine state. The model is described by a finite number of matrices whose elements are estimated on the basis of experimental data. We develop several methods of model identification and show their relationship to other methods of data analysis, such as spectral methods, autoregressive moving average (ARMA) approximations, and rational transfer function approximations.

We demonstrate that the model is convenient to use in applications by deriving closed-form expressions for various system performance characteristics (e.g., efficiency of forward error detection and error correction schemes and communication protocols). The model-related calculations are based on the methods of the matrix probability theory which is developed in this book. We illustrate the proposed methods by numerous examples and compare the results of analytical methods with experimental data and computer simulations.

The stochastic sequential machine model is popular in various applications, such as automatic control, queueing systems, speech and image recognition, telecommunications, and psychology. Depending on the application, the model has different names: hidden Markov model, probabilistic automaton, finite-state Markov model, and generalized Gilbert model. One of the reasons for the popularity of the model is its ability to approximate a large variety of stochastic processes; other reason is its relative simplicity. Because of the model's popularity, this book should be useful not only to specialists in telecommunications but to a much broader audience.

There are several books and many papers that deal with the stochastic sequential machine model. The majority of them are devoted to theoretical analysis of the model and intended for mathematicians. This book is intended for specialists in telecommunications and deals with practical aspects of model construction and application. It is based on the author's own experience in processing experimental data and analyzing real-world communication systems. Most of the methods considered have been used in practice. We provide numerous examples to aid the reader in understanding the proposed methods.

Several theoretical problems, such as various types of presentation and simplification of the model, which are directly related to creating the model from experimental data, are discussed in this book. Because some mathematical background is essential to understanding the proposed methods, to make the book self-contained we present a brief mathematical introduction related to matrix algebra, Markov chains, statistical inference, and error correcting codes. This introductory material is based mainly on a course taught by the author at AT&T Bell Laboratories.

The author is grateful to AT&T Bell Laboratories for providing the opportunity to write this book. Special thanks is due to Dan Leed without whose support and encouragement this book never would have appeared. Helpful comments on the manuscript made by Sam Boodaghians, Charlie Canada, Hector Corrales, Adam Irgon, Dan Jeske, Jay Padgett, Vasant Prabhu, and Richard Trenner are gratefully acknowledged.

This book was prepared using the UNIX† document preparation tools and facilities.

William Turin

† UNIX is a registered trademark of AT&T

NOTATION

$\mathbf{J}$	the set of all integers
$\mathbf{N}$	the set of all positive integers
$\sum\limits_{i_1,\ldots,i_n}$	sum over all possible values of $i_1,\ldots,i_n$
$\sum\limits_{i\in F}$	sum over all possible values of i that belong to a set F
$\mathbf{A} = [a_{ij}]_{m,n}$	a rectangular matrix with elements a_{ij} consisting of m rows and n columns
$\mathbf{A}+\mathbf{B}$	matrix sum: $\mathbf{A}+\mathbf{B} = [a_{ij}]_{m,n} + [b_{ij}]_{m,n} = [a_{ij}+b_{ij}]_{m,n}$
$\mathbf{AB}$	matrix product: $\mathbf{AB} = [a_{ij}]_{m,n} \cdot [b_{ij}]_{n,p} = [\sum\limits_{k=1}^{n} a_{ik}b_{kj}]_{m,n}$
$\lambda\mathbf{A}$	$= \lambda[a_{ij}]_{m,n} = [\lambda a_{ij}]_{m,n}$ a product of a number and a matrix
$\mathbf{A}' = [a'_{ij}]_{n,m}$	the transpose of the matrix $\mathbf{A} = [a_{ij}]_{m,n}$: $a_{ij} = a'_{ji}$
$diag\{a_i\}_n$	$= \begin{bmatrix} a_1 & 0 & \cdots & 0 \\ 0 & a_2 & \cdots & 0 \\ \cdots & \cdots & \cdots & \cdots \\ 0 & 0 & \cdots & a_n \end{bmatrix}$ a diagonal matrix
$\mathbf{I} = diag\{1\}$	a unit matrix
$\det \mathbf{A}$	matrix $\mathbf{A}$ determinant
$\mathbf{A}^{-1}$	matrix $\mathbf{A}$ inverse ($\mathbf{AA}^{-1} = \mathbf{A}^{-1}\mathbf{A} = \mathbf{I}$)
$\mathbf{a} = row\{a_i\}_n$	$= [a_1,a_2,\ldots,a_n]$ a row matrix
$\mathbf{a} = col\{a_i\}_n$	$= [a_1,a_2,\ldots,a_n]'$ a column matrix

$\mathbf{1}$	$= col\{1\}$ a column matrix consisting of ones
$\mathbf{A} = [\mathbf{A}_{ij}]_{m,n}$	a block matrix consisting of m rows and n columns of submatrices (matrix blocks) $\mathbf{A}_{ij}$
$block\ row\{\mathbf{A}_i\}_n$	$= [\mathbf{A}_i]_{1,n} = [\mathbf{A}_1, \mathbf{A}_2, \ldots, \mathbf{A}_n]$ a block row matrix
$block\ col\{\mathbf{A}_i\}_n$	$= [\mathbf{A}_i]_{n,1} = [\mathbf{A}_1, \mathbf{A}_2, \ldots, \mathbf{A}_n]'$ a block column matrix
$block\ diag\{\mathbf{A}_i\}_n$	$= \begin{bmatrix} \mathbf{A}_1 & 0 & \cdots & 0 \\ 0 & \mathbf{A}_2 & \cdots & 0 \\ \cdots & \cdots & \cdots & \cdots \\ 0 & 0 & \cdots & \mathbf{A}_n \end{bmatrix}$ a block diagonal matrix
$rank\ \mathbf{A}$	matrix $\mathbf{A}$ rank: the highest order of nonzero determinants which are composed of the elements of rows and columns of the matrix
$\mathbf{A} \otimes \mathbf{B}$	$= [a_{ij}]_{m,n} \otimes \mathbf{B} = [a_{ij}\mathbf{B}]_{m,n}$ Kronecker product of the matrices
$\mathbf{A} > \mathbf{B}$	denotes that all the elements of the matrix $\mathbf{A}$ are greater than the corresponding elements of the matrix $\mathbf{B}$: $a_{ij} > b_{ij}$ for all i and j
$\mathbf{A} \le \mathbf{B}$	denotes that all the elements of the matrix $\mathbf{A}$ are not greater than the corresponding elements of the matrix $\mathbf{B}$: $a_{ij} \le b_{ij}$ for all i and j
$\mathbf{e}^m$	denotes a sequence of m identical symbols $\mathbf{e}$: $\mathbf{e}^m = \mathbf{e}, \mathbf{e}, \ldots, \mathbf{e}$
i. i. d.	independent identically distributed (variables)
$Pr\{A\}$	probability of A
$Pr\{A \mid B\}$	conditional probability of A
$\mathbf{Pr}\{A\}$	matrix probability of A
$\mathbf{x} \in D$	$\mathbf{x}$ belongs to D
$\mathbf{x} \notin D$	$\mathbf{x}$ does not belong to D
$\bar{x} = \mathbf{E}\{x\}$	an expected value (mean) of x
$i \bmod p$	i modulo p: a remainder of division of i by p
$erf(x)$	error function $\frac{2}{\sqrt{\pi}} \int_0^x e^{-x^2} dx$
$erfc(x)$	co-error function $\frac{2}{\sqrt{\pi}} \int_x^\infty e^{-x^2} dx = 1 - erf(x)$
ACK	positive acknowledgement

ARQ	Automatic Repeat-reQuest (system)
AWGN	additive white Gaussian noise
BCH	Bose-Chaudhuri-Hocquenghem (code)
BSC	binary symmetric channel
CSOC	convolutional self-orthogonal code
DFT	discrete Fourier transform
EFS	error-free second
FEC	forward error corection
FFT	fast Fourier transform
GF(p)	size p Galois field
LS	least squares (method)
MDS	maximum-distance-separable (code)
ML	maximum likelihood
MLD	maximum likelihood decoder
NAK	negative acknowledgement
RS	Reed-Solomon (code)
SSM	stochastic sequential machine

INTRODUCTION

During the past two decades, digital transmission networks have become ubiquitous and complex. Many analytical tools and software packages have been created to analyze and evaluate the behavior of these systems. However, these tools often use simplistic models of the system elements' behavior such as exponential distribution of the intervals between failures, Poisson distribution of packets arriving at a node, binomial distribution of errors in a block. These models are selected to simplify the analysis, but they often differ from the actual networks sufficiently to make the computed results of doubtful value.

Computer simulation, on the other hand, can usually model real systems quite closely. But performance studies take a significant amount of computer time, so that in many cases it is not feasible to carry out complete system optimization when test cases are run in ordinary computer centers. Therefore, it is useful to develop mathematical tools and models that perform analytical calculations of the performance of real networks without incurring inaccuracies due to unrealistic assumptions. This book develops general methods for modeling digital transmission systems and applies these methods to practical performance analysis.

In Chapter 1 we discuss the rationale underlying choice of models for error source in digital transmission channels. In order to be able to analyze and predict a system's performance, it is necessary to have some simplified theoretical description of the system's operation—the system model. Usually, a communication system has several channels that may be statistically dependent. Therefore, the channel model should be able to describe the transmission distortion in a set of dependent channels. For example, in a full duplex system one station sends messages over the direct channel and receives acknowledgements over the return channel. If these channels are statistically dependent, we need a model that describes errors in the set of two channels. In general, we have a set of h digital channels with input and output defined by the vectors $\mathbf{a} = (a^{(1)}, \ldots, a^{(h)})$ and $\mathbf{b} = (b^{(1)}, \ldots, b^{(h)})$, whose k-th coordinates represent the k-th channel input and output symbols respectively. We call this set of channels a vector channel. The channel is said to be an *ideal* or *noiseless* if its input and output are the same, except for time delay.

Channel impairments distort the transmitted symbols, and therefore the received sequence contains errors whose values are defined by some measure of difference between transmitted and received symbols. This distortion is usually characterized by the conditional probabilities of

$$Pr(\mathbf{b}_i,\mathbf{b}_{i+1},\ldots,\mathbf{b}_{i+m} \mid \mathbf{a}_i,\mathbf{a}_{i+1},\ldots,\mathbf{a}_{i+m})$$

receiving the sequence of symbols $\mathbf{b}_i,\mathbf{b}_{i+1},\ldots,\mathbf{b}_{i+m}$ when the sequence $\mathbf{a}_i,\mathbf{a}_{i+1},\ldots,\mathbf{a}_{i+m}$ was sent. Here i is an integer ($i \in \mathbf{J}$) and m is a non-negative integer ($m \in \mathbf{N}$). The channel may also be described by the conditional probabilities of errors $Pr(\mathbf{e}_i,\mathbf{e}_{i+1},\ldots,\mathbf{e}_{i+m} \mid \mathbf{a}_i,\mathbf{a}_{i+1},\ldots,\mathbf{a}_{i+m})$. If, in addition, the error sequence does not depend on the transmitted sequence

$$Pr(\mathbf{e}_i,\mathbf{e}_{i+1},\ldots,\mathbf{e}_{i+m} \mid \mathbf{a}_i,\mathbf{a}_{i+1},\ldots,\mathbf{a}_{i+m}) = Pr(\mathbf{e}_i,\mathbf{e}_{i+1},\ldots,\mathbf{e}_{i+m})$$

for all $i \in \mathbf{J}$ and $m \in \mathbf{N}$, then the channel is said to be *symmetric*. The channel is called *stationary* if these probabilities do not depend on i.

We assume in the sequel that the channel is stationary and symmetric. The complete description of this channel is achieved by the multidimensional distributions of errors $Pr(\mathbf{e}_0,\mathbf{e}_1,\ldots,\mathbf{e}_m)$. This description assumes that the channels are ideally synchronized, that is, the number of received symbols is equal to the number of transmitted symbols. Small changes in the above description account for the synchronization loss. Even though the multidimensional distributions give us a complete error source description, their practical applications are quite limited because of their computational complexity. This complexity is one of the reasons for using an approximate channel error sequence description based on an *error source model*.

The most important requirement for an error source model is its agreement with the experimental data. Other important goals are model simplicity and convenience of use. Yet another feature is also desirable: system model generality. The model should be a particular case of a broad family of models with wide applicability. The menu should be rich enough to enable an investigator to choose a model that will agree with experimental data, so that model prediction will be credible.

The family of models that we use for the system description is the family of *stochastic sequential machines* (SSMs).[1] As is shown in Chapter 1, this family is general enough to approximate any finite-order multidimensional distribution. The model is capable of describing both the correlation between errors in a channel and the correlation between errors in different channels. Many error source models described in technical journals belong to this family. [2—19]

As a specific example we consider physical causes of error bursts in satellite channels and show that the SSM model is due to differential encoding, scrambling, forward error correction, intersymbol interference, nonlinear signal distortion, adjacent channel interference, modem defects, and so on. In addition, we analyze the relationship between the SSM model and some stochastic processes frequently used for dynamic system modeling, speech processing, image processing and queueing theory (autoregressive processes, renewal processes, Markov functions, semi-Markov processes, and matrix processes).

Because of the model's generality, the number of parameters that define an SSM may be large. Thus, we consider methods of minimizing their number. The problem of simplifying the model description is closely related to the mathematical problem of lumping the states of some class of semi-Markov processes. We derive an algorithm to minimize the number of model parameters.

The next step is to develop mathematical tools to use with the SSM model for system performance evaluation. Chapter 2 describes the needed tools, which are based on the notion of matrix probability. These tools include extension of basic probability theorems to matrix probabilities, closed-form expressions of matrix generating functions, recursive equations, signal-flow-graph methods, and factorization methods. These methods are then applied to the calculation of some basic performance measures (bit error rate, error-free-second probability, average error-burst length, error-free run distribution, number of errors in a block distribution, etc.).

Statistical methods of estimating model parameters are developed in Chapter 3. The problem of SSM model parameter estimation is reduced to the simpler problem of estimating the coefficients of a finite difference equation. This problem is closely related to some modern techniques of linear system identification, to order and parameter estimation of autoregressive moving average (ARMA) models, and to Padé approximation theory. The proposed methods are of greater interest to statistical inference relating to semi-Markov processes.

In Chapter 4 we calculate several basic parameters that characterize the effectiveness of forward error control (FEC) codes. We determine matrix probabilities of errors and of erasures in a received message for several commonly used block codes. The error matrix probability is expressed as a function of the parity check matrix through the matrix generating function for the general linear block code. The Fourier transform is applied to the probability evaluation. For the channel with independent errors these equations express the probability of incorrect decoding through the code-weight generator polynomial. We also analyze the performance of some convolutional codes, in particular convolutional self-orthogonal codes popular in satellite communications. The application of the Viterbi algorithm to channels with memory is proposed, and we derive the upper union bound on the algorithm performance. The results of Chapter 4 also serve as a basis for Chapter 5.

In Chapter 5 we compare several communication protocols and automatic-repeat-request (ARQ) schemes such as stop-and-wait and go-back-N. We determine message error probability, message erasure probability, message delay distribution, the throughput performance of the protocol, message duplication probability, message loss probability, synchronization loss probability, and the average time of synchronization recovery. We describe a communication protocol as an SSM and then analyze its performance using matrix probability methods developed in previous chapters.

To avoid impeding the development of concepts in the text, complex derivations are relegated to the appendices. They include the proofs of the conditions of Markov and semi-Markov lumpability, methods of matrix series summation, asymptotic approximation of matrix probabilities, block graph applications, derivation of the union bound on Viterbi algorithm performance, and derivation of the Reed-Solomon code weight generator. Appendix 5 provides a review of matrix properties relevant to the material discussed in this book. Appendix 6 provides a brief description of properties of Markov chains. Readers needing an introduction to matrix algebra and Markov chains are advised to read these appendices.

CHAPTER 1

ERROR SOURCE MODELS

1.1 DESCRIPTION OF THE SOURCE OF ERRORS BY STOCHASTIC SEQUENTIAL MACHINES

1.1.1 Stochastic Sequential Machines. The analysis of experimental data on real channels reveals error dependence that is characterized by the formation of error clusters (bursts). Let us assume that different types of clusters correspond to different states $s=1,2,\ldots,u$ of some source of errors. Denote

$$Pr(\mathbf{e}_1, s_1 \mid \mathbf{a}_{-n+1},\ldots,\mathbf{a}_0,\mathbf{a}_1;\, s_{-n},\ldots,s_{-1},s_0)$$

the conditional probability that the error in receiving the symbol $\mathbf{a}_1$ is equal to $\mathbf{e}_1$ and the source of errors passes into the state s_1, given that the symbols $\mathbf{a}_{-n+1},\ldots,\mathbf{a}_0,\mathbf{a}_1$ were transmitted in succession and the source of errors was in the states $s_{-n},\ldots,s_{-1},s_0$.

Now consider the model of the source of errors in which this probability does not depend on $\mathbf{a}_{-n+1},\ldots,\mathbf{a}_0,\mathbf{a}_1$, $s_{-n},\ldots,s_{-1},s_0$ for any n, $\mathbf{a}_{-n+1},\ldots$, $\mathbf{a}_0$, $\mathbf{a}_1$, $\mathbf{e}_1$, $s_{-n},\ldots$, s_{-1}, s_0, s_1:

$$Pr(\mathbf{e}_1, s_1 \mid \mathbf{a}_{-n+1},\ldots,\mathbf{a}_0,\mathbf{a}_1;\, s_{-n},\ldots,s_{-1},s_0) = Pr(\mathbf{e}_1, s_1 \mid \mathbf{a}_1; s_0) \qquad (1.1.1)$$

Physically this assumption means that, given the state s_0 of the source of errors during the transmission of the symbol $\mathbf{a}_0$, the following course of the process does not depend on the way the source of errors arrived into s_0 (in other words, it does not depend on the prehistory of the process). A model of the source of errors defined in this way is called a *stochastic sequential machine* (SSM) or probabilistic automaton.[1,20] Equation (1.1.1) expresses the so-called Markov property of the SSM.†

Suppose that an SSM is in the state s_0 with the probability $Pr(s_0)$ at the beginning of the transmission, and the input of the vector channel is the sequence $\mathbf{a}_1,\mathbf{a}_2,...,\mathbf{a}_m$ then the sequence of errors $\mathbf{e}_1,\mathbf{e}_2,...,\mathbf{e}_m$ appears with the probability $Pr(\mathbf{e}_1,...,\mathbf{e}_m \mid \mathbf{a}_1,...,\mathbf{a}_m)$, which can be computed according to the total probability formula

$$Pr(\mathbf{e}_1,...,\mathbf{e}_m \mid \mathbf{a}_1,...,\mathbf{a}_m) = \sum_{s_0,s_1,...,s_m} Pr(s_0)\prod_{k=1}^{m} Pr(\mathbf{e}_k,s_k \mid \mathbf{a}_k,s_{k-1}) \qquad (1.1.2)$$

It is convenient to rewrite this formula in a matrix form. Let $Pr(\mathbf{e}_1 \mid \mathbf{a}_1)$ be the matrix whose (s_0,s_1)-th element is the probability (1.1.1)

$$\mathbf{P}(\mathbf{e}_1 \mid \mathbf{a}_1) = [\, Pr(\mathbf{e}_1,s_1 \mid \mathbf{a}_1,s_0)\,]_{u,u}$$

Then (1.1.2) can be written as

$$Pr(\mathbf{e}_1,...,\mathbf{e}_m \mid \mathbf{a}_1,...,\mathbf{a}_m) = \mathbf{p}\prod_{k=1}^{m}\mathbf{P}(\mathbf{e}_k \mid \mathbf{a}_k)\mathbf{1} \qquad (1.1.3)$$

where $\mathbf{p} = row\{Pr(s_0)\}$ is the matrix row of the initial state probabilities and $\mathbf{1} = col\{1\}$ is the matrix column of ones. Indeed, according to the rule of the multiplication of matrices (see Ref [24] and Appendix 5), we have

$$\mathbf{pP}(\mathbf{e}_1 \mid \mathbf{a}_1) = \mathbf{p}_1$$

where $\mathbf{p}_1$ is the matrix row, whose s_1-th element is

$$\sum_{s_0=1}^{u} Pr(s_0)Pr(\mathbf{e}_1,s_1 \mid \mathbf{a}_1,s_0)$$

If we now multiply the matrices $\mathbf{p}_1$ and $\mathbf{1}$, we obtain

$$\mathbf{pP}(\mathbf{e}_1 \mid \mathbf{a}_1)\mathbf{1} = \sum_{s_0=1}^{u}\sum_{s_1=1}^{u} Pr(s_0)Pr(\mathbf{e}_1,s_1 \mid \mathbf{a}_1,s_0) \qquad (1.1.4)$$

The right-hand part of this equation coincides with (1.1.2) when $m=1$. We prove the correctness of formula (1.1.3) for every m similarly.

† Readers needing an introduction to Markov chains are advised to read Appendix 6 of this book.

The proposed model permits us to take into consideration the error source asymmetry, since error symbol probability depends on the transmitted symbol. Unfortunately, if the source of errors is asymmetric, the computations required by the model become far more complicated, since it is necessary to specify the process of the symbol's arrival at the channel input.

If the matrices $\mathbf{P}(\mathbf{e} \mid \mathbf{a})$ do not depend on $\mathbf{a}$, then the channel is symmetric and formula (1.1.2) can be written as

$$Pr(\mathbf{e}_1,...,\mathbf{e}_m \mid \mathbf{a}_1,...,\mathbf{a}_m) = \mathbf{p}\prod_{k=1}^{m}\mathbf{P}(\mathbf{e}_k)\mathbf{1} \tag{1.1.5}$$

In this particular case the SSM is called *autonomous*, since its states do not depend on the input. In the following discussion we consider only symmetric channels, and therefore formula (1.1.5) plays the main role.

Note that if the sequence of symbols on the vector channel input can be represented as the output of a certain autonomous SSM with the matrices $\mathbf{P}_1(\mathbf{a})=[Pr(\mathbf{a},i \mid j)]$ and with the matrix of the initial probabilities $\mathbf{p}_1=row\{Pr(i)\}$, then by increasing the number of states of the SSM it is possible to describe the channel as symmetric. Indeed, consider a source of errors the states of which take the form $\mu=(i,s_0)$, where i is the state of the SSM that "generates" the symbol $\mathbf{a}$ and s_0 is the state of the original asymmetric source of errors. Then according to the theorem of total probability we have

$$p_{\mu,\nu} = \sum_{\mathbf{a}} Pr(\mathbf{a},j \mid i)Pr(\mathbf{e},s_1 \mid \mathbf{a},s_0) \quad \mu=(i,s_0) \quad \nu=(j,s_1)$$

and the model of the symmetric channel with the matrices $\mathbf{P}_2(\mathbf{e}) = [p_{\mu\nu}]$ and with the matrices of the initial probabilities $\mathbf{p}_2 = row\{Pr(i)Pr(s_0)\}$ is equivalent to the original model. The matrices $\mathbf{P}_2(\mathbf{e})$ and $\mathbf{p}_2$ can be written as follows:

$$\mathbf{P}_2(\mathbf{e}) = \sum_{\mathbf{a}} \mathbf{P}_1(\mathbf{a}) \otimes \mathbf{P}(\mathbf{e} \mid \mathbf{a}) \quad \mathbf{p}_2 = \mathbf{p}_1 \otimes \mathbf{p}$$

where $\otimes$ is the Kronecker product of matrices. [21,22]

1.1.2 Stationary SSM. Obviously, the sequence of the states of the SSM is a Markov chain with the matrix

$$\mathbf{P} = [p_{ij}]_{u,u} = [\sum_{\mathbf{e}} Pr(\mathbf{e},j \mid i)]_{u,u} = \sum_{\mathbf{e}} \mathbf{P}(\mathbf{e}) \tag{1.1.6}$$

From now on we assume that the matrix $\mathbf{P}$ is *regular* [24] (see also Appendix 6). In this case the probability $p_{ij}^{(m)}$ that the chain moves from the state i to the state j in m steps has a limit when the number of steps tends to infinity, and this limit does not depend on the initial state i:

$$\lim_{m\to\infty} p_{ij}^{(m)} = \pi_j \tag{1.1.7}$$

These limits are called *equilibrium* or *stationary* probabilities of states of the Markov chain.

The matrix $\boldsymbol{\pi}=[\ \pi_1,\pi_2,...,\pi_u\ ]$ of the stationary probabilities can be found as a solution of the system:

$$\boldsymbol{\pi}\mathbf{P} = \boldsymbol{\pi} \qquad \boldsymbol{\pi}\mathbf{1} = 1 \tag{1.1.8}$$

(see Appendix 6).

If the matrix of the initial probabilities $\mathbf{p}$ of the Markov chain states coincides with the matrix of the stationary probabilities $\boldsymbol{\pi}$, the chain is called *stationary*. The above discussion shows that the stationary SSM can be defined by the collection of the matrices $\mathbf{P}(\mathbf{e})$ only, since the initial matrix in this case is determined as the unique solution of the system (1.1.8).

We often consider the binary SSM for which the matrices $\mathbf{P}(\mathbf{e})$ have the form

$$\mathbf{P}(\mathbf{e}) = \mathbf{PF}(\mathbf{e}) \quad \mathbf{F}(\mathbf{e}) = \prod_{i=1}^{h}(\mathbf{I} - \mathbf{E}_i)^{1-e^{(i)}}\mathbf{E}_i^{e^{(i)}} \tag{1.1.9}$$

where $\mathbf{e} = (e^{(1)},...,e^{(h)})$, $\mathbf{E}_i = diag\{\varepsilon_j^{(i)}\}$, $\mathbf{I} = diag\{1\}$ is the identity matrix. This case is interesting because it permits a simple interpretation of the model. The error source state transition is described as a Markov chain with the matrix $\mathbf{P}$. If the source of errors is in the state j, then the conditional probability of an error in the i-th channel is $\varepsilon_j^{(i)}$. This interpretation permits us to partition the states into "good" where the probability $\varepsilon_j^{(i)}$ is small and "bad" where this probability is not small. We illustrate this model in the next section.

1.1.3 Gilbert's Model. Gilbert [2] initiated the study of the Markov chain error source models. According to his model, the source of errors has two states: G (for good) and B (for bad or burst), which constitute a Markov chain with the matrix

$$\mathbf{P} = \begin{bmatrix} Q & P \\ p & q \end{bmatrix}$$

The model state transition diagram is shown in Fig. 1.1. In state G the probability of error is equal to zero ($\varepsilon_1^{(1)}$=0); in state B the probability of error is equal to $\varepsilon_2^{(1)}=1-h$. Thus,

$$\mathbf{F}(0) = \begin{bmatrix} 1 & 0 \\ 0 & h \end{bmatrix} \quad \mathbf{F}(1) = \begin{bmatrix} 0 & 0 \\ 0 & 1-h \end{bmatrix}$$

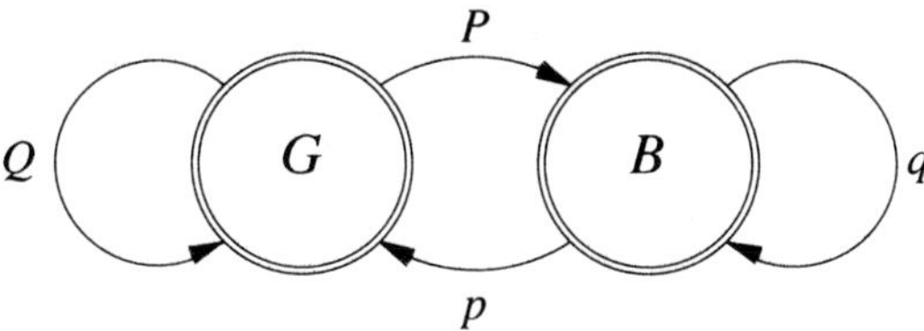

Figure 1.1. Gilbert's model state diagram.

According to formula (1.1.9), we obtain

$$\mathbf{P}(0) = \mathbf{PF}(0) = \begin{bmatrix} Q & Ph \\ p & qh \end{bmatrix} \qquad \mathbf{P}(1) = \mathbf{PF}(1) = \begin{bmatrix} 0 & P(1-h) \\ 0 & q(1-h) \end{bmatrix}$$

We will use this model for illustration throughout the book.

The bit-error probability can be found using equation (1.1.5)

$$p_{\text{err}} = \boldsymbol{\pi}\mathbf{P}(1)\mathbf{1} \tag{1.1.10}$$

where the stationary probabilities satisfy the system (1.1.8), which in our case has the form

$$\begin{aligned} \pi_1 Q + \pi_2 p &= \pi_1 \\ \pi_1 P + \pi_2 q &= \pi_2 \\ \pi_1 + \pi_2 &= 1 \end{aligned}$$

Solving this system we obtain

$$\pi_1 = p/(P+p) \qquad \pi_2 = P/(P+p)$$

so that

$$\boldsymbol{\pi} = \begin{bmatrix} \dfrac{p}{P+p} & \dfrac{P}{P+p} \end{bmatrix} = \frac{1}{P+p}[\,p \;\; P\,]$$

Equation (1.1.10) takes the form

$$p_{\text{err}} = \frac{1}{P+p}[\,p \;\; P\,]\begin{bmatrix}0 & P(1-h)\\ 0 & q(1-h)\end{bmatrix}\begin{bmatrix}1\\1\end{bmatrix} = \frac{(1-h)P}{P+p}$$

Let us now calculate the probability of the 01 error combination. For this purpose we use formula (1.1.5):

$$Pr(01) = \boldsymbol{\pi}\mathbf{P}(0)\mathbf{P}(1)\mathbf{1}$$

We have

$$\boldsymbol{\pi}\mathbf{P}(0) = \frac{1}{P+p}[\,p \;\; P\,]\begin{bmatrix}Q & Ph\\ p & qh\end{bmatrix} = \frac{1}{P+p}[\,p \;\; Ph\,]$$

$$\boldsymbol{\pi}\mathbf{P}(0)\mathbf{P}(1) = \frac{1}{P+p}[\,p \;\; Ph\,]\begin{bmatrix}0 & P(1-h)\\ 0 & q(1-h)\end{bmatrix} = \frac{1}{P+p}[0 \;\; P(p+qh)(1-h)]$$

And, finally,

$$Pr(01) = \boldsymbol{\pi}\mathbf{P}(0)\mathbf{P}(1)\mathbf{1} = \frac{1}{P+p}[0 \;\; P(p+qh)(1-h)]\begin{bmatrix}1\\1\end{bmatrix} = \frac{P(p+qh)(1-h)}{P+p}$$

1.1.4 Equivalent Models. In order to simplify the SSM model applications, we must consider a set of equivalent models and select a model whose description is simpler (a smaller order of the matrices or a simpler structure). We say that two models are equivalent if the processes of occurrence of errors described by these models are equivalent. [23] In other words, the models are equivalent if all their corresponding multidimensional distributions of errors are identical:

$$p(\mathbf{e}_1,\mathbf{e}_2,\ldots,\mathbf{e}_n) = p_1(\mathbf{e}_1,\mathbf{e}_2,\ldots,\mathbf{e}_n) \tag{1.1.11}$$

for all $n = 1,2,\ldots$ and all $\mathbf{e}_k$, $k = 1,2,\ldots,n$. According to this definition, two models with matrices $\mathbf{P}(\mathbf{e})$ and $\mathbf{P}_1(\mathbf{e})$ are equivalent if

$$\boldsymbol{\pi}\mathbf{P}(\mathbf{e}_1)\mathbf{P}(\mathbf{e}_2)\cdots\mathbf{P}(\mathbf{e}_n)\mathbf{1} = \boldsymbol{\pi}_1\mathbf{P}_1(\mathbf{e}_1)\mathbf{P}_1(\mathbf{e}_2)\cdots\mathbf{P}_1(\mathbf{e}_n)\mathbf{1} \tag{1.1.12}$$

for all $n = 1,2,\ldots$ and all $\mathbf{e}_k$, $k = 1,2,\ldots,n$.

It is convenient to use the following sufficient condition of equivalency:[1] If there exists a matrix $\mathbf{T}$ such that

$$\mathbf{T}\,\mathbf{P}_1(\mathbf{e}) = \mathbf{P}(\mathbf{e})\,\mathbf{T} \qquad \mathbf{T}\,\mathbf{1} = \mathbf{1} \tag{1.1.13}$$

for all $\mathbf{e}$, then the models with matrices $\mathbf{P}(\mathbf{e})$ and $\mathbf{P}_1(\mathbf{e})$ are equivalent.

Indeed, let $\boldsymbol{\pi}$ be the matrix of the stationary probabilities of the Markov chain with the matrix

$$\mathbf{P} = \sum \mathbf{P}(\mathbf{e})$$

Then $\boldsymbol{\pi}_1 = \boldsymbol{\pi}\mathbf{T}$ is the stationary vector of the Markov chain with the matrix

$$\mathbf{P}_1 = \sum \mathbf{P}_1(\mathbf{e})$$

as is seen from the following equations:

$$\boldsymbol{\pi}_1\,\mathbf{P}_1 = \boldsymbol{\pi}\,\mathbf{T}\,\mathbf{P}_1 = \boldsymbol{\pi}\,\mathbf{P}\,\mathbf{T} = \boldsymbol{\pi}\,\mathbf{T} = \boldsymbol{\pi}_1$$

$$\boldsymbol{\pi}_1\,\mathbf{1} = \boldsymbol{\pi}\,\mathbf{T}\,\mathbf{1} = \boldsymbol{\pi}\,\mathbf{1} = 1$$

Since

$$\boldsymbol{\pi}_1\,\mathbf{P}_1(\mathbf{e}_1)\,\mathbf{P}_1(\mathbf{e}_2)\cdots\,\mathbf{P}_1(\mathbf{e}_n)\,\mathbf{1} = \boldsymbol{\pi}\,\mathbf{T}\,\mathbf{P}_1(\mathbf{e}_1)\,\mathbf{P}_1(\mathbf{e}_2)\cdots\,\mathbf{P}_1(\mathbf{e}_n)\,\mathbf{1} =$$

$$\boldsymbol{\pi}\,\mathbf{P}(\mathbf{e}_1)\,\mathbf{P}(\mathbf{e}_2)\cdots\,\mathbf{P}(\mathbf{e}_n)\,\mathbf{T}\,\mathbf{1} = \boldsymbol{\pi}\,\mathbf{P}(\mathbf{e}_1)\,\mathbf{P}(\mathbf{e}_2)\cdots\,\mathbf{P}(\mathbf{e}_n)\,\mathbf{1}$$

equations (1.1.12) are satisfied and therefore the models are equivalent.

In the particular case when the matrix $\mathbf{T}$ is nonsingular, equation (1.1.12) takes on the form

$$\mathbf{P}_1(\mathbf{e}) = \mathbf{T}^{-1}\,\mathbf{P}(\mathbf{e})\,\mathbf{T} \qquad \mathbf{T}\mathbf{1} = \mathbf{1}$$

Thus, the matrices $\mathbf{P}_1(\mathbf{e})$ and $\mathbf{P}(\mathbf{e})$ are *similar* (see Appendix 5.1.5). They may be reduced to their simplest (normal) form.

1.1.5 The SSM Model Generality. We justify the generality of the considered model by mentioning that the SSM permits us to model various types of error dependence in different channels. To demonstrate this let us describe an SSM model in which errors in different channels occur independently. Suppose that the model of the source of errors in the k–th channel is described by the matrices

$$\mathbf{P}_k(e^{(k)}) = [Pr(e^{(k)}, i_k, j_k)]$$

The sources of errors in different channels are statistically independent if

$$Pr(\mathbf{e}, j \mid i) = \prod_{m=1}^{h} Pr(e^{(m)}, j_m \mid i_m) \tag{1.1.14}$$

where $\mathbf{e} = (e^{(1)}, e^{(2)}, ..., e^{(h)})$, $i = (i_1, i_2, ..., i_h)$, and $j = (j_1, j_2, ..., j_h)$. Consequently, if the elements of the matrix $\mathbf{P}(\mathbf{e})$ are determined according to formulas (1.1.14), then the sources of errors in different channels are statistically independent. Note that the matrix $\mathbf{P}(\mathbf{e})$, whose elements are determined by equations (1.1.14), is called a Kronecker product of the matrices[21,22] and is denoted as

$$\mathbf{P}(\mathbf{e}) = \mathbf{P}_1(e^{(1)}) \otimes \mathbf{P}_2(e^{(2)}) \otimes \cdots \otimes \mathbf{P}_h(e^{(h)}) \tag{1.1.15}$$

In many practical cases it is possible to justify the channel independence if they are physically isolated. In these cases we use formula (1.1.15) to describe the error source model. We will use it in Chapter 5 to model the error source in the two-way systems.

The other extreme is the channel deterministic dependence. We can describe a vector channel in which errors occur simultaneously in all its subchannels. Indeed, if the conditional error probability $\varepsilon_j^{(i)}$ equals to 0 for all the channels $(i=1,2,...,h)$ and some subset of states (say j=1,2,...,m) and equals to 1 otherwise, then

$$\mathbf{E}_i = \begin{bmatrix} 0 & 0 \\ 0 & \mathbf{I} \end{bmatrix} \qquad \mathbf{I} - \mathbf{E}_i = \begin{bmatrix} \mathbf{I} & 0 \\ 0 & 0 \end{bmatrix}$$

According to representation (1.1.9), $\mathbf{F}(\mathbf{1}) = \mathbf{E}_1$, $\mathbf{F}(\mathbf{0}) = \mathbf{I} - \mathbf{E}_1$, and $\mathbf{F}(\mathbf{e}) = 0$ otherwise. Therefore, in this case errors occur (or do not occur) simultaneously in all channels.

Thus, the SSM model can describe different degrees of error dependence in the channels.

This model can also reflect different degrees of dependence between errors in the same channel. Suppose that error sequence is an order k Markov chain. Then for any $\mathbf{e}_0, \mathbf{e}_1, ..., \mathbf{e}_n$ and $n \geq k$, the conditional probability of an error at the moment n

$$Pr(\mathbf{e}_n \mid \mathbf{e}_0, ..., \mathbf{e}_{n-1}) = Pr(\mathbf{e}_n \mid \mathbf{e}_{n-k}, ..., \mathbf{e}_{n-1}) \tag{1.1.16}$$

depends only on k previous errors $\mathbf{e}_{n-k}, ..., \mathbf{e}_{n-1}$. This Markov chain may be described as an SSM.

Consider a process $S_n = (\mathbf{e}_{n-k+1}, \mathbf{e}_{n-k+2}, ..., \mathbf{e}_n)$ whose states are defined as sequences of length k of the order k Markov chain. According to (1.1.16), the sequence of states S_n is a simple Markov chain. It is obvious also that state error probability $p_{S_n}(\mathbf{e}) = 1$ if $\mathbf{e} = \mathbf{e}_n$ and is equal to 0 otherwise. Therefore, the order k Markov chain is expressed as an SSM. Consequently, we could construct an SSM model of any error source whose order k multidimensional distributions are identical to those of the model. Since we always have only finite sets of experimental data, we may construct an SSM model for practically all statistical data. However, the size of the matrix, which defines the SSM, grows exponentially with k. This growth creates a problem of model simplification.

EXAMPLE 1.1.1: Let us illustrate the discussed technique of model building in the case of the second-order Markov chain ($k = 2$) and one binary symmetric channel ($h = 1$). In this case the Markov chain S_n has four states: (0,0), (1,0), (0,1), and (1,1). The chain transition probability matrix is

$$\mathbf{P} = \begin{bmatrix} Pr(0 \mid 0,0) & 0 & Pr(1 \mid 0,0) & 0 \\ Pr(0 \mid 1,0) & 0 & Pr(1 \mid 1,0) & 0 \\ 0 & Pr(0 \mid 0,1) & 0 & Pr(1 \mid 0,1) \\ 0 & Pr(0 \mid 1,1) & 0 & Pr(1 \mid 1,1) \end{bmatrix}$$

The state conditional probabilities of errors are equal to 1 when the chain is in the state (0,1) or (1,1) and equal to 0 otherwise. Therefore,

$$\mathbf{F}(0) = \begin{bmatrix} 1 & 0 & 0 & 0 \\ 0 & 1 & 0 & 0 \\ 0 & 0 & 0 & 0 \\ 0 & 0 & 0 & 0 \end{bmatrix} \qquad \mathbf{F}(1) = \begin{bmatrix} 0 & 0 & 0 & 0 \\ 0 & 0 & 0 & 0 \\ 0 & 0 & 1 & 0 \\ 0 & 0 & 0 & 1 \end{bmatrix}$$

According to (1.1.9),

$$\mathbf{P}(0) = \mathbf{PF}(0) = \begin{bmatrix} Pr(0 \mid 0,0) & 0 & 0 & 0 \\ Pr(0 \mid 1,0) & 0 & 0 & 0 \\ 0 & Pr(0 \mid 0,1) & 0 & 0 \\ 0 & Pr(0 \mid 1,1) & 0 & 0 \end{bmatrix}$$

$$\mathbf{P}(1) = \mathbf{P}\,\mathbf{F}(1) = \begin{bmatrix} 0 & 0 & Pr(1 \mid 0,0) & 0 \\ 0 & 0 & Pr(1 \mid 1,0) & 0 \\ 0 & 0 & 0 & Pr(1 \mid 0,1) \\ 0 & 0 & 0 & Pr(1 \mid 1,1) \end{bmatrix}$$

■

It is usually said that impulse noise is the major contributor to the channel error bursts. However, other sources also produce error bursts. We analyze some of these sources in the following section.

1.1.6 Satellite Channel Model. The major results of the classical information transmission theory are based on the assumption that channel noise is *additive white Gaussian noise* (AWGN) and channel errors are *independent and identically distributed* (i. i. d.). A satellite channel is considered an example for which this theory is good enough.†

However, the real satellite channel error statistics showed that these assumptions are far from reality. Even if we assume that the space noise is the AWGN, the channel output errors are not i. i. d. It may be explained by the various types of channel impairments (such as intersymbol interference, nonlinearities, adjacent channel interference, modem defects, etc.), and different digital signal transformations (such as differential encoding, scrambling, error detection and error correction coding, etc.). Consider, for example, a typical satellite communication system as shown in Fig. 1.2. The transmitter model consists of a differential encoder, transmit shaping filters, and quadrature phase-shift keying (QPSK) modulator. The transponder model consists of a channel input filter, traveling wave tube (TWT) amplifier, and the transponder output filter. The receiver model includes a receive filter, symbol timing and carrier recovery circuitry, coherent demodulator, data detector, and differential decoder.

Two independent data streams (Data-P and Data-Q in Fig. 1.2) appear on the differential encoder input and two data streams (Data-G and Data-H) appear on the differential decoder output, thereby forming the set of two binary channels. We will show that in the presence AWGN the error sequence on the differential decoder output can be described using the SSM model.

It is convenient to use complex number notations to describe signal transformations in QPSK-modulated systems. We therefore denote $z_n = p_n + \mathrm{j}\,q_n$ ($p_n = \pm 1$, $q_n = \pm 1$, $-\infty < n < \infty$) the differential encoder input data and $w_n = c_n + \mathrm{j}\,d_n$ ($c_n = \pm 1$, $d_n = \pm 1$, $-\infty < n < \infty$) its output data. It is easy to verify that the differential encoder operation [25] may be defined by

† The major part of this section is reprinted from the IEEE *Journal on Selected Areas in Communications*, vol. 6, no. 1, January 1988, with the permission of the IEEE.

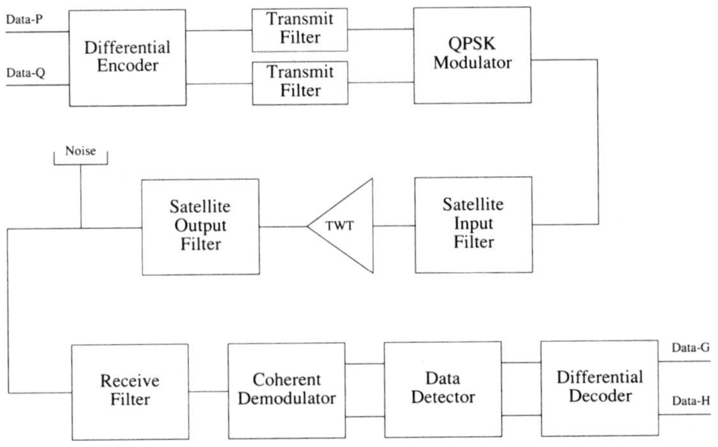

Figure 1.2.. Functional block diagram of satellite channel.

$$w_n = -0.5(1+j)w_{n-1}z_n^* \tag{1.1.17}$$

where z_n^* denotes the complex conjugate of z_n.

The input to the satellite nonlinear amplifier is equal to

$$s(t) = \text{Re}\, u(t)e^{j\omega_c t} \tag{1.1.18}$$

where ω_c is the carrier frequency,

$$u(t) = \sum_{n=-\infty}^{\infty} w_n p(t - nT) \tag{1.1.19}$$

$p(t)$ is a pulse shape which is the convolution of the original rectangular pulse with the transmit filter impulse response and the satellite input filter low-pass equivalent, [26] and T is the rectangular pulse duration.

The satellite TWT amplifier operation could be described by the equation [25]

$$s_1(t) = \text{Re}\, u_1(t)e^{j\omega_c t} \tag{1.1.20}$$

where

$$u_1(t) = v(R(t))u(t) \tag{1.1.21}$$

with $v(R)=(f(R)/R)e^{jg(R)}$. Functions $f(R)$ and $g(R)$ characterize the TWT AM/AM and AM/PM conversion, $R(t)=|u(t)|$ is the TWT input signal amplitude.

Assume that the satellite output signal is distorted by the AWGN. Uplink noise may be compounded with that of the downlink by using the method described in Ref. [27]. Assume also that the receiver carrier phase tracking is ideal and that the variances of the recovered carrier and of the symbol clock produce insignificant performance loss, given the transmission of purely random data bits. Then the receiver data sampler output is

$$u_3(nT)=u_2(nT)+N(nT) \tag{1.1.22}$$

where $u_2(t)$ and $N(t)$ are the low-pass representations of the signal and noise [26] on the output of the receiver filter. The complex signal $u_2(t)$ is a convolution of the TWT output signal $u_1(t)$ with the low-pass equivalents of the satellite output filter and the receiver input filter. The inphase and quadrature components of the received signal represent the real and imaginary parts of $u_3(nT)=s_{I,n}+\mathrm{j}\,s_{Q,n}$.

The data detector makes a decision about the transmitted bit, depending on the difference between the sampled level and the decision threshold. The data detector outputs the sequence

$$\omega_n=e_n+\mathrm{j}\,f_n=sign\,(s_{I,n})+\mathrm{j}\,sign\,(s_{Q,n}) \tag{1.1.23}$$

where $sign(x)=1$ for $x\geq 0$ and $sign(x)=0$ for $x<0$.

Finally, the differential decoder performs the operation which is inverse to that of the differential encoder:

$$\zeta_n=-0.5(1+\mathrm{j})\omega_n^*\omega_{n-1} \tag{1.1.24}$$

Equations (1.1.17) through (1.1.24) may be treated as a satellite channel physical model. According to this model, error sequences may be expressed as the output of the SSM.

Consider first the channel without the differential encoding. In this case, w_n is the channel input and ω_n its output. An error occurs if $w_n\neq\omega_n$. In the absence of noise and TWT nonlinearity, the only source of signal degradation is channel misequalization, which results in intersymbol interference. That is, the signal amplitude at the sampling instant nT depends not only on w_n but also on w_i for $i=n-l,\ldots,n+l$. If the channel state is defined as a vector $S_n=(w_{n-l},\ldots,w_{n+l})$ and the input bit sequence is purely random, then the sequence S_n is the Markov chain. In a more general case, when the input sequence is a Markovian function we will construct the Markov chain by combining S_n with the states of the source Markovian function.

Satellite nonlinearity increases the intersymbol interference and, according to (1.1.20) and (1.1.21), adds the cross-coupling between the inphase and quadrature components. Therefore, more neighboring samples must be taken into account in the channel state S_n definition. The actual number of samples depends on the parameters of the TWT and also on the offset from the saturation.

According to equation (1.1.22), the state bit-error probabilities may be determined from the following equations:

$$p_{S_n}(\mathbf{e}) = p_{S_n}(e_I, e_Q) = p_{S_n}(e_I) p_{S_n}(e_Q) \tag{1.1.25}$$

where

$$p_{S_n}(e_I) = 0.5 erfc(|\mathrm{Re} u_2(nT)|/\sigma\sqrt{2}) \quad p_{S_n}(e_Q) = 0.5 erfc(|\mathrm{Im} u_2(nT)|/\sigma\sqrt{2})$$

and σ^2 is the power of the noise on the output of the receiver filter and

$$erfc(x) = \frac{2}{\sqrt{\pi}} \int_x^{\infty} e^{-x^2} dx$$

The average error probabilities for the inphase and quadrature channels are equal to

$$p(e_I) = \sum_{S_n} p(S_n) p_{S_n}(e_I) \tag{1.1.26}$$

and

$$p(e_Q) = \sum_{S_n} p(S_n) p_{S_n}(e_Q) \tag{1.1.27}$$

respectively. In these equations $p(S_n)$ denote stationary probabilities of the SSM states which may be found from equations (1.1.19). In the particular case when purely random data bits are transmitted, all these stationary probabilities are equal to 4^{-2l-1}.

We have shown that in the absence of the differential coding the error sequence may be described as an SSM. The differential coding increases the model's complexity, but still the model may be described as an SSM. According to equation (1.1.24), the differential encoder output depends on two consecutive samples on the data detector output. If we add ω_{n-1} to the state definition by introducing states $S_{d,n} = (S_n, \omega_{n-1})$, then error sequences on the differential decoder output will also be modeled by SSM.

1.1.7 SSM Channel Combination. Suppose that the link between receiving and transmitting stations is composed of two independent SSM channels. Our goal is to describe the error source in the combination of the channels. We assume that the channels are connected through a digital repeater that recovers digital sequences from the first channel and then transmits them over the second channel.

If we assume that the models are described by the matrices of the form (1.1.9)

$$\mathbf{P}_1(\mathbf{e}) = \mathbf{P}_1\mathbf{F}_1(\mathbf{e}) \quad \mathbf{P}_2(\mathbf{e}) = \mathbf{P}_2\mathbf{F}_2(\mathbf{e})$$

for the first and second channels in the link, then the composite channel can be described by the product states with the transition probability matrix

$$\mathbf{P} = \mathbf{P}_1 \otimes \mathbf{P}_2$$

and the matrix of the conditional probabilities of errors

$$\mathbf{F}(\mathbf{e}) = \sum_{\mathbf{e}_1} \mathbf{F}_1(\mathbf{e}_1) \otimes \mathbf{F}_2(\mathbf{e}-\mathbf{e}_1)$$

Using the Kronecker product properties,[21,22] we can write the matrices $\mathbf{P}(\mathbf{e})$ that describe the composite channel as

$$\mathbf{P}(\mathbf{e}) = \mathbf{P}\mathbf{F}(\mathbf{e}) = \sum_{\mathbf{e}_1} \mathbf{P}_1(\mathbf{e}_1) \otimes \mathbf{P}_2(\mathbf{e}-\mathbf{e}_1).$$

EXAMPLE 1.1.2: Suppose that a channel is composed of a terrestrial channel whose error source is described by the Gilbert model (see Sec. 1.1.3) and a satellite channel with independent errors that occur with the probability p_0. The composite channel is described by the SSM model with the matrices

$$\mathbf{P}(0) = \mathbf{P}_1(0) \otimes \mathbf{P}_2(0) + \mathbf{P}_1(1) \otimes \mathbf{P}_2(1)$$
$$\mathbf{P}(1) = \mathbf{P}_1(1) \otimes \mathbf{P}_2(0) + \mathbf{P}_1(0) \otimes \mathbf{P}_2(1)$$

where

$$\mathbf{P}_1(0) = \begin{bmatrix} Q & Ph \\ p & qh \end{bmatrix} \quad \mathbf{P}_1(1) = \begin{bmatrix} 0 & P(1-h) \\ 0 & q(1-h) \end{bmatrix}$$

$$\mathbf{P}_2(0) = q_0 = 1-p_0 \quad \mathbf{P}_2(1) = p_0$$

Performing the matrix multiplication, we obtain the matrices describing the composite channel

$$\mathbf{P}(0) = \begin{bmatrix} Qq_0 & P(p_0+hq_0-hp_0) \\ pq_0 & q(p_0+hq_0-hp_0) \end{bmatrix} \qquad \mathbf{P}(1) = \begin{bmatrix} Qp_0 & P(q_0+hp_0-hq_0) \\ pp_0 & q(q_0+hp_0-hq_0) \end{bmatrix}$$

■

1.2 BINARY SYMMETRIC STATIONARY CHANNEL

1.2.1 Model Description. The single channel case ($h=1$) plays the most important role in applications. In this case the probability $Pr(e_1,e_2,...,e_m)$ of occurrence of a sequence of errors $e_1,e_2,...,e_m$ can be computed by the formula

$$Pr(e_1,e_2,...,e_m) = \boldsymbol{\pi}\prod_{i=1}^{m}\mathbf{P}(e_i)\mathbf{1} \tag{1.2.1}$$

Multidimensional error-free run distributions have the form

$$f_\lambda(\lambda_1,\lambda_2,...,\lambda_m) = Pr(0^{\lambda_1}10^{\lambda_2}\cdots 0^{\lambda_m}1 \mid 1) = \boldsymbol{\pi}\mathbf{P}(1)\prod_{i=1}^{m}\{\mathbf{P}(0)^{\lambda_i}\mathbf{P}(1)\}\mathbf{1} \,/\, \boldsymbol{\pi}\mathbf{P}(1)\mathbf{1} \tag{1.2.2}$$

The distributions of the intervals between the correct symbols is given by

$$f_l(l_1,l_2,...,l_m) = Pr(1^{l_1}01^{l_2}\cdots 1^{l_m}0 \mid 0) = \boldsymbol{\pi}\mathbf{P}(0)\prod_{i=1}^{m}\{\mathbf{P}(1)^{l_i}\mathbf{P}(0)\}\mathbf{1} \,/\, \boldsymbol{\pi}\mathbf{P}(0)\mathbf{1} \tag{1.2.3}$$

It is possible that in these equations $\lambda_i=0$ and $l_i=0$ so that $f_\lambda(0,\ 0,...,\ 0) = Pr(11\cdots 1 \mid 1) = Pr(1^m \mid 1)$ and $f_l(m) = Pr(1^m0 \mid 0)$ are linearly dependent. This dependency creates an ambiguity in experimental data processing. To eliminate the ambiguity we consider the distributions that do not permit $\lambda_i=0$ or $l_i=0$:

$$p_\lambda(\lambda_1,\lambda_2,...,\lambda_m) = Pr(0^{\lambda_1}10^{\lambda_2}\cdots 0^{\lambda_m}1 \mid 1,\lambda_1>0,\ \lambda_2>0,...,\lambda_m>0) = \boldsymbol{\pi}\mathbf{P}(1)\prod_{i=1}^{m}\{\mathbf{P}(0)^{\lambda_i}\mathbf{P}(1)\}\mathbf{1} \,/\, \boldsymbol{\pi}\mathbf{P}(1)[\mathbf{P}(0)(\mathbf{I}-\mathbf{P}(0))^{-1}\mathbf{P}(1)]^m\mathbf{1} \tag{1.2.4}$$

$$p_l(l_1,l_2,...,l_m) = Pr(1^{l_1}01^{l_2}\cdots 1^{l_m}0 \mid 0,\ l_1>0,\ l_2>0,...,l_m>0) = \boldsymbol{\pi}\mathbf{P}(0)\prod_{i=1}^{m}\{\mathbf{P}(1)^{l_i}\mathbf{P}(0)\}\mathbf{1} \,/\, \boldsymbol{\pi}\mathbf{P}(0)[\mathbf{P}(1)(\mathbf{I}-\mathbf{P}(1))^{-1}\mathbf{P}(0)]^m\mathbf{1} \tag{1.2.5}$$

These distributions are connected with the distributions (1.2.2) and (1.2.3):

$$p_\lambda(\lambda_1,\lambda_2,\ldots,\lambda_m) = f_\lambda(\lambda_1,\lambda_2,\ldots,\lambda_m) / c_\lambda$$
$$c_\lambda = 1-\sum_{\lambda_1=0}^{1}\sum_{\lambda_2=0}^{1}\cdots\sum_{\lambda_m=0}^{1} f_\lambda(\lambda_1,\lambda_2,\ldots,\lambda_m)+f_\lambda(1,1,\ldots,1)$$

$$p_l(l_1,l_2,\ldots,l_m) = f_l(l_1,l_2,\ldots,l_m) / c_l$$
$$c_l = 1-\sum_{l_1=0}^{1}\sum_{l_2=0}^{1}\cdots\sum_{l_m=0}^{1} f_l(l_1,l_2,\ldots,l_m)+f_l(1,1,\ldots,1)$$

In the case of a single variable these equations become

$$p_\lambda(\lambda) = f_\lambda(\lambda) / [1-f_\lambda(0)] \tag{1.2.6}$$

$$p_l(l) = f_l(l) / [1-f_l(0)] \tag{1.2.7}$$

where

$$f_\lambda(\lambda) = \boldsymbol{\pi}\mathbf{P}(1)\mathbf{P}^\lambda(0)\mathbf{P}(1)\mathbf{1} / \boldsymbol{\pi}\mathbf{P}(1)\mathbf{1} \qquad \lambda=0,1,\ldots$$

$$f_l(l) = \boldsymbol{\pi}\mathbf{P}(0)\mathbf{P}^l(1)\mathbf{P}(0)\mathbf{1} / \boldsymbol{\pi}\mathbf{P}(0)\mathbf{1} \qquad l=0,1,\ldots$$

EXAMPLE 1.2.1: Let us illustrate the distributions calculation for Gilbert's model, described in Sec. 1.1.3. The error-free run distribution (1.2.2)

$$f_\lambda(\lambda) = \boldsymbol{\pi}\mathbf{P}(1)\mathbf{P}^\lambda(0)\mathbf{P}(1)\mathbf{1} / \boldsymbol{\pi}\mathbf{P}(1)\mathbf{1} \tag{1.2.8}$$

can be calculated using spectral representation (A.5.21) of **P**(0)

$$\mathbf{P}(0) = \mathbf{T}^{-1}\begin{bmatrix} g_1 & 0 \\ 0 & g_2 \end{bmatrix}\mathbf{T} \tag{1.2.9}$$

assuming that g_1 and g_2 are *different* eigenvalues of the matrix **P**(0), which are found from the characteristic equation

$$\det(\mathbf{P}(0) - g\mathbf{I}) = \det \begin{bmatrix} Q-g & Ph \\ p & qh-g \end{bmatrix} =$$

$$g^2 - (Q+qh)g + h(Q-p) = 0$$

The roots of this equation are given by

$$g_{1,2} = b \pm \sqrt{b^2+h(p-Q)} \qquad b=(Q+hq)/2$$

The transform matrix $\mathbf{T}$ is composed from the matrix $\mathbf{P}(0)$ eigenvectors

$$\mathbf{T} = \begin{bmatrix} p & g_1-Q \\ p & g_2-Q \end{bmatrix} \quad \mathbf{T}^{-1} = [p(g_2-g_1)]^{-1} \begin{bmatrix} g_2-Q & -g_1+Q \\ -p & p \end{bmatrix} \tag{1.2.10}$$

Using the spectral representation (1.2.9) we obtain according to (A.5.23)

$$\mathbf{P}^{\lambda}(0) = \mathbf{T}^{-1} \begin{bmatrix} g_1^{\lambda} & 0 \\ 0 & g_2^{\lambda} \end{bmatrix} \mathbf{T}$$

so that equation (1.2.8) becomes

$$f_{\lambda}(\lambda) = a_1(1-g_1)g_1^{\lambda} + a_2(1-g_2)g_2^{\lambda}$$

where

$$a_1 = \frac{(1-h)(p-Q+qg_1)}{(1-g_1)(g_1-g_2)} \qquad a_2 = 1-a_1$$

Quite analogously, we find the error series distribution

$$f_l(l) = \boldsymbol{\pi}\mathbf{P}(0)\mathbf{P}^l(1)\mathbf{P}(0)\mathbf{1} \,/\, \boldsymbol{\pi}\mathbf{P}(0)\mathbf{1}$$

which, after some algebra, can be reduced to

$$f_l(l) = \begin{cases} 1-a & \text{for } l=0 \\ a(p+qh)(1-h)^{l-1}q^{l-1} & \text{for } l>0 \end{cases}$$

where $a = P(1-h)/(p+Ph)$ is the bit-error probability (see Sec. 1.1.3). Typically, this distribution is not geometric, but the distribution $p_l(l)$, which is defined by equation (1.2.7), is geometric

$$p_l(l) = (p+qh)(1-h)^{l-1}q^{l-1} \quad l=1,2,...$$

The distribution $p_\lambda(\lambda)$ is bigeometric:

$$p_\lambda(\lambda) = b_1(1-g_1)g_1^\lambda + b_2(1-g_2)g_2^\lambda$$

where $b_i = a_i g_i/(p+qh)$. Thus, the model may be characterized by the geometric distribution of the series of errors and the bigeometric distribution of the error-free runs.

■

As we mentioned in Sec. 1.1.4, it is possible to make certain transformations of the matrices $\mathbf{P}(0)$ and $\mathbf{P}(1)$ to obtain an SSM model that is equivalent to the original one. In particular, it was shown that the similarity transformations (1.1.13)

$$\mathbf{P}_1(0) = \mathbf{T}^{-1}\mathbf{P}(0)\mathbf{T} \quad \mathbf{P}_1(1) = \mathbf{T}^{-1}\mathbf{P}(1)\mathbf{T} \tag{1.2.11}$$

where the matrix $\mathbf{T}$ satisfies the conditions

$$\mathbf{T}^{-1}\mathbf{P}(0)\mathbf{T} \geq 0 \quad \mathbf{T}^{-1}\mathbf{P}(1)\mathbf{T} \geq 0 \quad \mathbf{T1} = \mathbf{1} \tag{1.2.12}$$

yield an equivalent model with the matrices $\mathbf{P}_1(e)$. Using this fact we can select the matrix $\mathbf{T}$ so as to simplify the description of the model. Let us consider several useful descriptions.

1.2.2 Diagonalization of the Matrices. Assume that the roots $g_1, g_2, ..., g_n$ of the characteristic polynomial det $(\mathbf{P}(0) - g\mathbf{I})$ are different. Then there is a matrix $\mathbf{T}$ (see Appendix 5.1.5) such that

$$\mathbf{P}_1(0) = \mathbf{T}^{-1}\mathbf{P}(0)\mathbf{T} = diag\{g_i\} \tag{1.2.13}$$

If all $g_i \geq 0$, the first of the conditions in (1.2.12) is satisfied. If the last condition in (1.2.12) does not hold, then it is possible to satisfy it by an appropriate normalization. To do so we choose the matrix $\mathbf{T}_1 = \mathbf{TB}$ with $\mathbf{B} = diag\{b_i\}$, $\mathbf{T}^{-1}\mathbf{1} = col\{b_i\}$, $b_1 \neq 0, ..., b_n \neq 0$ as a transforming matrix: $\mathbf{T}_1\mathbf{P}(0)\mathbf{T} = \mathbf{P}_1(0)$ and $\mathbf{T}_1\mathbf{1} = \mathbf{T}^{-1}\mathbf{T1} = \mathbf{1}$. If at the same time $\mathbf{P}_1(1) = \mathbf{T}_1^{-1}\mathbf{P}(1)\mathbf{T}_1 \geq 0$, then we have a model with a diagonal matrix $\mathbf{P}_1(0)$ which is equivalent to the original one. Similar transformations can be performed with the matrix $\mathbf{P}(1)$. The convenience of using a diagonal matrix in the calculations is obvious. For example, according to (1.2.4)

$$p_\lambda(\lambda) = \sum_{i=1}^{n} A_i g_i^\lambda \tag{1.2.14}$$

In this case the distribution of lengths of the intervals between the errors is polygeometric, and the parameters of the progressions are the eigenvalues of the matrix $\mathbf{P}(0)$.

EXAMPLE 1.2.2: Let us illustrate the above transformations for Gilbert's model. Using the matrix $\mathbf{T}$ of Example 1.2.1, we obtain the equivalent representation of the model

$$\mathbf{P}_1(0) = \begin{bmatrix} g_1 & 0 \\ 0 & g_2 \end{bmatrix} \qquad \mathbf{P}_1(1) = \begin{bmatrix} a_1(1-g_1) & -a_1(1-g_1) \\ -a_2(1-g_2) & a_2(1-g_2) \end{bmatrix}$$

in which the values a_1, a_2, g_1, and g_2 are defined in Example 1.2.1. The matrix $\mathbf{P}_1(1)$ has negative elements, conditions (1.2.5) are not satisfied, and therefore this matrix does not represent an SSM. In Sec. 1.3 we will generalize the SSM definition so that only conditions (1.2.4) will be important.

■

1.2.3 Models with Two Sets of States. In certain applications of the SSM model with the matrices $\mathbf{P}(0) = \mathbf{P}(\mathbf{I} - \mathbf{E})$ and $\mathbf{P}(1) = \mathbf{PE}$, it is convenient to have two sets of states: "good" states where error probability is small and "bad" states where this probability is large. It is possible to find an equivalent model in which the probabilities of errors for some states take the values $\varepsilon_g = \min \varepsilon_i$ and for other states $\varepsilon_b = \max \varepsilon_i$. To achieve it we decompose a state for which $\varepsilon_g < \varepsilon_j < \varepsilon_b$ into two substates j_1 and j_2. Let ε_g be the error probability in the first substate and ε_b in the second substate. Define the transition probabilities

$$p_{i_1 j_1} = p_{i_2 j_1} = \alpha p_{ij} \qquad p_{i_1 j_2} = p_{i_2 j_2} = (1 - \alpha) p_{ij} \tag{1.2.15}$$

We choose α in such a way that the error probability in the combined state is

$$\varepsilon_j = \alpha \varepsilon_g + (1 - \alpha)\varepsilon_b \tag{1.2.16}$$

Thus, $\alpha = (\varepsilon_b - \varepsilon_j)/(\varepsilon_b - \varepsilon_g)$.

Let the states of the initial chain be ordered in such a way that the first are those in which the conditional probabilities of errors are ε_g; the next are those that correspond to ε_j, which is neither equal to ε_g nor ε_b; and the last states are those in which the probabilities of errors are ε_b. Then the matrices $\mathbf{P}$ and $\mathbf{E}$ can be written in the block form

$$\mathbf{P} = \begin{bmatrix} \mathbf{P}_{11} & \mathbf{P}_{12} & \mathbf{P}_{13} \\ \mathbf{P}_{21} & \mathbf{P}_{22} & \mathbf{P}_{23} \\ \mathbf{P}_{31} & \mathbf{P}_{32} & \mathbf{P}_{33} \end{bmatrix} \qquad \mathbf{E} = \begin{bmatrix} \mathbf{E}_g & 0 & 0 \\ 0 & \mathbf{E}_i & 0 \\ 0 & 0 & \mathbf{E}_b \end{bmatrix} \tag{1.2.17}$$

where $\mathbf{E}_g = \varepsilon_g \mathbf{I}_g$, $\mathbf{E}_b = \varepsilon_b \mathbf{I}_b$, $\mathbf{E}_i = diag\{\varepsilon_j\}$ for $\varepsilon_g < \varepsilon_j < \varepsilon_b$.

The matrices $\mathbf{P}_1$ and $\mathbf{E}_1$ of the chain transformed according to (1.2.15) and (1.2.16) have the form

$$\mathbf{P}_1 = \begin{bmatrix} \mathbf{P}_{11} & \mathbf{P}_{12}\mathbf{E}_p & \mathbf{P}_{12}\mathbf{E}_q & \mathbf{P}_{13} \\ \mathbf{P}_{11} & \mathbf{P}_{12}\mathbf{E}_p & \mathbf{P}_{12}\mathbf{E}_q & \mathbf{P}_{13} \\ \mathbf{P}_{11} & \mathbf{P}_{12}\mathbf{E}_p & \mathbf{P}_{12}\mathbf{E}_q & \mathbf{P}_{13} \\ \mathbf{P}_{11} & \mathbf{P}_{12}\mathbf{E}_p & \mathbf{P}_{12}\mathbf{E}_q & \mathbf{P}_{13} \end{bmatrix} \qquad \mathbf{E}_1 = \begin{bmatrix} \mathbf{E}_g & 0 & 0 & 0 \\ 0 & \mathbf{E}_g^{(1)} & 0 & 0 \\ 0 & 0 & \mathbf{E}_b^{(1)} & 0 \\ 0 & 0 & 0 & \mathbf{E}_b \end{bmatrix} \tag{1.2.18}$$

where $\mathbf{E}_p = diag\{(\varepsilon_b - \varepsilon_j)/(\varepsilon_b - \varepsilon_g)\}$, $\mathbf{E}_q = diag\{(\varepsilon_j - \varepsilon_g)/(\varepsilon_b - \varepsilon_g)\}$, $\mathbf{E}_g^{(1)} = \varepsilon_g \mathbf{I}_i$, and $\mathbf{E}_b^{(1)} = \varepsilon_b \mathbf{I}_i$. Of course, all the previous arguments are heuristic. Let us give a formal proof of equivalency of the representations (1.2.17) and (1.2.18). The matrix $\mathbf{P}_1$ can be rewritten as

$$\mathbf{P}_1 = \mathbf{TP} \begin{bmatrix} \mathbf{I}_g & 0 & 0 & 0 \\ 0 & \mathbf{E}_p & \mathbf{E}_q & 0 \\ 0 & 0 & 0 & \mathbf{I}_b \end{bmatrix} \quad \text{where} \quad \mathbf{T} = \begin{bmatrix} \mathbf{I}_g & 0 & 0 \\ 0 & \mathbf{I}_i & 0 \\ 0 & \mathbf{I}_i & 0 \\ 0 & 0 & \mathbf{I}_b \end{bmatrix} \tag{1.2.19}$$

But then $\mathbf{P}_1 \mathbf{E}_1 \mathbf{T} = \mathbf{TPE}$, $\mathbf{P}_1(\mathbf{I} - \mathbf{E}_1)\mathbf{T} = \mathbf{TP}(\mathbf{I} - \mathbf{E})$, so that (1.1.13) is satisfied, thereby proving the equivalency of (1.2.17) and (1.2.18).

It is easy to verify that all the proofs are also valid in the case where ε_g is replaced by any number ε_x such that $0 \le \varepsilon_x \le \varepsilon_g$, and ε_b is replaced by ε_y such that $\varepsilon_b \le \varepsilon_y \le 1$, and the decomposition is performed only for the states in a certain subset. In particular, every SSM model is equivalent to the *canonic SSM model* with $\varepsilon_x = 0$ and $\varepsilon_y = 1$.

EXAMPLE 1.2.3: Let us find a model with $\varepsilon_x = 0$ and $\varepsilon_y = 1$ that is equivalent to Gilbert's model with the matrices

$$\mathbf{P} = \begin{bmatrix} Q & P \\ p & q \end{bmatrix} \qquad \mathbf{E} = \begin{bmatrix} 0 & 0 \\ 0 & 1-h \end{bmatrix}$$

In this case $\varepsilon_g = 0$ and there are no states with the probability of errors $\varepsilon_x = 1$ if $h \neq 0$. Therefore $\mathbf{E}_p = h$, $\mathbf{E}_q = 1-h$, and from (1.2.19) we obtain

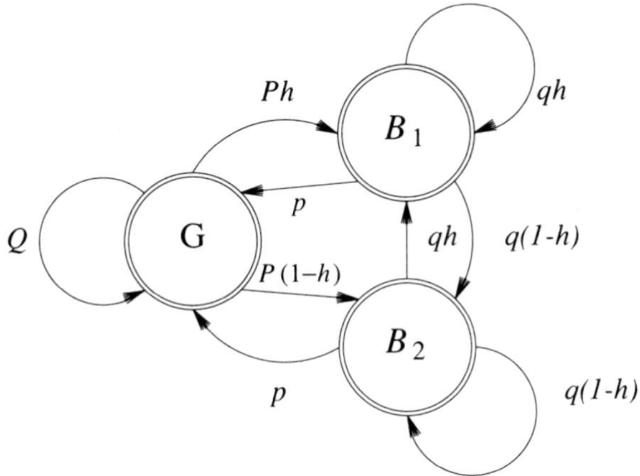

Figure 1.3. Gilbert's model expanded state diagram.

$$\mathbf{P}_1 = \begin{bmatrix} 1 & 0 & 0 \\ 0 & 1 & 0 \\ 0 & 1 & 0 \end{bmatrix} \begin{bmatrix} Q & P \\ p & q \end{bmatrix} \begin{bmatrix} 1 & 0 & 0 \\ 0 & h & 1-h \end{bmatrix} = \begin{bmatrix} Q & Ph & P(1-h) \\ p & qh & q(1-h) \\ p & qh & q(1-h) \end{bmatrix}$$

The matrix $\mathbf{E}_1$ is found from (1.2.18) and takes the form

$$\mathbf{E}_1 = \begin{bmatrix} 0 & 0 & 0 \\ 0 & 0 & 0 \\ 0 & 0 & 1 \end{bmatrix}$$

Therefore, the two-state model in which errors occur only in the second state, with the probability $1-h$, is equivalent to the three-state model in which errors never occur in the first two states and always happen in the third state. Flow graphs of the model are depicted in Figs. 1.1 and 1.3.

■

1.2.4 Block Matrix Representation. In the SSM model applications it is convenient to represent the matrices in the block form because this form simplifies some calculations.

Consider a Markov chain with the matrix $\mathbf{P}$, and partition the set of its states into two subsets, A_0 and A_1. For simplicity we attribute all the states with the initial numbers of $i=1,2,...,R$ to the subset A_0, and the remaining states with the numbers of $i=R+1,...,n$ we attribute to the subset A_1. In correspondence with this partition, we rewrite the matrix $\mathbf{P}$ in the form of blocks

$$\mathbf{P} = \begin{bmatrix} \mathbf{P}_{00} & \mathbf{P}_{01} \\ \mathbf{P}_{10} & \mathbf{P}_{11} \end{bmatrix} \tag{1.2.20}$$

We assume that when the Markov chain is in the states of the set A_1 an error occurs, and that no error occurs in the states of the set A_0. It is clear that this model is a particular case of the SSM when

$$\mathbf{P}(0) = \mathbf{P}(\mathbf{I} - \mathbf{E}) = \begin{bmatrix} \mathbf{P}_{00} & 0 \\ \mathbf{P}_{10} & 0 \end{bmatrix} \quad \mathbf{P}(1) = \mathbf{PE} = \begin{bmatrix} 0 & \mathbf{P}_{01} \\ 0 & \mathbf{P}_{11} \end{bmatrix} \tag{1.2.21}$$

where

$$\mathbf{E} = \begin{bmatrix} 0 & 0 \\ 0 & \mathbf{I} \end{bmatrix}$$

The advantage of representation (1.2.20) is that the occurrence of errors is deterministically connected to the state of the Markov chain: $e_i = 0$ if the chain state $s_i \le R$ and $e_i = 0$ otherwise. This representation is called a *Markov function.*

In calculations involving block matrices the formulas are not usually as compact as in calculations using the matrices $\mathbf{P}(e)$, but the orders of the matrices are smaller which leads to faster computations. Probability distribution (1.2.1) takes the form

$$Pr(e_1,e_2,...,e_m) = \boldsymbol{\pi}_{e_1}\mathbf{P}_{e_1e_2}\cdots\mathbf{P}_{e_{m-1}e_m}\mathbf{1} = \boldsymbol{\pi}_{e_1}\prod_{i=1}^{m-1}\mathbf{P}_{e_ie_{i+1}}\mathbf{1} \tag{1.2.22}$$

Distribution (1.2.2) of lengths of intervals between the errors has the form

$$\begin{gathered} f_\lambda(\lambda_1,...,\lambda_m) = Pr(10^{\lambda_1}10^{\lambda_2}\cdots 10^{\lambda_m}1 \mid 1) = \\ \boldsymbol{\pi}_1\prod_{i=1}^{m}\{\mathbf{P}_{10}\mathbf{P}_{00}^{\lambda_i-1}\mathbf{P}_{01}\}\mathbf{1} / \boldsymbol{\pi}_1\mathbf{1} \end{gathered} \tag{1.2.23}$$

and distribution (1.2.3) of the lengths of intervals between correct symbols can be written as

$$\begin{gathered} f_l(l_1,...,l_m) = Pr(01^{l_1}01^{l_2}\cdots 01^{l_m}0 \mid 0) = \\ \boldsymbol{\pi}_0\prod_{i=1}^{m}\{\mathbf{P}_{01}\mathbf{P}_{11}^{l_i-1}\mathbf{P}_{10}\}\mathbf{1} / \boldsymbol{\pi}_0\mathbf{1} \end{gathered} \tag{1.2.24}$$

Symbols $\boldsymbol{\pi}_e$ in formulas (1.2.22) and (1.2.23) denote the blocks of the matrix

$\boldsymbol{\pi} = [\boldsymbol{\pi}_0 \ \ \boldsymbol{\pi}_1]$. The matrices $\boldsymbol{\pi}_e$ satisfy the equations

$$\boldsymbol{\pi}_0\mathbf{P}_{00}+\boldsymbol{\pi}_1\mathbf{P}_{10}=\boldsymbol{\pi}_0 \quad \boldsymbol{\pi}_0\mathbf{P}_{01}+\boldsymbol{\pi}_1\mathbf{P}_{11}=\boldsymbol{\pi}_1 \quad \boldsymbol{\pi}_0\mathbf{1}+\boldsymbol{\pi}_1\mathbf{1}=1 \tag{1.2.25}$$

The transformation of similarity (1.2.4) with the matrix

$$\mathbf{T} = block\ diag\{\mathbf{T}_e\} = \begin{bmatrix} \mathbf{T}_0 & 0 \\ 0 & \mathbf{T}_1 \end{bmatrix} \tag{1.2.26}$$

leads to

$$\mathbf{R}_{ij} = \mathbf{T}_i^{-1}\mathbf{P}_{ij}\mathbf{T}_j \qquad (i,j = 0,1) \tag{1.2.27}$$

If the conditions

$$\mathbf{T}^{-1}\mathbf{PT} \geq 0 \qquad \mathbf{T1} = \mathbf{1} \tag{1.2.28}$$

hold, then the model with the matrix

$$\mathbf{P}_1 = \begin{bmatrix} \mathbf{R}_{00} & \mathbf{R}_{01} \\ \mathbf{R}_{10} & \mathbf{R}_{11} \end{bmatrix}$$

is equivalent to the original model with the matrix **P**.

If all the eigenvalues of the matrices $\mathbf{P}_{00}$ and $\mathbf{P}_{11}$ are positive and different, then there exists a matrix **T** of the form (1.2.26), which diagonalizes the matrices $\mathbf{P}_{00}$ and $\mathbf{P}_{11}$. If conditions (1.2.28) hold as well, then the matrix **P** can be transformed to the form

$$\mathbf{P}_1 = \begin{bmatrix} diag\{\gamma_i\} & \mathbf{R}_{01} \\ \mathbf{R}_{01} & diag\{g_i\} \end{bmatrix} \tag{1.2.29}$$

We showed above that the transformation

$$\mathbf{TP}_1 = \mathbf{PT} \qquad \mathbf{T1} = \mathbf{1} \tag{1.2.30}$$

with the rectangular matrix **T** as in (1.2.26) is more general. In this case the blocks of the matrices $\mathbf{P}_1$ and **P** are related by

$$\mathbf{T}_{i0}\mathbf{R}_{0j}+\mathbf{T}_{i1}\mathbf{R}_{1j} = \mathbf{P}_{i0}\mathbf{T}_{0j}+\mathbf{P}_{i1}\mathbf{T}_{1j} \qquad (i,j = 0,1) \tag{1.2.31}$$

This type of transformation may not only simplify the matrix structure but also may decrease the matrix size.

EXAMPLE 1.2.4: In Example 1.2.3 we showed that Gilbert's model can be described by the Markov function with the matrix

$$\begin{bmatrix} Q & Ph & P(1-h) \\ p & qh & q(1-h) \\ p & qh & q(1-h) \end{bmatrix}$$

Let us apply the transformation (1.2.27) to simplify the model description.

Choosing

$$\mathbf{T}_0 = [p(g_2-g_1)]^{-1} \begin{bmatrix} g_2-Q & -g_1+Q \\ -p & p \end{bmatrix} \qquad \mathbf{T}_1 = 1$$

(see Example 1.2.1), we obtain

$$\mathbf{M}_1 = \begin{bmatrix} g_1 & 0 & a_1(1-g_1)(g_2-g_1) \\ 0 & g_2 & a_2(1-g_2)(g_2-g_1) \\ d_1 & d_2 & q(1-h) \end{bmatrix}$$

where $d_1 = (g_2-Q-qh)/(g_2-g_1)$, $d_2 = (g_1-Q-qh)/(g_1-g_2)$, a_1, a_2, g_1, and g_2 are defined in Example 1.2.1. This matrix is not stochastic and cannot be used to describe the Markov chain; however, the additional transformation

$$\mathbf{P}_1 = \mathbf{T}_2^{-1}\mathbf{M}_1\mathbf{T}_2$$

with the matrix

$$\mathbf{T}_2 = \begin{bmatrix} a_1(g_2-g_1) & 0 & 0 \\ 0 & a_2(g_2-g_1) & 0 \\ 0 & 0 & 1 \end{bmatrix}$$

yields the desired result

$$\mathbf{P}_1 = \begin{bmatrix} g_1 & 0 & 1-g_1 \\ 0 & g_2 & 1-g_2 \\ b_1 & b_2 & p_{33} \end{bmatrix}$$

where $p_{33} = q(1-h)$. Several modifications of Gilbert's model use this matrix structure.[5,11,12] We will consider these models in Sec. 1.5.1.

One can restore the original Gilbert model parameters using the relations of Example 1.2.1.

$$h = \frac{g_1 g_2}{g_1+g_2+p_3-1} \qquad q = \frac{p_3}{1-h} \qquad Q = g_1+g_2-qh$$ ■

1.3 ERROR SOURCE DESCRIPTION BY MATRIX PROCESSES

1.3.1 Matrix Process Definition. The requirement of non-negativity of the elements of matrices describing a Markov function limits the class of transformations that simplify these matrices. It is obvious, however, that if the matrices $\mathbf{P}(e)$ undergo an arbitrary transformation of similarity $\mathbf{T}^{-1}\mathbf{P}(e)\mathbf{T}$ with a matrix $\mathbf{T}$ whose elements are complex numbers (see Example 1.2.2), then the probability $Pr(e_1,e_2,...,e_n)$ of a sequence of errors can be found by a formula which is quite analogous to (1.2.1):

$$Pr(e_1,e_2,...,e_n) = \mathbf{y}\prod_{i=1}^{n}\mathbf{M}(e_i)\mathbf{z}$$

where

$$\mathbf{M}(e_i) = \mathbf{T}^{-1}\mathbf{P}(e_i)\mathbf{T} \quad \mathbf{y} = \boldsymbol{\pi}\mathbf{T} \quad \mathbf{z} = \mathbf{T}^{-1}\mathbf{1}$$

which can easily be verified:

$$Pr(e_1,e_2,...,e_n) = \boldsymbol{\pi}\prod_{i=1}^{n}\mathbf{P}(e_i)\mathbf{1} =$$

$$\mathbf{y}\mathbf{T}^{-1}\prod_{i=1}^{n}\mathbf{P}(e_i)\mathbf{1} = \mathbf{y}\prod_{i=1}^{n}\mathbf{M}(e_i)\mathbf{T}^{-1}\mathbf{1} = \mathbf{y}\prod_{i=1}^{n}\mathbf{M}(e_i)\mathbf{z}$$

The matrices $\mathbf{M}(e)$, $\mathbf{y}$, and $\mathbf{z}$ can have arbitrary complex elements.

Processes with distributions defined by previous relations represent a particular case of matrix or pseudo-Markov processes. According to,[28] we say that a process $Y = \{y_t : t \in T, y_t \in \Omega\}$ is called a matrix process if there exist matrices

a, **R**(t), $\mathbf{B}_\alpha$, **b** with complex elements such that

$$Pr(y_{t_1} = \alpha_1, y_{t_1+t_2} = \alpha_2, ..., y_{t_1+t_2+...+t_m} = \alpha_m) = \mathbf{a}\prod_{i=1}^{m}\mathbf{R}(t_i)\mathbf{B}_{\alpha_i}\mathbf{b} \qquad (1.3.1)$$

for every $\alpha \in \Omega$. Here **a** is a matrix row and **b** is a matrix column. In the following discussion we consider only matrix processes of a particular form where $y_j = e_j$ is binary, and the square matrix $\mathbf{R}(t) = \mathbf{R}$ does not depend on t and has a finite order n. Evidently, the SSM is a particular case of the matrix process model with $\mathbf{a} = \boldsymbol{\pi}, \mathbf{R} = \mathbf{P}$, $\mathbf{B}_0 = \mathbf{I} - \mathbf{E}$, $\mathbf{B}_1 = \mathbf{E}$, $\mathbf{b} = \mathbf{1}$. However, not every matrix process can be represented as a function of a Markov chain with a finite number of states. [28—36]

Matrix processes have many properties that make them similar to Markov functions. Indeed, denoting $\mathbf{M}(e) = \mathbf{R}\mathbf{B}_e$, we obtain by (1.3.1)

$$Pr(e_1, e_2, ..., e_m) = \mathbf{a}\prod_{i=1}^{m}\mathbf{M}(e_i)\mathbf{b} \qquad (1.3.2)$$

which coincides formally with (1.2.1). Although $\mathbf{M} = \mathbf{M}(0) + \mathbf{M}(1)$ may not be a stochastic matrix, it possesses certain properties of stochastic matrices. Like them, it has an eigenvalue $\lambda = 1$. Indeed, it follows from the relation

$$\sum_{e_1,...,e_m} Pr(e_1, ..., e_m) = \mathbf{a}\mathbf{M}^m\mathbf{b} = 1$$

that

$$\sum_{m=0}^{\infty}\mathbf{a}\mathbf{M}^m\mathbf{b} = \mathbf{a}(\mathbf{I} - \mathbf{M}z)^{-1}\mathbf{b} = 1/(1-z)$$

The inverse matrix $(\mathbf{I} - \mathbf{M}z)^{-1}$ can be expressed using formula (A.5.2) as as

$$(\mathbf{I} - \mathbf{M}z)^{-1} = \mathbf{B}(z)/\Delta(z)$$

where $\Delta(z) = \det(\mathbf{I} - \mathbf{M}z)$ is the *characteristic polynomial* and $\mathbf{B}(z)$ is the matrix polynomial called the *adjoint matrix*. From the previous equation we obtain

$$\mathbf{a}\mathbf{B}(z)\mathbf{b}/\Delta(z) = 1/(1-z)$$

or $\Delta(z) = \mathbf{a}\mathbf{B}(z)\mathbf{b}(1-z)$. But then $\Delta(1) = 0$, and therefore the matrix **M** has an eigenvalue $\lambda = 1$.

The collection { **a**, **M**(0), **M**(1), **b** } is called the matrix process representation. In the regular case, when the matrix **M** has a simple eigenvalue $\lambda = 1$ and the absolute values of the rest of the eigenvalues are smaller than 1 one can prove (similarly to Theorem A.6.1) that

$$\lim_{m\to\infty} \mathbf{M}^m = \mathbf{zy}$$

exists. Vectors **z** and **y** are the right and left eigenvectors of the matrix **M**, corresponding to the eigenvalue $\lambda = 1$

$$\mathbf{Mz} = \mathbf{z} \qquad \mathbf{yM} = \mathbf{y} \tag{1.3.3}$$

and satisfying the condition

$$\mathbf{yz} = 1 \tag{1.3.4}$$

Passing to the limit when $s\to\infty$, $j\to\infty$ in the equation

$$Pr(\Omega^s, e_1, e_2, \ldots, e_m, \Omega^j) = \mathbf{aM}^s \prod_{i=1}^{m} \mathbf{M}(e_i)\mathbf{M}^j\mathbf{b}$$

we obtain

$$Pr(e_1, e_2, \ldots, e_m) = \mathbf{y}\prod_{i=1}^{m} \mathbf{M}(e_i)\mathbf{z} \tag{1.3.5}$$

so that the initial and final matrices of a stationary matrix process satisfy conditions (1.3.3) and (1.3.4).

Under our assumptions the value (1.3.5) does not depend on the choice of **y** and **z** satisfying conditions (1.3.3) and (1.3.4), since the matrices **y** and **z** are determined uniquely up to numerical factors by the conditions (1.3.3): $\mathbf{y} = \mu\mathbf{y}_0$, $\mathbf{z} = \nu\mathbf{z}_0$. The condition (1.3.4) implies that $\mathbf{yz} = \mu\nu\mathbf{y}_0\mathbf{z}_0 = 1$ and $\mu\nu = 1$ if $\mathbf{y}_0\mathbf{z}_0 = 1$. Substituting $\mathbf{y} = \mu\mathbf{y}_0$, $\mathbf{z} = \nu\mathbf{z}_0$ in (1.3.5), we observe that the value of the right-hand side of (1.3.5) does not change when we replace **y** by $\mathbf{y}_0$ and **z** by $\mathbf{z}_0$.

Obviously, the distribution of the lengths of intervals between errors and between correct symbols in the matrix process model can be found by formulas (1.2.2) and (1.2.3), in which we replace the matrices $\boldsymbol{\pi}$, $\mathbf{P}(e)$, and **1** by **y**, $\mathbf{M}(e)$, and **z** respectively.

Thus, a matrix process is a natural generalization of a Markov function. By recognizing that the matrices $\mathbf{M}(e)$ that define a matrix process may have complex elements, we find more ways to transform these matrices while replacing a matrix process by its equivalent than we do in the case of a Markov function. Repeating the transformations of Sec. 1.1.4, we can prove that if there exists a matrix **T** such that

$$\mathbf{T}\mathbf{M}_1(e) = \mathbf{M}(e)\mathbf{T} \quad e=0,1 \tag{1.3.6}$$

then the matrix processes $\{\mathbf{yT},\mathbf{M}_1(0),\mathbf{M}_1(1),\mathbf{z}_1\}$ and $\{\mathbf{y},\mathbf{M}(0),\mathbf{M}(1),\mathbf{Tz}_1\}$ are equivalent.

Obviously, if a matrix **T** is square and nonsingular, then (1.3.6) is a transformation of the similarity. Therefore it is possible to reduce any matrix $\mathbf{M}(e)$ to its Jordan normal form (A.5.22). In particular, if the matrix $\mathbf{M}(e)$ has a simple structure, then it can be reduced to a diagonal form (A.5.21). It is conceivable that in certain transformations of types (1.3.6), we obtain non-negative matrices $\mathbf{M}_1(0) \geq 0$ and $\mathbf{M}_1(1) \geq 0$, such that the matrix $\mathbf{M}_1 = \mathbf{M}_1(0) + \mathbf{M}_1(1)$ is stochastic and $\mathbf{z}_1 = \mathbf{1}$. In this case the matrix process is equivalent to a Markov function.

Thus, the matrices representing a matrix process can be reduced to a simpler form than the matrices representing an SSM model, since the restrictions of type (1.2.12) are absent. It is more convenient to use them in processing experimental data. However, matrix processes are not convenient for probabilistic interpretations of the process and for error source modeling. We use the matrix processes to represent the SSM model in analytic computations transforming the matrices $\mathbf{P}(e)$ by formulas (1.2.11) to their simplest forms. In such transformations we use, of course, formulas of type (1.3.5) and not formulas (1.2.1).

1.3.2 Block Matrix Representation. Let us consider a matrix process with matrices

$$\mathbf{M}(0) = \begin{bmatrix} \mathbf{M}_{00} & 0 \\ \mathbf{M}_{10} & 0 \end{bmatrix} \qquad \mathbf{M}(1) = \begin{bmatrix} 0 & \mathbf{M}_{01} \\ 0 & \mathbf{M}_{11} \end{bmatrix} \tag{1.3.7}$$

$$\mathbf{y} = [\ \mathbf{y}_0 \ \ \mathbf{y}_1\ ] \qquad \mathbf{z} = \begin{bmatrix} \mathbf{z}_0 \\ \mathbf{z}_1 \end{bmatrix}$$

We can reduce the orders of the matrices involved in calculations by expressing the distributions with the help of the matrix blocks. The distribution (1.3.5) takes the form

$$Pr(e_1,e_2,\ldots,e_m) = \mathbf{y}_{e_1} \prod_{i=1}^{m-1} \mathbf{M}_{e_i e_{i+1}} \mathbf{z}_{e_m} \tag{1.3.8}$$

and the matrices $\mathbf{y}_e$ and $\mathbf{z}_e$ in the stationary case satisfy the equations

$$\sum_{i=0}^{1} \mathbf{y}_i \mathbf{M}_{ij} = \mathbf{y}_j \quad \sum_{j=0}^{1} \mathbf{M}_{ij} \mathbf{z}_j = \mathbf{z}_i \quad \sum_{i=1}^{1} \mathbf{y}_i \mathbf{z}_i = 1 \quad i,j=0,1 \tag{1.3.9}$$

Notice that because of (1.3.7) the probabilities of error and lack of error are given by

$$p_{\text{err}} = Pr(1) = \mathbf{y}_1 \mathbf{z}_1 \quad p_{\text{cor}} = Pr(0) = \mathbf{y}_0 \mathbf{z}_0 \tag{1.3.10}$$

The multidimensional distributions of the intervals between the errors and correct symbols have the form

$$p_0(\lambda_1, \lambda_2, ..., \lambda_m) = \mathbf{y}_1 \prod_{i=1}^{m} \{\mathbf{M}_{10} \mathbf{M}_{00}^{\lambda_i - 1} \mathbf{M}_{01}\} \mathbf{z}_1 / \mathbf{y}_1 \mathbf{M}_{10} \mathbf{z}_0 \tag{1.3.11}$$

$$p_1(l_1, l_2, ..., l_m) = \mathbf{y}_0 \prod_{i=1}^{m} \{\mathbf{M}_{01} \mathbf{M}_{11}^{l_i - 1} \mathbf{M}_{10}\} \mathbf{z}_0 / \mathbf{y}_0 \mathbf{M}_{01} \mathbf{z}_1 \tag{1.3.12}$$

We can easily show that an arbitrary matrix process is equivalent to some matrix process, the matrices of which have the form (1.2.21). Indeed, it is obvious that the process with the matrices

$$\mathbf{M}_1(0) = \begin{bmatrix} \mathbf{M}(0) & 0 \\ \mathbf{M}(0) & 0 \end{bmatrix} \quad \mathbf{M}_1(1) = \begin{bmatrix} 0 & \mathbf{M}(1) \\ 0 & \mathbf{M}(1) \end{bmatrix}$$

is equivalent to the process with the matrices $\mathbf{M}(e)$. The formal proof follows immediately from theorem 1.3.1 because of the relation $\mathbf{TM}_1(e) = \mathbf{M}(e)\mathbf{T}$, where $\mathbf{T} = [\ \mathbf{I}\ \ \mathbf{0}\ ]$.

1.3.3 Matrix Processes and Difference Equations. The probabilities of various combinations of symbols governed by a matrix process can be expressed by difference equations. For the sake of simplifying the discussion, we consider the binary matrix process $\{e_t\}$, (e_t=0,1). For a matrix process the probability of a series $e,e,...,e = e^k$ can be found from equation (1.3.8):

$$p_e(k) = Pr(e^k) = \mathbf{y}_e \mathbf{M}_{ee}^{k-1} \mathbf{z}_e \tag{1.3.13}$$

Let

$$\psi_e(\lambda) = \lambda^{\nu_e} - \sum_{i=1}^{\nu_e} a_{ei} \lambda^{\nu_e - i} \tag{1.3.14}$$

be any annulling polynomial of the matrix $\mathbf{M}_{ee}$, i.e.,

$$\psi_e(\mathbf{M}_{ee}) = \mathbf{M}_{ee}^{\nu_e} - \sum_{i=1}^{\nu_e} a_{ei}\mathbf{M}_{ee}^{\nu_e - i} = 0 \qquad (1.3.15)$$

One such polynomial, according to the Cayley-Hamilton theorem (A.5.29) is the characteristic polynomial $\Delta(\lambda) = \det(\lambda\mathbf{I} - \mathbf{M}_{ee})$ of the matrix $\mathbf{M}_{ee}$.

After multiplying equation (1.3.15) by $\mathbf{M}_{ee}^{k-1-\nu_e}$ we obtain

$$\mathbf{M}_{ee}^{k-1} = \sum_{i=1}^{\nu_e} a_{ei}\mathbf{M}_{ee}^{k-i-1}$$

Multiplying this equation from the left by $\mathbf{y}_e$ and on the right by $\mathbf{z}_e$, we obtain, because of (1.3.13),

$$p_e(k) = \sum_{i=1}^{\nu_e} a_{ei}p_e(k-i) \qquad (1.3.16)$$

Thus, the distribution $p_e(k)$ satisfies a linear homogeneous difference equation with constant coefficients. The coefficients of this equation are not uniquely defined. However, the coefficients of the equation with the minimal ν_e are uniquely determined. If assume the opposite that there is an equation

$$p_e(k) = \sum_{i=1}^{\nu_e} b_{ei}p_e(k-i)$$

with at least one coefficient b_i different from a_i, then, subtracting this equation from (1.3.16), we obtain the equation whose order is smaller than ν_e. This contradicts the ν_e minimality and, therefore, our assumption was incorrect. Thus, the coefficients of the equation with the minimal ν_e are uniquely defined. Obviously, the minimal ν_e does not exceed the order of the matrix $\mathbf{M}_{ee}$.

If the order of the matrix $\mathbf{M}_{ee}$ is greater than the minimal ν_e, it is possible to find an equation of the form of (1.3.13) with matrices of the minimal order ν_e. Indeed, let us consider a vector

$$\mathbf{p}_e(k) = [\, p_e(k) \;\; p_e(k+1) \ldots \; p_e(k+\nu_e-1) \,]$$

Its coordinates satisfy equation (1.3.16) which can be expressed in the following matrix form

$$\mathbf{p}_e(k) = \mathbf{p}_e(k-1)\mathbf{A}_{ee}$$

where

$$\mathbf{A}_{ee} = \begin{bmatrix} 0 & 0 & \ldots & 0 & a_{e\nu_e} \\ 1 & 0 & \ldots & 0 & a_{e\nu_e-1} \\ \cdot & \cdot & \cdot & \cdot & \cdot \\ 0 & 0 & \ldots & 1 & a_{e1} \end{bmatrix}$$

This special structure of a matrix is called the matrix *echelon form*. Repeating the above recursion, we obtain

$$\mathbf{p}_e(k) = \mathbf{p}_e(k-2)\mathbf{A}_{ee}^2 = \cdots = \mathbf{p}_e(1)\mathbf{A}_{ee}^{k-1} = \mathbf{y}_e^{(1)}\mathbf{A}_{ee}^{k-1}$$

where

$$\mathbf{y}_e^{(1)} = [\, p_e(1) \;\; p_e(2) \ldots \; p_e(\nu_e) \,]$$

Multiplying of both sides of this equation from the right by the matrix column $\mathbf{z}_e^{(1)} = [\ 1 \ \ 0 \ \ldots \ 0\ ]'$, we obtain

$$p_e(k) = \mathbf{y}_e^{(1)}\mathbf{A}_{ee}^{k-1}\mathbf{z}_e^{(1)}$$

This equation is analogous to (1.3.13) with the matrix $\mathbf{M}_{ee}$ of the minimal order.

EXAMPLE 1.3.1: Let us find the difference equation for the error-free run distribution according to Gilbert's model.

As we saw in Example 1.2.1, the error-free run distribution is presented by equation (1.2.8) which has the form of (1.3.13) with $\mathbf{M}_{00} = \mathbf{P}(0)$. The characteristic polynomial of $\mathbf{P}(0)$ equals

$$\Delta(g) = g^2 - (Q+qh)g + h(Q-p)$$

Thus the difference equation is given by

$$f_\lambda(k) = (Q+qh)f_\lambda(k-1) + h(p-Q)f_\lambda(k-2)$$

and its matrix has the following form

$$\mathbf{A}_{00} = \begin{bmatrix} 0 & h(p-Q) \\ 1 & Q+qh \end{bmatrix} \qquad \blacksquare$$

Similarly, we can show that the two-dimensional distributions $p(0^k 1^m)$, $p(1^k 0^m)$ satisfy two-dimensional difference equations. Indeed, it follows from (1.3.15) that

$$\begin{aligned} p(0^k 1^m) &= \sum_i a_{0i} p(0^{k-i} 1^m) \\ p(0^k 1^m) &= \sum_i a_{1i} p(0^k 0^{m-i}) \end{aligned} \qquad (1.3.17)$$

To represent these equations in the matrix form we introduce the matrix

$$\mathbf{P}_0(k,m) = \left[p(0^{k+i-1} 1^{m+j-1}) \right]_{v_0 v_1}$$

which, according to (1.3.17), can be expressed as

$$\mathbf{P}_0(k,m) = \mathbf{A'}_{00} \mathbf{P}_0(k-1,m) \qquad \mathbf{P}_0(k,m) = \mathbf{P}_0(k,m-1)\mathbf{A}_{11}$$

and therefore

$$\mathbf{P}_0(k,m) = (\mathbf{A'}_{00})^{m-1} \mathbf{P}_0(1,1) \mathbf{A}_{11}^{k-1}$$

where $\mathbf{A'}$ is a transposed matrix $\mathbf{A}$.

Multiplying the matrix $\mathbf{P}_0(k,m)$ from the left by $\mathbf{y}_0^{(2)} = [\ 1\ 0\ 0 \ldots 0\ ]$ and from the right by $\mathbf{z}_1^{(1)}$ we obtain

$$p(0^k 1^m) = \mathbf{y}_0^{(2)} (\mathbf{A'}_{00})^{k-1} \mathbf{P}_0(1,1) \mathbf{A}_{11}^{m-1} \mathbf{z}_1^{(1)}$$

The probability $p(1^k 0^m)$ can be expressed similarly:

$$p(1^k 0^m) = \mathbf{y}_1^{(2)} (\mathbf{A'}_{11})^{k-1} \mathbf{P}_1(1,1) \mathbf{A}_{00}^{m-1} \mathbf{z}_0^{(1)}$$

with

$$\mathbf{P}_1(k,m) = \left[p(1^{k+i-1} 0^{m+j-1}) \right]_{v_1 v_0}$$

Since matrices $\mathbf{A'}_{ee}$ and $\mathbf{A}_{ee}$ are similar (i.e., $\mathbf{A'}_{ee} = \mathbf{T}_e^{-1} \mathbf{A}_{ee} \mathbf{T}_e$), the previous relations can be rewritten in the form

$$p(0^k 1^m) = \mathbf{y}_0^{(3)} \mathbf{A}_{00}^{k-1} \mathbf{A}_{01} \mathbf{A}_{11}^{m-1} \mathbf{z}_1^{(1)}$$
$$p(1^k 0^m) = \mathbf{y}_1^{(3)} \mathbf{A}_{11}^{k-1} \mathbf{A}_{10} \mathbf{A}_{00}^{m-1} \mathbf{z}_0^{(1)}$$

where

$$\mathbf{y}_e^{(3)} = \mathbf{y}_e^{(2)} \mathbf{T}_e^{-1} \qquad \mathbf{A}_{e(1-e)} = \mathbf{T}_e \mathbf{P}(0,0) \mathbf{A}_{(1-e)(1-e)}$$

The equations developed in this section establish connections between parameters of a matrix process and the coefficients of the corresponding difference equations. It is convenient to use these representations when estimating parameters of matrix processes.

1.3.4 Matrix Processes and Markov Functions. As mentioned above, the matrix processes are a direct generalization of Markov functions. The restrictions imposed on the matrices describing a Markov function complicate the solution of the problem of the equivalence of two Markov functions and the problem of simplifying the model description. However, since Markov functions are matrix processes, the conditions of their equivalence can be formulated in terms of matrix processes. For two binary SSM models with matrices $\mathbf{P}(e)$ and $\mathbf{P}_1(e)$ to be equivalent it is necessary and sufficient that they be equivalent as matrix processes. In other words, it is necessary and sufficient that the models' ranks, canonic basis sequences, and their probabilities $p(s_i e t_j)$ coincide (see Appendix 1).

Any matrix process can be replaced by an equivalent process with the matrices $\mathbf{A}(e)$ of the minimal order, equal to the rank of the process (see Appendix 1). The similar replacement is impossible in the case of Markov functions.[33] Not every matrix process is equivalent to a function of a Markov chain with a finite number of states,[28,34] but every matrix process is equivalent to a function of a Markov chain with a countable number of states.[28]

The constraints that separate the class of Markov functions from the class of the matrix processes complicate the minimizing of the sizes of the matrices that define the Markov functions. The question of minimization of the number of states is closely related to Markov chain state lumpability (see Appendix 1). However, if an SSM model is used for analytical computations, then minimization in the class of the matrix processes is possible, since the computation formulas have the same form for matrix processes as for Markov functions. If one uses the SSM model for the error source simulation, then it is convenient to use a Markov function description, thus achieving a simple probabilistic interpretation.

Occasionally, especially in the processing of experimental data, it is important to know whether there is a possibility of reducing the number of states in the description of the SSM model. A simple sufficient condition for the minimality of the number of states of the Markov chain can be obtained by analyzing the distributions of the lengths of the intervals between errors and the lengths of the intervals between correct symbols.

If the distributions of the length λ of the intervals between the errors and the lengths l of the intervals between correct symbols of a matrix process are polygeometric

$$p_0(\lambda) = \sum_{i=1}^{m_0} a_{0i} q_{0i}^{\lambda-1} \tag{1.3.18}$$

$$p_1(l) = \sum_{i=1}^{m_1} a_{1i} q_{1i}^{l-1} \tag{1.3.19}$$

$(\,|q_{e1}| < |q_{e2}| < \cdots < |q_{em_e}|\,)$, and the matrices $\mathbf{P}_{ee}$ have the orders m_e, then the number of the model states $m_0 + m_1$ is minimal (see Appendix 1). We will often use this simple argument.

1.4 ERROR SOURCE DESCRIPTION BY SEMI-MARKOV PROCESSES

1.4.1 Semi-Markov Processes. Consider now a generalization of the error source model based on semi-Markov processes: The source of errors has a finite number of states; transitions of the process from state to state are governed by a *semi-Markov* [37] or *Markov renewal* [38] process, which may be described as follows.

Suppose that the process was initially in the state c_0. It enters then the state c_1 with the probability $f^{(0)}_{c_0 c_1}(l_0)$ after l_0 steps. From the state c_1 it passes into the state c_2 with the probability $f_{c_1 c_2}(l_1)$ after l_1 steps, and so on.

Consider now a set of h binary channels and assume that errors in these channels are generated by the common source, which may be in one of k states forming a semi-Markov process. If the error source transfers from state i to state j, an error $\mathbf{e}=(e_1, e_2, \ldots, e_h)$ appears with probability $\varepsilon_{ij}(\mathbf{e})$, where $e_k=1$ if k–th channel bit is in error, and $e_k=0$ otherwise.

This model is obviously a generalization of the model based on Markov processes. To describe the model we need a matrix of initial state probabilities $\mathbf{p}_0 = [\, p_i \,]_{1,k}$, matrix of initial transition probabilities $\mathbf{F}_0(l) = [\, f^{(0)}_{ij}(l) \,]_{k,k}$, matrix of transition probabilities $\mathbf{F}(l) = [\, f_{ij}(l) \,]_{k,k}$, and matrices of the state-error probabilities $\mathbf{E}(\mathbf{e}) = [\, \varepsilon_{ij}(\mathbf{e}) \,]_{k,k}$. We showed above that the Markov model of the error source is general enough to describe a wide variety of error statistics. However, the size of matrices used for defining the model may be quite large. The semi-Markov approach often allows a significant decrease in the number of the model parameters.

An alternative description of a semi-Markov process is based on imbedded Markov chains. Let us consider the semi-Markov process only at the moments when it changes its states. The new process is the Markov chain with the transition probabilities

$$p^{(0)}_{ij} = \sum_{l=1}^{\infty} f^{(0)}_{ij}(l)$$

for the first transition from the initial state and

$$p_{ij} = \sum_{l=1}^{\infty} f_{ij}(l)$$

for subsequent transitions. The intervals between the two consecutive transitions have the distributions

$$w_{ij}^{(0)}(l) = f_{ij}^{(0)}(l) \,/\, p_{ij}^{(0)} \qquad w_{ij}(l) = f_{ij}(l)/p_{ij} \tag{1.4.1}$$

which are called the *state-holding time* probabilities.[37] The successive state transitions are governed by the Markov chain with transition matrices $\mathbf{P}^{(0)} = [\, p_{ij}^{(0)} \,]_{k,k}$, $\mathbf{P} = [\, p_{ij} \,]_{k,k}$. After a transition $i \to j$ has been selected, the process still holds the previous state i for l steps, according to transition distributions (1.4.1).

Thus, the proposed model can be described by the matrix of the state initial probabilities $\mathbf{p}_0 = [\, p_i \,]_{1,k}$, matrices of transition probabilities $\mathbf{P}^{(0)} = [\, p_{ij}^{(0)} \,]_{k,k}$, $\mathbf{P} = [\, p_{ij} \,]_{k,k}$, matrices of the interval distributions $\mathbf{W}^{(0)}(l) = [\, w_{ij}^{(0)}(l) \,]_{k,k}$ and $\mathbf{W}(l) = [\, w_{ij}(l) \,]_{k,k}$, and matrices of the probabilities of errors $\mathbf{E}(\mathbf{e}) = [\, \varepsilon_{ij}(\mathbf{e}) \,]_{k,k}$. We call the described error source model the *semi-Markov model*, and the corresponding channel is called a *symmetric semi-Markov channel.*

If the interval distributions depend only on the current state of the process

$$w_{ij}(l) = w_i(l) \qquad i,j=1,2,...,k$$

the process is called *autonomous*. In this case the matrix $\mathbf{W}(l)$ has identical columns. Any channel model based on a semi-Markov process is equivalent to some model based on an autonomous semi-Markov process. Indeed, consider a process whose states are defined as pairs (i,j) of the above-defined Markov process. The new process is a Markov process with the transition probabilities

$$p_{(ij),(m,n)} = \begin{cases} p_{jn} & \text{for} \quad m=j \\ 0 & \text{for} \quad m \neq j \end{cases}$$

For a process thus defined, the transition distribution $w_{ij}(l)$ depends only on the first state (i,j) if we assume that impossible transitions from the state (i,j) also have the same interval distribution.

Clearly, a Markov chain is a particular case of the semi-Markov process (with the same transition probability matrix and $\mathbf{W}(1)=\mathbf{I}$). Since triplets (i,j,l) constitute a Markov chain (with an infinite number of states), a semi-Markov process may be considered as a function of some Markov chain. We use this result to apply the statistical methods developed for Markov chains to semi-Markov models.

In general, it is possible for a semi-Markov process to transfer from some state i into the same state (with the probability $f_{ii}(l)$) so that no *real* transition of the process occurs. These types of transitions are called *virtual* transitions.[37] If we assume that the conditional probabilities $\varepsilon_{ij} = \varepsilon_i$ depend only on the current state of the process, then we can replace the semi-Markov process with an equivalent process whose virtual transitions are real. In order to do so, we have to replace the interval transition distribution $f_{ij}(l)$ with the real transition distribution

$$\begin{aligned} f_{ij}^{(1)}(l) &= \sum_{m=1}^{\infty} f_{ij}^{*m}(l) \quad \text{for } i \neq j \\ f_{ii}^{(1)}(l) &= 0 \end{aligned} \tag{1.4.2}$$

where $f_{ij}^{*m}(l)$ denotes the convolution

$$f_{ij}^{*m}(l) = \sum_{l_1+\ldots+l_m=l} f_{ii}(l_1) \cdots f_{ii}(l_{m-1}) f_{ij}(l_m)$$

Since the z-transform of a convolution is equal to the product of the z-transforms of its components,[39] the z-transform of the distribution (1.4.2)

$$\rho_{ij}^{(1)}(z) = \sum_{l=1}^{\infty} f_{ij}^{*m}(l) z^{-l} = \sum_{l=1}^{\infty} \rho_{ii}^{m-1}(z) \rho_{ij}(z)$$

has the form

$$\rho_{ij}^{(1)}(z) = \rho_{ij}(z)/[1-\rho_{ii}(z)] \tag{1.4.3}$$

where

$$\rho_{ij}(z) = \sum_{l} f_{ij}(l) z^{-l}$$

is the z-transform of the original distribution. Distribution (1.4.2) can be found by decomposing its z-transform (1.4.3) into the Laurent series or by using the inverse Fourier transform.

EXAMPLE 1.4.1: For example, a Markov chain may be treated as a particular case of the semi-Markov process whose interval distributions have the form $f_{ij}(1) = p_{ij}$. The z-transform $\rho_{ij}(z) = p_{ij}/z$ and equation (1.4.3) then takes the form

$$\rho_{ij}^{(1)}(z) = p_{ij}/(z-p_{ii}) = \sum_{l=1}^{\infty} p_{ii}^{l-1} p_{ij} z^{-l}$$

Therefore the real transition interval distribution $f_{ij}^{(1)}(l) = (1-p_{ii})p_{ii}^{l-1}$ is geometric, and the Markov chain can also be described as a semi-Markov process with geometric interval distributions.

The semi-Markov process with the interval distribution in the form of (1.4.2) is equivalent to the original process. The transitional probabilities of the modified process are equal to

$$p_{ij}^{(1)} = \sum_l f_{ij}^{(1)}(l) = \rho_{ij}^{(1)}(1) = p_{ij}/(1-p_{ii})$$

for $i \neq j$. Therefore, the i-th row of the modified matrix of transition probabilities has the form

$$(1-p_{ii})^{-1}[\, p_{i1} \;\; p_{i2} \cdots p_{i,i-1} \;\; 0 \;\; p_{i,i+1} \cdots p_{ik} \,] \tag{1.4.4}$$

In the above derivations we assume that the state i is not an initial state of the process. For the initial state the transformations can be performed quite analogously.

Thus, if $p_{ij} \neq 0$, the error source model description is not unique. Therefore, we assume that the transformations (1.4.2) and (1.4.4) are performed for every $i=1,2,...,k$ and the diagonal elements of the matrix **P** are equal to zero. We call such a representation of a semi-Markov process *canonical*.

In a particular case, when the source can be in only one of the two states and the transition matrix has the form

$$\mathbf{P} = \begin{bmatrix} 0 & 1 \\ 1 & 0 \end{bmatrix}$$

we have an *alternating renewal process* with different probabilities of errors in different states. In this case, one of the lengths is often regarded as the length of a cluster of errors, the other as the time between clusters.

■

In practice, models based on semi-Markov processes or alternating processes are used relatively rarely, because of the difficult calculations they entail. However, it is often convenient to represent the SSM model in the form of the semi-Markov process for processing experimental data and especially for simulating an error flow. Let us study, then, under what conditions it is possible to describe the SSM model by the semi-Markov process.

1.4.2 Semi-Markov Lumping of Markov Chain. Suppose we are given a homogeneous Markov chain with matrix $\mathbf{P} = [\, p_{ij} \,]_{u,u}$ and a matrix of initial probabilities $\mathbf{p}_0 = row\{p_i\} = [\, p_i \,]_{1,u}$. Let $\{A_1, A_2, ..., A_s\}$ be a partition of the set of all states into nonintersecting subsets:

$$\bigcup_{i=1}^{s} A_i = A \quad A_i \cap A_j = \varnothing \quad i \neq j$$

Let us consider a process y_t whose states are defined by the condition $y_t = i$ if at time t the Markov chain is in some state from the set A_i. The process y_t is usually called a *Markov function* (function of a Markov chain) or, in the terminology of Ref. [40], a *lumped process.*

To simplify notations let us assume that the Markov chain states are numbered in such a way that

$$A_1 = \{1,2,...,m_1\},\ A_2 = \{m_1+1,...,m_2\},...,\ A_s = \{m_{s-1}+1,...,m_s\}$$

If the assumption is not valid, we renumber the states (see Appendix 6). In correspondence with this partition we can write the matrices $\mathbf{P}$ and $\mathbf{p}_0$ in the block form

$$\mathbf{P} = block[\mathbf{P}_{ij}]_{s,s} = \begin{bmatrix} \mathbf{P}_{11} & \mathbf{P}_{12} & \cdots & \mathbf{P}_{1s} \\ \mathbf{P}_{21} & \mathbf{P}_{22} & \cdots & \mathbf{P}_{2s} \\ \cdots & \cdots & \cdots & \cdots \\ \mathbf{P}_{s1} & \mathbf{P}_{s2} & \cdots & \mathbf{P}_{ss} \end{bmatrix}$$

$$\mathbf{p}_0 = block\ row\{\mathbf{p}_i\} = [\ \mathbf{p}_1 \ \mathbf{p}_2 \ \cdots \ \mathbf{p}_s]$$

where matrix block $\mathbf{P}_{ij}$ consists of all transition probabilities $p_{\alpha\beta}$, $\alpha \in A_i$, $\beta \in A_j$.

The lumped process y_t is not usually a Markov or semi-Markov process. Several necessary and sufficient conditions for the lumped process to be a Markov chain may be found in Ref. [40] (see also Appendix 1). We find conditions for the lumped process to be a semi-Markov process.

DEFINITION 1.4.1: We say that a Markov chain with the initial distribution $\mathbf{p}_0 = [\ \mathbf{p}_1 \ \mathbf{p}_2 \ \cdots \ \mathbf{p}_s\ ]$ is $\mathbf{p}_0$-semi-Markov *lumpable* by a partition $\{A_1, A_2, ..., A_s\}$, if the lumped process y_t is a homogeneous semi-Markov process.

■

Theorem 1.4.1: A necessary and sufficient condition for a Markov chain to be semi-Markov lumpable by a partition $\{A_1, A_2, ..., A_s\}$ is that there exist a matrix row of the initial probabilities $\hat{\mathbf{p}}_0 = row\{\hat{p}_i\}$ and transition matrix distributions $\mathbf{F}^{(0)}(l) = [\ f_{ij}^{(0)}(l)\ ]_{s,s}$ $\mathbf{F}(l) = [\ f_{ij}(l)\ ]_{s,s}$ such that

$$\mathbf{p}_{y_0} \prod_{i=0}^{m} \mathbf{P}_{y_i y_i}^{l_i - 1} \mathbf{P}_{y_i y_{i+1}} \mathbf{1} = \hat{p}_{y_0} f_{y_0 y_1}^{(0)}(l_0) \prod_{i=1}^{m} f_{y_i y_{i+1}}(l_i) \qquad (1.4.5)$$

for all $y_i = 1,2,...,s$, $l_i \in \mathbf{N}$, $i=0,1,...,m$, $m \in \mathbf{N}$.

The proof of the theorem is obvious, since the previous relation is actually a paraphrase of the semi-Markov process definition. The left-hand side of the equation is the probability that the Markov function starts at y_0, stays in this state for l_0 steps, then enters y_1, stays in this state for l_1 steps, enters y_2, and so on and finally enters y_{m+1}. The right-hand side of the equation is the probability of the same trajectory if one assumes that y_t is a semi-Markov process.

■

Let us consider the Markov process in the moments of its transitions from a state of one subset A_i into a state of another subset A_j, assuming that the sets A_i are not absorbing. We call a process built in this way a *transition process*, which is obviously a Markov chain. Denote its matrix $\mathbf{Q} = [\ \mathbf{Q}_{ij}\ ]$. Let α and β be two states of the initial Markov chain and $\alpha \in A_i$, $\beta \in A_j$. The probability $q_{\alpha\beta}$ of transition from the state α into state β is a sum of probability $p_{\alpha\beta}$ of transition of the initial process from the state α into the state β in one step, in two steps:

$$\sum_{\gamma \in A_i} p_{\alpha\gamma} p_{\gamma\beta}$$

and so on. Noting that these probabilities are elements of block matrices, $\mathbf{P}_{ij}$, $\mathbf{P}_{ii}\mathbf{P}_{ij}$, and so on; we obtain

$$\mathbf{Q}_{ij} = \sum_{m=0}^{\infty} \mathbf{P}_{ii}^{m} \mathbf{P}_{ij}$$

Since the set A_i is not absorbing, this series converges. Its sum is

$$\mathbf{Q}_{ij} = (\mathbf{I} - \mathbf{P}_{ii})^{-1} \mathbf{P}_{ij} \tag{1.4.6}$$

It is obvious also that $\mathbf{Q}_{ii} = 0$. Thus, if the initial process has the matrix $\mathbf{P} = [\ \mathbf{P}_{ij}\ ]$, then the transition process has the matrix

$$\mathbf{Q} = \begin{bmatrix} 0 & (\mathbf{I}-\mathbf{P}_{11})^{-1}\mathbf{P}_{12} & \cdots & (\mathbf{I}-\mathbf{P}_{11})^{-1}\mathbf{P}_{1s} \\ (\mathbf{I}-\mathbf{P}_{22})^{-1}\mathbf{P}_{21} & 0 & \cdots & (\mathbf{I}-\mathbf{P}_{22})^{-1}\mathbf{P}_{2s} \\ \cdots & \cdots & \cdots & \cdots \\ (\mathbf{I}-\mathbf{P}_{ss})^{-1}\mathbf{P}_{s1} & (\mathbf{I}-\mathbf{P}_{ss})^{-1}\mathbf{P}_{s2} & \cdots & 0 \end{bmatrix} \tag{1.4.7}$$

If we denote as $p_{\alpha\beta}(l)$ the probabilities that the chain remains in the set A_i during l steps, under the condition that the initial state belongs to this set ($\alpha \in A_i$) and the final state does not belong to the set ($\beta \notin A_i$), then we can verify that the probabilities $p_{\alpha\beta}(l)$ are the elements of the matrix

$$\mathbf{Q}_i(l) = \mathbf{P}_{ii}^{l-1}(\mathbf{I} - \mathbf{P}_{ii}) \tag{1.4.8}$$

Using the transition process notations, we rewrite the multidimensional distributions in the left-hand side of (1.4.5) as

$$\mathbf{p}_{y_0}\prod_{i=0}^{m}\mathbf{P}_{y_iy_i}^{l_i-1}\mathbf{P}_{y_iy_{i+1}}\mathbf{1} = \mathbf{p}_{y_0}\prod_{i=0}^{m}\mathbf{Q}_{y_i}(l_i)\mathbf{Q}_{y_iy_{i+1}}\mathbf{1} \tag{1.4.9}$$

and equation (1.4.5) becomes

$$\mathbf{p}_{y_0}\prod_{i=0}^{m}\mathbf{Q}_{y_i}(l_i)\mathbf{Q}_{y_iy_{i+1}}\mathbf{1} = \hat{p}_{y_0}f_{y_0y_1}^{(0)}(l_0)\prod_{i=1}^{m}f_{y_iy_{i+1}}(l_i) \tag{1.4.10}$$

If the Markov chain with the matrix $\mathbf{P}$ is regular, the transition process is not necessarily regular as well. This fact becomes obvious if we consider a Markov chain with two states and matrix

$$\mathbf{P} = \begin{bmatrix} p_{11} & p_{12} \\ p_{21} & p_{22} \end{bmatrix}$$

The matrix of the transition process, according to (1.4.7), is equal to

$$\mathbf{Q} = \begin{bmatrix} 0 & 1 \\ 1 & 0 \end{bmatrix}$$

and is not regular. However, if the matrix $\mathbf{P}$ is regular and its invariant vector is $\boldsymbol{\pi} = block\ row\{\boldsymbol{\pi}_i\}$, then it is easy to verify that the matrix $\mathbf{Q}$ has the invariant vector $\mathbf{q} = block\ row\{\mathbf{q}_i\}$ of the form

$$\mathbf{q}_i = \boldsymbol{\pi}_i(\mathbf{I} - \mathbf{P}_{ii}) / \sum_{i=1}^{s}\boldsymbol{\pi}_i(\mathbf{I} - \mathbf{P}_{ii})$$

which is the unique solution of the system

$$\mathbf{qQ} = \mathbf{q} \quad \mathbf{q1} = 1 \tag{1.4.11}$$

(see Appendix 1.3).

If the process starts at a random moment, it is necessary to use the initial transition distribution $f_{ij}^{(0)}(l)$. However, if the process starts at the moment of the set change (or, equivalently, the Markov function state change), then $f_{ij}^{(0)}(l) = f_{ij}(l)$

[37] and equation (1.4.10) takes on the form

$$\mathbf{q}_{y_1}\prod_{i=1}^{m}\mathbf{Q}_{y_i}(l_i)\mathbf{Q}_{y_iy_{i+1}}\mathbf{1} = \hat{q}_{y_1}\prod_{i=1}^{m}f_{y_iy_{i+1}}(l_i) \tag{1.4.12}$$

where $\mathbf{q} = [\mathbf{q}_1 \ \mathbf{q}_2 \ \ldots \ \mathbf{q}_s \ ]$ is the initial distribution of the chain states at the moment of a transition from one of the sets A_i. Equation (1.4.12) can also be obtained from (1.4.10) by performing summation with respect to y_0 and l_0. In this case

$$\mathbf{q}_j = \sum_{i\neq j}\mathbf{p}_i\mathbf{Q}_{ij}$$

or $\mathbf{q} = \mathbf{pQ}$.

Thus, if a Markov chain is semi-Markov **p**-lumpable with respect to $\{A_1,A_2,\ldots,A_s\}$ and sets A_i are not absorbing, then it is **pQ**-lumpable with the initial transition distribution $f_{ij}^{(0)}(l)$ coinciding with the intermediate transition distribution $f_{ij}(l)$. We assume in the sequel that all the necessary conditions are met and use (1.4.12) as a criterion of a semi-Markov lumpability.

Consider now some properties of lumpable processes. For m=1, equation (1.4.12) takes on the form

$$\mathbf{q}_i\mathbf{Q}_i(l)\mathbf{Q}_{ij}\mathbf{1} = \hat{q}_i f_{ij}(l) \tag{1.4.13}$$

If we perform summation with respect to l, we obtain

$$\mathbf{q}_i\mathbf{Q}_{ij}\mathbf{1} = \hat{q}_i\hat{p}_{ij} \tag{1.4.14}$$

The summation with respect to j gives

$$\mathbf{q}_i\mathbf{1} = \hat{q}_i$$

since the sum of the elements of each row of a stochastic matrix is equal to one:

$$\sum_{j=1}^{s}\mathbf{Q}_{ij}\mathbf{1} = \mathbf{1}$$

The z-transform of the distribution

$$f_{ij}(l) = \mathbf{q}_i\mathbf{P}_{ii}^{l-1}\mathbf{P}_{ij}\mathbf{1}\,/\,\mathbf{q}_i\mathbf{1} \tag{1.4.15}$$

is a rational function

$$\rho_{ij}(z) = \sum_{l=1}^{\infty} f_{ij}(l) z^{-l} = \mathbf{q}_i (z\mathbf{I} - \mathbf{P}_{ii})^{-1} \mathbf{P}_{ij} \mathbf{1} \,/\, \mathbf{q}_i \mathbf{1} \tag{1.4.16}$$

so the interval distribution $f_{ij}(l)$ can be found by carrying out a partial-fraction expansion and identifying the inverse z-transform of the partial fractions (see Appendices 2.1 and 5.1.3):

$$f_{ij}(l) = \sum_{k} [A_{ijk}(l)\cos(a_{ik}l) + B_{ijk}(l)\sin(a_{ik}l)] q_{ik}^{l-1} \tag{1.4.17}$$

where $A_{ijk}(l)$ and $B_{ijk}(l)$ are polynomials of l, and $q_{ik}[\cos(a_{ik}) + \mathrm{j}\sin(a_{ik})]$ are the roots of the characteristic polynomial $\Delta(z) = \det(z\mathbf{I} - \mathbf{P}_{ii})$ or, in other words, eigenvalues of the matrix $\mathbf{P}_{ii}$.

The previous equations are the *necessary* conditions of semi-Markov lumpability. Consequently, if a semi-Markov process has an interval transition distribution that does not have the form of (1.4.17), then one cannot represent this process as a function of a Markov chain with a finite number of states.

Equations (1.4.14) and (1.4.15) lead us to the following conclusion. If a Markov chain is semi-Markov lumpable by the partition $\{A_1, A_2, \ldots, A_s\}$, then the initial distribution and the interval transition distributions are uniquely defined. Thus, to check whether the Markov chain is lumpable, one may first determine the semi-Markov process parameters from equations (1.4.14) and (1.4.15) and then check if they satisfy (1.4.10). However, the system (1.4.10) contains an infinite number of equations, hence it is necessary to develop some methods to verify the process lumpability in a finite number of steps. These methods are presented in Appendix 1.2. We discuss here the simplest and, from the practical point of view, the most important sufficient conditions of lumpability.

Theorem 1.4.2: If the sets of states $\{A_1, A_2, \ldots, A_s\}$ of a Markov chain are not absorbing, and the transition matrices have the form

$$\mathbf{P}_{ij} = \mu_{ij} \mathbf{M}_i \mathbf{N}_j \quad i \neq j \quad (i, j = 1, 2, \ldots, s) \tag{1.4.18}$$

where $\mathbf{N}_j$ is a matrix row, $\mathbf{M}_i$ is a matrix column, and μ_{ij} is a number, then the Markov chain is semi-Markov lumpable.

The proof of the theorem is given in Appendix 1. ■

One can easily verify that rows of the product $\mathbf{M}_i \mathbf{N}_j$ are proportional so that condition (1.4.18) is equivalent to proportionality in the rows of the matrices

$$[\mathbf{P}_{i1} \;\; \mathbf{P}_{i2} \; \ldots \; \mathbf{P}_{i-1,i} \;\; \mathbf{P}_{i+1,i} \;\; \mathbf{P}_{is}]$$

Let us consider a particular case when the set of states is partitioned into two subsets $\{A_1, A_2\}$ and conditions (1.4.18) hold. The matrix of the Markov chain has the form

$$\mathbf{P} = \begin{bmatrix} \mathbf{P}_{11} & \mu_{12}\mathbf{M}_1\mathbf{N}_2 \\ \mu_{21}\mathbf{M}_2\mathbf{N}_1 & \mathbf{P}_{22} \end{bmatrix}$$

We obtain an equivalent semi-Markov process with the matrix

$$\hat{\mathbf{P}} = \begin{bmatrix} 0 & 1 \\ 1 & 0 \end{bmatrix}$$

and with the interval transition distributions $f_{12}(l) = \mathbf{N}_1\mathbf{P}_{11}^{l-1}(\mathbf{I} - \mathbf{P}_{11})\mathbf{1}$ and $f_{21}(l) = \mathbf{N}_2\mathbf{P}_{22}^{l-1}(\mathbf{I} - \mathbf{P}_{22})\mathbf{1}$.

Thus, in this case the Markov process is equivalent to an alternating renewal process. This result coincides with the result presented in Ref. [13]. Other conditions for equivalency to an alternating renewal process presented in Ref. [13] are in effect conditions of strong and weak lumpability [40] of a Markov chain states by the partition $\{A_1, A_2\}$.

In practice the opposite problem often arises: to determine whether a model described by of a semi-Markov process (and an alternating renewal process in particular) can be replaced by an equivalent SSM model. The solution of this problem is especially important in the processing of experimental data on errors in real communication channels, since the parameters of the semi-Markov model are usually more easily estimated from experimental data, but analytic calculations are simpler for the SSM model. As shown above, the equivalence of a semi-Markov model to a Markov model requires that the interval distributions $f_{ij}(l)$ have the form (1.4.15). These conditions are sufficient when additional constraints related to stochasticity are imposed.

Theorem 1.4.3: If the interval distributions of the semi-Markov process with matrices

$$\hat{\mathbf{P}} = [\, \hat{p}_{ij} \,]_{s,s} \quad \hat{\mathbf{P}}_0 = [\, \hat{p}_i \,]_{1,s} \tag{1.4.19}$$

have the form (1.4.15), where the matrices $\mathbf{p}_i$, $\mathbf{P}_{ii}$ satisfy the conditions

$$\mathbf{P}_{ii} > 0 \quad \mathbf{p}_i \geq 0 \quad \mathbf{p}_i \neq 0 \quad \mathbf{P}_{ii}\mathbf{1} < \mathbf{1} \quad \mathbf{p}_i\mathbf{P}_{ii} \leq \mathbf{p}_i \tag{1.4.20}$$

then this semi-Markov process is equivalent to the Markov function with the matrix

$$\mathbf{P} = \begin{bmatrix} \mathbf{P}_{11} & \hat{p}_{12}\mathbf{M}_1\mathbf{N}_2 & \cdots & \hat{p}_{1s}\mathbf{M}_1\mathbf{N}_s \\ \hat{p}_{21}\mathbf{M}_2\mathbf{N}_1 & \mathbf{P}_{22} & \cdots & \hat{p}_{2s}\mathbf{M}_2\mathbf{N}_s \\ \cdots & \cdots & \cdots & \cdots \\ \hat{p}_{s1}\mathbf{M}_s\mathbf{N}_1 & \hat{p}_{s2}\mathbf{M}_s\mathbf{N}_2 & \cdots & \mathbf{P}_{ss} \end{bmatrix} \tag{1.4.21}$$

where $\mathbf{M}_i = (\mathbf{I} - \mathbf{P}_{ii})\mathbf{1}$, $\mathbf{N}_j = \mathbf{p}_j(\mathbf{I} - \mathbf{P}_{jj})/\mathbf{P}_j(\mathbf{I} - \mathbf{P}_{jj})\mathbf{1}$, and the vectors of the initial probabilities are bound by the conditions

$$\mathbf{p}_0 = block\ row\{\mu_i \mathbf{p}_i\} \tag{1.4.22}$$

where

$$\mu_i = \frac{\hat{p}_i[\mathbf{p}_i(\mathbf{I}-\mathbf{P}_{ii})\mathbf{1}]^{-1}}{\sum_j \hat{p}_j[\mathbf{p}_j(\mathbf{I}-\mathbf{P}_{jj})\mathbf{1}]^{-1}\mathbf{p}_j\mathbf{1}}$$

Proof: It is sufficient to show that the matrix $\mathbf{P}$ is a stochastic matrix, since the equivalence of a Markov function to a semi-Markov process follows from Theorem 1.4.2. Because of (1.4.26) we have $\mathbf{P}_{ii} \geq 0$, but then $\mathbf{P}_{ij} = \hat{p}_{ij}\mathbf{M}_i\mathbf{N}_j \geq 0$, and since

$$\sum_{j=1}^{s}\mathbf{P}_{ii}\mathbf{1} = \mathbf{P}_{ii}\mathbf{1} + \sum_{j\neq i}\hat{p}_{ij}(\mathbf{I} - \mathbf{P}_{ii})\mathbf{1} = \mathbf{1}$$

the matrix $\mathbf{P}$ is stochastic.

∎

If the matrix $\mathbf{P}$ is regular and $\hat{\mathbf{p}}_0$ is an invariant vector of the matrix $\hat{\mathbf{P}}$, then the vector $\mathbf{p}_0$ defined by formula (1.4.22) specifies the stationary distribution. Indeed,

$$\sum_{i=1}^{s}\mu_i\mathbf{p}_i\mathbf{P}_{ij} = \mu_j\mathbf{p}_j\mathbf{P}_{jj} + \sum_{i\neq j}\hat{p}_{ij}\mu_i\mathbf{p}_i\mathbf{M}_i\mathbf{N}_j =$$

$$\mu_j\mathbf{p}_j\mathbf{P}_{jj} + \sum_{i\neq j}\hat{p}_{ij}\hat{p}_i\mu_j\mathbf{p}_j(\mathbf{I} - \mathbf{P}_{jj})/\hat{p}_j = \mu_j\mathbf{p}_j$$

1.4.2.1 Polygeometric Distribution. The particular case in which the transition distribution is polygeometric,

$$w_i(l) = \sum_{j=1}^{J_i} a_{ij}(1-q_{ij})q_{ij}^{l-1} \tag{1.4.23}$$

$a_{ij} \geq 0$, $0 \leq q_{ij} < q_{i,j+1} < 1$, $(i=1,2,...,s,\ j=1,2,...,J_i)$, plays a special role in practice. These distributions can be represented in the form of (1.4.14) where

$$\mathbf{p}_i = row\{a_{i1}/(1-q_{i1})\} = \mathbf{N}_i(\mathbf{I} - \mathbf{Q}_i)^{-1} \tag{1.4.24}$$

$$\mathbf{N}_i = row\{a_{ij}\} \quad \mathbf{Q}_i = \mathbf{P}_{ii} = diag\{q_{ij}\} \tag{1.4.25}$$

It is easy to check that under the imposed constraints these matrices satisfy the conditions of Theorem 1.4.3, and therefore the representation in the form of the Markov function with the matrix (1.4.21) is possible.

The particular case of Theorem 1.4.3 considered above has a simple probabilistic interpretation. We consider formula (1.4.23) as the formula of total probability:

$$w_i(l) = \sum_j a_{ij} w_{ij}(l)$$

where a_{ij} is regarded as the probability of choice of a random variable l_{ij} with geometric distribution $w_{ij}(l) = (1-q_{ij})q_{ij}^{l-1}$. Now we can consider the process $\{i,j\}$ obtained from the initial process by decomposing the state i in J_i categories of states $(i, 1)$, $(i, 2),...,(i,J_i)$. Since the time spent in each category of states has geometric distribution, the process $\{c,j\}$ is a Markov chain. The transition probabilities of this chain have the form

$$Pr((k,m) \mid (i,j)) = \begin{cases} q_{ij} & \text{for } (k,m) = (i,j) \\ (1-q_{ij})\hat{p}_{ik}a_{km} & \text{for } (k,m) \neq (i,j) \end{cases}$$

Note that the matrix consisting of these probabilities coincides with (1.4.21) under conditions (1.4.24) and (1.4.25).

Consider now the case when the model of the source of errors is described by an alternating renewal process with the polygeometric distributions of the lengths of clusters and intervals between them of the type (1.4.23). In this case,

$$\hat{\mathbf{P}} = \begin{bmatrix} 0 & 1 \\ 1 & 0 \end{bmatrix}$$

and therefore the equivalent Markov function, according to (1.4.21), has the matrix

$$\mathbf{P} = \begin{bmatrix} \mathbf{Q}_1 & (\mathbf{I} - \mathbf{Q}_1)\mathbf{1}\mathbf{N}_2 \\ (\mathbf{I} - \mathbf{Q}_2)\mathbf{1}\mathbf{N}_1 & \mathbf{Q}_2 \end{bmatrix} \tag{1.4.26}$$

1.5 SOME PARTICULAR ERROR SOURCE MODELS

In this section we consider several error source models based on experimental data for a large variety of real channels. We show that all of them can be described as an SSM.

1.5.1 Single-Error-State Models. Let us consider one of the popular SSM models in which errors occur only in one state.[5,11,12] If the conditional probability of error in this state differs from 1, the model can be replaced with the equivalent model in which this probability is equal to 1 (see Example 1.2.3).

In this case the Markov chain matrix has the form

$$\mathbf{P} = \begin{bmatrix} \mathbf{P}_{11} & \mathbf{P}_{12} \\ \mathbf{P}_{21} & p_{nn} \end{bmatrix} \tag{1.5.1}$$

where $\mathbf{P}_{12}$ is a matrix row and $\mathbf{P}_{21}$ is a matrix column. The conditions (1.4.18) are satisfied with

$$\mathbf{M}_1 = \mathbf{P}_{12} \quad \mathbf{N}_1 = \mathbf{P}_{21} \quad \mu_{12}=\mu_{21}=1 \quad \mathbf{N}_2=1 \quad \mathbf{M}_2=1$$

The interval transition distributions are given by

$$f_{12}(l) = \mathbf{P}_{21}\mathbf{P}_{11}^{l-1}\mathbf{P}_{12}\mathbf{1} \quad f_{21}(l) = p_{nn}^{l-1}(1-p_{nn})$$

This Markov chain is equivalent to a renewal process with the so-called *phase type* [41] interval distribution

$$f(0)=p_{nn} \quad f(l) = (1-p_{nn})f_{12}(l) \quad \text{for } l>0 \tag{1.5.2}$$

If the distribution $f_{12}(l)$ is polygeometric,

$$f_{12}(l) = \sum_{i=1}^{n-1} a_i(1-g_i)g_i^{l-1} \quad a_i>0 \quad g_i>0 \quad i=1,2,...,n-1$$

the matrix $\mathbf{P}_{11}$ can be diagonalized. Then the Markov chain matrix (1.5.1) takes the form

$$\mathbf{P} = \begin{bmatrix} g_1 & 0 & \dots & 0 & 1-g_1 \\ 0 & g_2 & \dots & 0 & 1-g_2 \\ \cdots & \cdots & \cdots & \cdots & \cdots \\ 0 & 0 & \dots & g_{n-1} & 1-g_{n-1} \\ b_1 & b_2 & \dots & b_{n-1} & p_{nn} \end{bmatrix} \tag{1.5.3}$$

where $b_j=(1-p_{nn})a_j$. The model whose matrix has this form is called Fritchman's partitioned model and is often used in practice. [5,11,12,19] It is easy to verify that the model's stationary distribution can be expressed as

$$\boldsymbol{\pi} = [p_e b_1(1-g_1)^{-1},\ p_e b_2(1-g_2)^{-1}, \dots,\ p_e b_{n-1}(1-g_{n-1})^{-1},\ p_e\ ]$$

where

$$p_e = (\ 1+\sum_{i=1}^{n-1} b_i/(1-g_i)\ )^{-1}$$

is the bit error probability.

Several models, not equivalent to Fritchman's partitioned model, are based on the random walk between the states [7,10]. In the case of pure random walk the transition probability matrix (1.5.1) is given by [10,39]

$$\mathbf{P} = \begin{bmatrix} 0.5 & 0.5 & 0 & 0 & \cdots & 0 & 0 & 0 \\ 0.5 & 0 & 0.5 & 0 & \cdots & 0 & 0 & 0 \\ 0 & 0.5 & 0 & 0.5 & \cdots & 0 & 0 & 0 \\ 0 & 0 & 0.5 & 0 & \cdots & 0 & 0 & 0 \\ \cdots & \cdots & \cdots & \cdots & \cdots & \cdots & \cdots & \cdots \\ 0 & 0 & 0 & 0 & \cdots & 0.5 & 0 & 0.5 \\ 0 & 0 & 0 & 0 & \cdots & 0 & 0.5 & 0.5 \end{bmatrix}$$

The eigenvalues of this matrix are equal to [39]

$$g_i = \cos(\frac{2j-1}{2n-1}\pi) \qquad i=1,2,\dots,n-1$$

Some of these eigenvalues are negative, and therefore the model's matrix is not diagonalizable in the class of Markov functions. It may be transferred to its diagonal form in the class of matrix processes (see Sec. 1.3).

It follows from Theorem 1.4.3 that a renewal process with phase-type interval distribution is equivalent to a Markov function. Thus, if intervals between consecutive errors are independent, the model can be constructed by approximating the interval distribution with the phase-type distribution. In particular, an alternating process model[42] with polygeometric interval distribution can be expressed in Fritchman's partitioned form (1.5.3). This result is often used in practice for estimating model parameters. Some of examples are given below.

Since Gilbert's model is a single-error-state model, the previous results are applicable to it. Therefore, the model matrix may be presented in the form of (1.5.3).

$$\mathbf{P} = \begin{bmatrix} g_1 & 0 & 1-g_1 \\ 0 & g_2 & 1-g_2 \\ b_1 & b_2 & q(1-h) \end{bmatrix} \tag{1.5.4}$$

We obtained this result independently in Example 1.2.4.

Because the model has only three independent parameters, it is sufficient to approximate the error-free run distribution with the bigeometric function

$$f_\lambda(\lambda) = a_1(1-g_1)g_1^\lambda + a_2(1-g_2)g_2^\lambda$$

and then obtain the model parameters by solving some algebraic equations (see Example 1.2.4 and Ref. [2]). For statistical measurements of one of the real telephone channels,[43] the following estimates have been found[2] $a_1 = 0.184$, $a_2 = 0.816$, $g_1 = 0.99743$, $g_2 = 0.81$. These values give the model parameter estimates $h = 0.84$, $P = 0.003$, $p = 0.034$. Matrix (1.5.4) is estimated by

$$\mathbf{P} = \begin{bmatrix} 0.99743 & 0 & 0.00257 \\ 0 & 0.81 & 0.19 \\ 0.15556 & 0.68988 & 0.15456 \end{bmatrix} \tag{1.5.5}$$

Using similar reasoning, the following models have been constructed for a high-frequency radio link[5]

$$\mathbf{P} = \begin{bmatrix} 0.66 & 0 & 0.34 \\ 0 & 0.9991 & 0.0009 \\ 0.44 & 0.34 & 0.22 \end{bmatrix} \tag{1.5.6}$$

and for a troposcatter channel:[11]

$$\mathbf{P} = \begin{bmatrix} 0.99952 & 0 & 0 & 0.00048 \\ 0 & 0.99534 & 0 & 0.00466 \\ 0 & 0 & 0.94196 & 0.05804 \\ 0.06902 & 0.25289 & 0.63308 & 0.04501 \end{bmatrix} \tag{1.5.7}$$

A typical model for the T1 digital repeatered line [122] has the matrix

$$\mathbf{P} = \begin{bmatrix} 0.949282615949 & 0 & 0.050717384051 \\ 0 & 0.999999997789 & 0.000000002211 \\ 0.144624167460 & 0.768791627022 & 0.086584205518 \end{bmatrix} \tag{1.5.8}$$

1.5.2 Alternating Renewal Process Models. An alternating renewal process model is a particular case of semi-Markov model with two states. These states alternate and have different interval transition distributions (see Sec. 1.4.1.). Several models satisfy this description.

One of the models of this type was proposed by Bennet and Froelich.[44] According to this model, bursts of errors occur independently with the fixed probability p_{b}; length l burst appears with the probability $P_{\text{b}}(l)$. Inside bursts, errors occur with some probability ε; there are no errors outside bursts.

Strictly speaking, this model permits the overlapping of several bursts, but, in practice it is difficult to identify such a case, so overlapping bursts are treated as a single burst. If we assume that bursts do not overlap, the model can be described as a semi-Markov process with the transition probability matrix

$$\mathbf{P} = \begin{bmatrix} q_{\text{b}} & p_{\text{b}} \\ q_{\text{b}} & p_{\text{b}} \end{bmatrix}$$

and the state-holding time distributions

$$w_{11}(1) = 1 \quad w_{22}(l) = P_{\text{b}}(l)$$

If we transform this description to its canonical form (see Sec. 1.4.1), we obtain an alternating renewal process with the matrix

$$\mathbf{P} = \begin{bmatrix} 0 & 1 \\ 1 & 0 \end{bmatrix}$$

and transition distributions

$$f_{12}(l) = p_b q_b^{l-1} \quad f_{21}(l) = \sum_{m=1}^{\infty} q_b p_b^{m-1} P_b^{*m}(l)$$

The z-transforms of these transition distributions are defined by equations (1.4.3), which in our case take the form

$$\rho_{12}(z) = \frac{p_b}{z-q_b} \quad \rho_{21}(z) = \frac{q_b \rho_b(z)}{z-p_b \rho_b(z)}$$

If the burst-length distribution is matrix geometric, this model, according to Theorem 1.4.3, can be described as an SSM. If the burst-length distribution is polygeometric,

$$P_b(l) = \sum_{i=1}^{n-1} a_i(1-g_i)g_i^{l-1} \quad a_i > 0 \quad g_i > 0 \quad i=1,2,...,n-1$$

then the model matrix can be written in the form of (1.5.3). However, in contrast to the single-error-state models considered in the previous section, this model has only one state in which errors do not occur. In all the remaining states they occur with the probability ε.

If the burst-length distribution is geometric

$$P_b(l) = (1-g)g^{l-1}$$

then

$$\rho_{21}(z) = \frac{(1-g)p_b}{z-gq_b-p_b}$$

and therefore the transition distributions are also geometric:

$$f_{12}(l) = p_b q_b^{l-1} \quad f_{21}(l) = (1-q)q^{l-1}$$

where $q = gq_b + p_b$. In this case this model is Gilbert's model with the matrix

$$\mathbf{P} = \begin{bmatrix} q_b & p_b \\ 1-q & q \end{bmatrix}$$

The error-free runs in this case have the bigeometric distribution

$$f_\lambda(\lambda) = a_1(1-g_1)g_1^\lambda + a_2(1-g_2)g_2^\lambda$$

in which g_1 and g_2 are different positive roots of the quadratic equation (see Example 1.2.1)

$$g^2 - (q_b+q(1-\varepsilon))g + (1-\varepsilon)(q_b+q-1) = 0$$

Analysis of experimental data showed that the assumption that error bursts are independent is not valid in some cases. In many cases the assumption that error bursts occur independently inside larger error clusters agreed with the experimental data.[45] It has been discovered that, for a large variety of telephone channels, error clusters occurred independently with the fixed probability p_c. The error-cluster length distribution was geometric

$$P_c(l) = (1-g_c)g_c^{l-1}$$

and the burst-length distribution was trigeometric

$$P_b(l) = a_1(1-g_1)g_1^{l-1} + a_2(1-g_2)g_2^{l-1} + a_3(1-g_3)g_3^{l-1}$$

Ignoring the possibility of error-cluster overlapping and error bursts overlapping inside the clusters, we can easily conclude that this model can be described by an alternating renewal process with the transition distributions

$$f_{12}(l) = a_{11}(1-g_{11})g_{11}^{l-1} + a_{12}(1-g_{12})g_{12}^{l-1} \quad f_{21}(l) = P_b(l)$$

where g_{11} and g_{12} are the roots of the quadratic equation

$$g^2 - [q_c+q(1-p_{bc})]g + (1-p_{bc})(q_c+q-1) = 0$$

In this equation $q_c=1-p_c$, p_{bc} is the conditional probability that an error burst starts at any bit inside the cluster, and $q = g_c q_c + p_c$.

This model is a particular case of the SSM considered in Sec. 1.4.2.1 and has the matrix (1.4.26), in which

$$\mathbf{Q}_1 = \begin{bmatrix} g_{11} & 0 \\ 0 & g_{11} \end{bmatrix} \quad \mathbf{Q}_2 = \begin{bmatrix} g_1 & 0 & 0 \\ 0 & g_2 & 0 \\ 0 & 0 & g_3 \end{bmatrix}$$

$$\mathbf{N}_1 = [\,a_1 \;\; a_2 \;\; a_3\,] \qquad \mathbf{N}_2 = [\,a_{11} \;\; a_{12}\,]$$

A typical model, described in Ref. [13], has the parameter estimates

$$\mathbf{Q}_1 = \begin{bmatrix} 0.99789 & 0 \\ 0 & 0.9999951 \end{bmatrix} \qquad \mathbf{Q}_2 = \begin{bmatrix} 0.26 & 0 & 0 \\ 0 & 0.93 & 0 \\ 0 & 0 & 0.992 \end{bmatrix}$$

$$\mathbf{N}_1 = [\,0.416 \;\; 0.267 \;\; 0.317\,] \qquad \mathbf{N}_2 = [\,0.86 \;\; 0.14\,]$$

and the probability of errors inside bursts $\varepsilon = 0.67$.

1.6 CONCLUSION

We have considered various methods of modeling digital stochastic processes and proved that the SSM model is general enough to approximate wide variety of experimental data. The actual model building on the basis of the experimental data is presented in Chapter 3.

We discussed advantages and disadvantages of different descriptions of the model and presented several methods of simplifying the model. It turned out that to minimize the number of the model parameters it is necessary to generalize the model allowing its parameters to be complex numbers. This generalization preserves the matrix-train form of multidimensional distributions. The set of probabilities of all possible error sequences is a finite-dimensional linear space, therefore it is possible to express probability of any sequence as a linear combination of probabilities of some basis sequences. We showed in Appendix 1 that the basis selection can simplify the model description. The major part of this material is an adaptation of results published in mathematical papers and monographs.

In Appendix 1 we prove basic statements made in this chapter and show the relationships between the basis sequence selection and the problems of lumping the states of a Markov chain.

CHAPTER 2

MATRIX PROBABILITIES

2.1 MATRIX PROBABILITIES AND THEIR PROPERTIES

A communication system performance analysis includes the calculation of performance characteristics such as received symbol-error probability, the average interval between errors, the average length of an error burst, error number distribution in a message, etc. The convenience of calculating these and similar characteristics often determines the choice of an error source model. In this chapter we develop methods of calculation based on the SSM model.

With this model, the probabilities of various events depend on the SSM state. The conditional probabilities of an event for different SSM states form a matrix that we call the matrix probability of the event. Usage of matrix probabilities simplifies calculations and allows us to find solutions in the matrix form that are applicable to all SSM models.

2.1.1 Basic Definitions. The matrix probability of a sequence of errors $\mathbf{e}_1, \mathbf{e}_2, ..., \mathbf{e}_m$ is defined by the following equation

$$\mathbf{P}(\mathbf{e}_1, \mathbf{e}_2, ..., \mathbf{e}_m) = \prod_{i=1}^{m} \mathbf{P}(\mathbf{e}_i) \tag{2.1.1}$$

Using this notation we can rewrite equation (1.1.5) as

$$Pr(\mathbf{e}_1, \mathbf{e}_2, ..., \mathbf{e}_m) = \mathbf{p}\mathbf{P}(\mathbf{e}_1, \mathbf{e}_2, ..., \mathbf{e}_m)\mathbf{1} \tag{2.1.2}$$

It is convenient to use matrix probabilities, because they do not depend on the SSM initial state distribution. The transitive property of the multiplication of matrix probabilities permits the construction of a "matrix probability theory" whose methods, as we will show later, are useful in calculating SSM characteristics. The corresponding scalar probabilities are found from formula (2.1.2).

Consider now an event A, defined by a set of combinations of errors from the interval $(t_0,t_1]$: $\mathbf{e}_{t_0+1},\mathbf{e}_{t_0+2},...,\mathbf{e}_{t_1}$. Let $\Omega_{t_0}^m$ be the set of all possible ordered sequences of the type $(\mathbf{e}_{t_0+1},\mathbf{e}_{t_0+2},...,\mathbf{e}_{t_0+m})$. We say that the event A depends only on the values $\mathbf{e}_t$ in the interval $(t_0,t_1]$, if this event is a subset of the set $\Omega_{t_0}^m$.

Suppose that events A and B depend only on values $\mathbf{e}_t$ in the interval $(t_0,t_1]$. The union of these sets $A \cup B$ is called the events sum; the intersection of the sets $A \cap B$ is called the events product; the complement of the set A (relative to $\Omega_{t_0}^m$) is called the opposite event $\bar{A}$. We define the events σ-algebra as the collection $N_{t_0}^m$ of all possible subsets of the set $\Omega_{t_0}^m$ including also the impossible event $\varnothing$.

Let A be an event that depends only on values $\mathbf{e}_t$ in the interval $(t_{0A},t_{1A}]$ and B be an event that depends only on the values $\mathbf{e}_t$ in the interval $(t_{0B},t_{1B}]$. If these intervals do not coincide, then the events A and B belong to different σ-algebras. Denote $\tau_0 = \min(t_{0A},t_{0B})$, $\tau_1 = \max(t_{1A},t_{1B})$. Then the interval $(\tau_0,\tau_1]$ contains the intervals $(t_{0A},t_{1A}]$ and $(t_{0B},t_{1B}]$. Consider now cylindrical expansions A_1 and B_1 of the sets A and B to $\Omega_{\tau_0}^{\tau_1-\tau_0}$. The elements of these sets are formed from the corresponding elements of sets A and B by placing all possible symbols on the additional positions that do not belong to the original intervals. We associate the sets A and B with the corresponding cylindrical sets A_1 and B_1 from $\Omega_{\tau_0}^{\tau_1-\tau_0}$. Events A_1 and B_1 belong to the same σ-algebra; therefore, the operations with these events are defined. The operations with the events A and B are defined as operations with the corresponding events A_1 and B_1.

EXAMPLE 2.1.1: Let A be the occurrence of an error in the first position of a block consisting of two symbols and B be the occurrence of an error on the second position of this block. The event A depends only on the value $\mathbf{e}_t$ in the moment t_0+1 (i.e., in the interval $(t_0,t_0+1]$); the event B depends only on value $\mathbf{e}_t$ in the interval $(t_0+1,t_0+2]$. These events correspond to the cylindric sets $A_1 = \{(1,0),\ (1,1)\}$, $B_1 = \{(0,1),\ (1,1)\}$ belonging to the σ-algebra $N_{t_0}^2$. The event $A \cap B$ is the occurrence of two errors in the block which is defined as $A_1 \cap B_1 = (1,1)$.

■

Suppose that the event A depends only on the values $\mathbf{e}_t$ in the interval $(t_0,t_1]$. Then the matrix

$$\mathbf{P}(A;t_0,t_1) = \sum_{\mathbf{e}_1,...,\mathbf{e}_m} \prod_{i=1}^{m} \mathbf{P}(\mathbf{e}_i) \tag{2.1.3}$$

is called a matrix probability of the event A. In this equation the summation is performed over all sequences $\mathbf{e}_1,\mathbf{e}_2,...,\mathbf{e}_m$ that constitute the set A, and $\mathbf{P}(\mathbf{e}_i)$ are the parameters of the SSM (or matrix process). The ij-th element of this matrix is the conditional probability

$$p_{ij}(A;t_0,t_1) = Pr(A, y_{t_1} = j \mid y_{t_0} = i)$$

of transition of the Markov chain with the matrix

$$\mathbf{P} = \sum_{\Omega} \mathbf{P}(\mathbf{e}) = [\, p_{ij} \,]_{u,u}$$

from the state $y_{t_0} = i$ (at the moment t_0) into the state $y_{t_1} = j$ (at the moment t_1) and the event A occurrence:

$$\mathbf{P}(A;t_0,t_1) = [\, p_{ij}(A;t_0,t_1) \,]_{u,u}$$

This interpretation follows immediately from formula (2.1.3), according to which

$$p_{ij}(A;t_0,t_1) = \sum_{A} \sum_{c_1 \cdots c_{m-1}} Pr(\mathbf{e}_1, c_1 \mid i) Pr(\mathbf{e}_2, c_2 \mid c_1) \cdots Pr(\mathbf{e}_m, j \mid c_{m-1}) =$$

$$\sum_{c_1 \cdots c_{m-1}} p_{ic_1} p_{c_1 c_2} \cdots p_{c_{m-1} j} Pr(A \mid i, c_1, c_2, \ldots, c_{m-1}, j)$$

The last expression is $Pr(A, y_{t_1} = j \mid y_{t_0} = i)$ due to the theorem of total probability.

According to (2.1.2), the usual (scalar) probability of A is defined as

$$p(A;t_0,t_1) = \mathbf{p}\mathbf{P}(A;t_0,t_1)\mathbf{1} \tag{2.1.4}$$

For the sake of simplicity, we will often omit the symbols t_0 and t_1 from the matrix probability notation. If the Markov chain is stationary, then the probability of an event A is given by

$$p(A;t_0,t_1) = p(A) = \boldsymbol{\pi}\mathbf{P}(A)\mathbf{1} \tag{2.1.5}$$

where $\boldsymbol{\pi}$ is the chain stationary distribution.

EXAMPLE 2.1.2: Consider Gilbert's model (see Sec. 1.1.3) with the matrices

$$\mathbf{P}(0) = \begin{bmatrix} Q & Ph \\ p & qh \end{bmatrix} \qquad \mathbf{P}(1) = \begin{bmatrix} 0 & P(1-h) \\ 0 & q(1-h) \end{bmatrix}$$

Let us find the probability of two errors in a block of three symbols.

The event A is composed of the union of sequences (110), (101), (011). Therefore, according to formula (2.1.3),

$$\mathbf{P}(A) = \mathbf{P}(0)\mathbf{P}^2(1) + \mathbf{P}(1)\mathbf{P}(0)\mathbf{P}(1) + \mathbf{P}^2(1)\mathbf{P}(0)$$

After multiplication and addition of the matrices in the right-hand side of this equation we obtain

$$\mathbf{P}(A) = (1-h)^2 \begin{bmatrix} Ppq & (Pp+Qq)P+3Pq^2h \\ pq^2 & 2Ppq+3q^3h \end{bmatrix}$$

The corresponding scalar probability is defined by (2.1.5). Multiplying the matrices

$$p(A) = \boldsymbol{\pi}\mathbf{P}(A)\mathbf{1} = \frac{1}{P+p}[\,p,\ P\,]\,\mathbf{P}(A)\begin{bmatrix} 1 \\ 1 \end{bmatrix}$$

we find that

$$p(A) = (1-h)^2 P\,(Pp+2pq+3q^2h)/(P+p) \qquad \blacksquare$$

2.1.2 Properties of Matrix Probabilities. Consider now the basic properties of matrix probabilities that play the main role in applications.

1. *The Unity Theorem.* The following formula holds

$$\mathbf{P}(\Omega^m) = \mathbf{P}^m \tag{2.1.6}$$

where $\mathbf{P} = \sum_{\Omega}\mathbf{P}(\mathbf{e})$. The proof of this formula follows immediately from (2.1.3):

$$\mathbf{P}(\Omega^m) = \sum_{\Omega^m}\prod_{i=1}^{m}\mathbf{P}(\mathbf{e}_i) = \prod_{i=1}^{m}\sum_{\mathbf{e}_i\in\Omega}\mathbf{P}(\mathbf{e}_i) = \mathbf{P}^m$$

The matrix $\mathbf{P}^m$ plays the role of unity in the matrix theory of probabilities; no matrix probability for the events in the σ–algebra N^m can exceed this probability:

$$0 \le \mathbf{P}(A) \le \mathbf{P}^m \qquad A \in N^m$$

2. *Interval Expansion.* If an event A depends only on values $\mathbf{e}_t$ in the interval $(t_0,t_1]$, then the matrix probability of the cylindrical expansion of the event is given by

$$\mathbf{P}(A;t_0-\tau_0,t_1+\tau_1) = \mathbf{P}^{\tau_0}\mathbf{P}(A;t_0,t_1)\mathbf{P}^{\tau_1} \tag{2.1.7}$$

The proof of this property is similar to the previous proof.

If the Markov chain is regular (see Appendix 6), then, letting $\tau_0\to\infty$ and $\tau_1\to\infty$, we obtain

$$\mathbf{P}(A;-\infty,\infty) = \mathbf{1}\boldsymbol{\pi}\mathbf{P}(A;t_0,t_1)\mathbf{1}\boldsymbol{\pi}$$

Since

$$\boldsymbol{\pi}\mathbf{P}(A;t_0,t_1)\mathbf{1} = p(A;t_0,t_1)$$

is a number, we can rewrite the previous equation as

$$\mathbf{P}(A;-\infty,\infty) = p(A;t_0,t_1)\mathbf{1}\boldsymbol{\pi}$$

3. *The Theorem of Addition of Probabilities.* If $A\in N^m$ and $B\in N^m$, then

$$\mathbf{P}(A\cup B) = \mathbf{P}(A) + \mathbf{P}(B) - \mathbf{P}(A\cap B) \tag{2.1.8}$$

The proof of this theorem immediately follows from the theorem of addition of scalar probabilities and the rule of addition of matrices.

4. *The Theorem of Multiplication of Probabilities.* Let $A\in N^m$ and $B\in N^m$. Then, according to the theorem of multiplication of scalar probabilities, we have

$$Pr(A\cap B,y_{t_1}=j\,|\,y_{t_0}=i)=Pr(A,y_{t_1}=j\,|\,y_{t_0}=i)Pr(B\,|\,A,y_{t_1}=j,y_{t_0}=i) \tag{2.1.9}$$

This formula permits us to determine all the elements of the matrix probability $\mathbf{P}(A\cap B;t_0,t_1)$. Let us consider several particular cases of this equation.

If an event A depends only on $\mathbf{e}_t$ in the interval $(t_0,\tau]$ for $t_0<\tau<t_1$, then

$$p_{ij}(A\cap B;t_0,t_1) = \sum_{k=1}^{u} Pr(A\cap B,y_\tau=k,y_{t_1}=j\,|\,y_{t_0}=i) = \sum_{i=1}^{u} Pr(A,y_\tau\,|\,y_{t_0}=i)Pr(B,y_{t_1}=j\,|\,A,y_\tau=k,y_{t_0}=i) \tag{2.1.10}$$

If for any $i=1,2,\ldots,u$ the condition

$$Pr(B,y_{t_1}=j\,|\,A,y_\tau=k,y_{t_0}=i) = Pr(B,y_{t_1}=j\,|\,A,y_\tau=k)$$

holds, then we call the matrix

$$\mathbf{P}(B\,|\,A;\tau,t_1) = [\,Pr(B,y_{t_1}=j\,|\,A,y_\tau=k)\,]_{u,u}$$

the conditional matrix probability of event B given event A. In this case formula (2.1.10) can be rewritten in the matrix form

$$\mathbf{P}(A\cap B;t_0,t_1) = \mathbf{P}(A;t_0,\tau)\mathbf{P}(B\,|\,A;\tau,t_1) \tag{2.1.11}$$

If we assume that A and B are conditionally independent [23], then

$$\mathbf{P}(A\cap B;t_0,t_1) = \mathbf{P}(A;t_0,\tau)\mathbf{P}(B;\tau,t_1) \tag{2.1.12}$$

This formula is similar to the usual probability of a product of two independent events. However, in this case the events A and B are not independent (they depend on the states of the same Markov chain).

5. *The Formula of Total Probability.* The formula of total probability is an obvious corollary of the theorems of addition and multiplication of probabilities. If $A\in N^m$ and the events H_k are pairwise disjoint and form a full group of events, i.e.,

$$\bigcup_k H_k = \Omega^m \quad H_i\cap H_j = \varnothing \text{ for } i\neq j$$

then, obviously,

$$\mathbf{P}(A;t_0,t_1) = \sum_k \mathbf{P}(A\cap H_k;t_0,t_1)$$

The matrix probability $\mathbf{P}(A\cap H_k;t_0,t_1)$ in the last formula can be obtained from the theorem of multiplication of probabilities. In particular, if formula (2.1.11) is valid, then

$$\mathbf{P}(A;t_0,t_1) = \sum_k \mathbf{P}(H_k;t_0,\tau_k)\mathbf{P}(A\,|\,H_k;\tau_k,t_1) \tag{2.1.13}$$

EXAMPLE 2.1.3: Let $\mathbf{P}_n(t)$ be the matrix probability that t errors occur in a block of length n. Then according to the formula of total probability we have

$$\mathbf{P}_n(t) = \mathbf{P}(0)\mathbf{P}_{n-1}(t) + \mathbf{P}(1)\mathbf{P}_{n-1}(t-1)$$ ■

2.2 MATRIX TRANSFORMS

Let ξ be a non-negative random discrete variable. Suppose that an event $\xi=x$ depends only on $\mathbf{e}_t$ in the interval $(t_0,t_1]$. Then a matrix distribution of the random variable ξ is defined as a matrix function $\mathbf{P}_\xi(x;t_0,t_1)$, whose value for any fixed x is the matrix probability of $\xi=x$

$$\mathbf{P}_\xi(x;t_0,t_1) = \mathbf{P}(\xi=x;t_0,t_1)$$

The ij-th element of the matrix $\mathbf{P}_\xi(x;t_0,t_1)$ is the probability that $\xi=x$ and that the Markov chain transfers from the state $y_{t_0}=i$ into the state $y_{t_1}=j$.

For the sake of simplicity, we usually omit the symbols t_0 and t_1 in the matrix distribution notation.

Matrix distributions possess many properties of their scalar counterparts. We do not discuss them here, because their proof is obvious. However, distribution transforms play a very important role in applications, and we consider them in some detail.

2.2.1 Matrix Generating Functions. The generating function of a matrix distribution $\mathbf{P}_\xi(x)$ is defined as the power series

$$\mathbf{\Phi}_\xi(z) = \sum_x \mathbf{P}_\xi(x)z^x \tag{2.2.1}$$

which converges inside the unit disk ($|z|\leq 1$) of the complex z-plane. Its elements are defined by

$$\phi_{ij}(z) = \sum_x p_{ij}(x)z^x \tag{2.2.2}$$

If the distribution-generating function is known, then the matrix probability $\mathbf{P}_\xi(x)$, being a coefficient of z^x of the Taylor series (2.2.1), is given by the inversion formulas [46]

$$\mathbf{P}_\xi(x) = \frac{1}{x!}\frac{d^x}{dz^x}\mathbf{\Phi}_\xi(z)\,\Big|_{z=0} \tag{2.2.3}$$

or

$$\mathbf{P}_{\xi}(x) = \frac{1}{2\pi j} \oint_{\gamma} \mathbf{\Phi}_{\xi}(z) z^{-x-1} dz \tag{2.2.4}$$

where γ is a counterclockwise contour that encircles the origin and lies entirely in the region of convergence of $\mathbf{\Phi}_{\xi}(z)$, which contains $|z| \le 1$.

It is convenient to use generating functions of matrix distributions when dealing with SSM because, as we will see, they have a comparatively simple form (in contrast to matrix distributions). Besides, it is easier to find distribution moments and asymptotics using generating functions.

EXAMPLE 2.2.1: Let us calculate the matrix distribution of the number of errors $\mathbf{P}_3(t)$ in a block of length 3 for the Gilbert model considered in Sec. 1.1.3.

By the formula of total probability we have

$$\mathbf{P}_3(0) = \mathbf{P}^3(0)$$

$$\mathbf{P}_3(1) = \mathbf{P}^2(0)\mathbf{P}(1)+\mathbf{P}(0)\mathbf{P}(1)\mathbf{P}(0)+\mathbf{P}(1)\mathbf{P}^2(0)$$

$$\mathbf{P}_3(2) = \mathbf{P}^2(1)\mathbf{P}(0)+\mathbf{P}(1)\mathbf{P}(0)\mathbf{P}(1)+\mathbf{P}(0)\mathbf{P}^2(1)$$

$$\mathbf{P}_3(3) = \mathbf{P}^3(1)$$

The generating function of this distribution is

$$\mathbf{\Phi}_t(z) = \mathbf{P}_3(0) + \mathbf{P}_3(1)z + \mathbf{P}_3(2)z^2 + \mathbf{P}_3(3)z^3$$

Substituting the values of matrix probabilities in this formula and simplifying the result we obtain

$$\mathbf{\Phi}_{\xi}(z) = (\mathbf{P}(0) + \mathbf{P}(1)z)^3$$

This generating function has the same form as for independent events [39] occurring with the probability $p(1)$ [binomial distribution $\binom{3}{t} p^t(1) p^{3-t}(0)$]. We encounter this fact very often. However, the matrix distribution $\mathbf{P}_3(t)$ cannot generally be written in the form $\binom{3}{t}\mathbf{P}^t(1)\mathbf{P}^{3-t}(0)$, because the product of matrices is noncommutative.

The matrix distribution $\mathbf{P}_3(t)$ can be found from the generating function by using the inversion formula (2.2.3). For example, let us find $\mathbf{P}_3(1)$. We have

$$\mathbf{\Phi}'_t(z) = \frac{d}{dz}(\mathbf{P}(0)+z\mathbf{P}(1))^3 = \mathbf{P}(1)(\mathbf{P}(0)+z\mathbf{P}(1))^2 +$$
$$(\mathbf{P}(0)+z\mathbf{P}(1))\mathbf{P}(1)(\mathbf{P}(0)+z\mathbf{P}(1)) + (\mathbf{P}(0)+z\mathbf{P}(1))^2\mathbf{P}(1)$$

Therefore, †

$$\mathbf{P}_3(1) = \mathbf{\Phi}'_t(0) = \mathbf{P}(1)\mathbf{P}(0)^2 + \mathbf{P}(0)\mathbf{P}(1)\mathbf{P}(0) + \mathbf{P}^2(0)\mathbf{P}(1)$$

The expansion of the function $\mathbf{\Phi}_t(z)$ in the vicinity of the origin can be obtained by the matrix polynomial expansion. The simplest way of doing it is by performing matrix multiplication and gathering the same power terms:

$$\mathbf{\Phi}_t(z) = (\mathbf{P}(0) + \mathbf{P}(1)z)(\mathbf{P}(0) + \mathbf{P}(1)z)(\mathbf{P}(0) + \mathbf{P}(1)z) = \mathbf{P}^3(0) +$$
$$[\mathbf{P}^2(0)\mathbf{P}(1)+\mathbf{P}(0)\mathbf{P}(1)\mathbf{P}(0)+\mathbf{P}(1)\mathbf{P}^2(0)]z +$$
$$[\mathbf{P}^2(1)\mathbf{P}(0)+\mathbf{P}(1)\mathbf{P}(0)\mathbf{P}(1)+\mathbf{P}(0)\mathbf{P}^2(1)]z^2 + \mathbf{P}^3(1)z^3$$

The generating function of an ordinary distribution has the form

$$\phi_\xi(z) = \mathbf{p}\mathbf{\Phi}_\xi(z)\mathbf{1} \tag{2.2.5}$$

Several obvious properties of matrix distribution generating functions are listed below:

1. If $\mathbf{p}$ is an arbitrary probabilistic vector, then we obtain from (2.2.5)

$$\phi_\xi(1) = \mathbf{p}\mathbf{\Phi}_\xi(1)\mathbf{1} = 1 \tag{2.2.6}$$

2. If a random variable ξ depends on $\mathbf{e}_t$ in the interval $(t_0, t_1]$, then

$$\mathbf{\Phi}_\xi(1) = \sum_x \mathbf{P}(x;t_0,t_1) = \mathbf{P}^{t_1-t_0} \tag{2.2.7}$$

3. It is convenient to use generating functions to find matrix binomial moments

† We draw the reader's attention to the technique of differentiating the matrix function $(\mathbf{P}(0)+\mathbf{P}(1)z)^3$. Because the product of the matrices is noncommutative, generally speaking $[(\mathbf{P}(0)+\mathbf{P}(1)z)^3]' \neq 3(\mathbf{P}(0)+\mathbf{P}(1)z)^2\mathbf{P}(1)$.

$$\mathbf{B}_m = \sum_n \binom{n}{m} \mathbf{P}_\xi(n)$$

Indeed, it is easy to check that if we differentiate (2.2.1) m times and then let $z=1$, we obtain the binomial moment

$$\mathbf{B}_m = m!\mathbf{\Phi}_\xi^{(m)}(1)$$

In particular, for $m=1$ this formula gives the matrix mean $\mathbf{B}_1$, and consequently the scalar mean is given by

$$\bar{\xi} = \phi'_\xi(1) = \mathbf{pB}_1\mathbf{1} = \mathbf{p\Phi}'_\xi(1)\mathbf{1} \tag{2.2.8}$$

4. We say that the random variables ξ and η are conditionally independent if for any x and y from the region of their definition events $\xi=x$ and $\eta=y$ are conditionally independent. If the random variables ξ and η are conditionally independent, then according to the formula of total probability (2.1.13), the matrix distribution of their sum is the convolution

$$\mathbf{P}_{\xi+\eta}(y) = \sum_{x=0}^{y} \mathbf{P}_\xi(x)\mathbf{P}_\eta(y-x) \tag{2.2.9}$$

If we denote as $\mathbf{\Phi}_\xi(z)$ and $\mathbf{\Phi}_\eta(z)$ the generating functions of the ξ and η distributions, then we can easily verify that the generating function of their sum is

$$\mathbf{\Phi}_{\xi+\eta}(z) = \mathbf{\Phi}_\xi(z)\mathbf{\Phi}_\eta(z) \tag{2.2.10}$$

In the general case of n conditionally independent variables, the generating function of their sum distribution has the form

$$\mathbf{\Phi}_{\xi_1+\xi_2+\ldots+\xi_n}(z) = \prod_{i=1}^{n} \mathbf{\Phi}_{\xi_i}(z) \tag{2.2.11}$$

Since the distribution of the sum is the convolution of distributions of the addends, the previous equation is often called the *convolution theorem:* the generating function of a convolution is equal to the product of its terms' generating functions. The converse is also true, because of the uniqueness of the Taylor series.[46]

EXAMPLE 2.2.2: Let us find the matrix generating function $\mathbf{\Phi}_t^{(n)}(z)$ of the distribution of the number of errors in a block of length n according to Gilbert's model (see Example 2.2.1).

For a block of length $n=1$ we have

$$\mathbf{\Phi}_e(z) = \mathbf{P}(0) + \mathbf{P}(1)z$$

The number of errors in the block is equal to the sum $t = e_1+e_2+\cdots+e_n$ so that, according to formula (2.2.11),

$$\mathbf{\Phi}_t^{(n)}(z) = \mathbf{\Phi}_e^n(z) = (\mathbf{P}(0) + \mathbf{P}(1)z)^n$$

The average number of errors in the block can be found using equation (2.2.8). The generating function derivative with respect to z is

$$\phi'_t(z) = \sum_{i=0}^{n-1} \mathbf{p}(\mathbf{P}(0)+z\mathbf{P}(1))^i \mathbf{P}(1)(\mathbf{P}(0)+z\mathbf{P}(1))^{n-i-1}\mathbf{1}$$

With $z=1$ we have

$$\overline{t} = \sum_{i=0}^{n-1} \mathbf{p}\mathbf{P}^i \mathbf{P}(1)\mathbf{P}^{n-i-1}\mathbf{1}$$

Since $\mathbf{P}=\mathbf{P}(0)+\mathbf{P}(1)$ is a stochastic matrix, sums of the elements in each of its rows are equal to 1 (that is $\mathbf{P1}=1$), and therefore

$$\overline{t} = \sum_{i=0}^{n-1} \mathbf{p}\mathbf{P}^i \mathbf{P}(1)\mathbf{1}$$

If we also assume that the chain is stationary (that is $\mathbf{p}=\boldsymbol{\pi}=\boldsymbol{\pi}\mathbf{P}$), then this formula can be simplified:

$$\overline{t} = \sum_{i=0}^{n} \boldsymbol{\pi}\mathbf{P}(1)\mathbf{1} = np_e$$

where $p_e=\boldsymbol{\pi}\mathbf{P}(1)\mathbf{1}$ is the bit error probability.

■

The previous results can be generalized for multidimensional distributions. Let $\{\mathbf{A}_{m_1,m_2,\ldots,m_k}\}$ be a sequence of matrices. Then the matrix

$$\mathbf{\Phi}(z_1,z_2,\ldots,z_k) = \sum_{m_1,m_2,\ldots,m_k} \mathbf{A}_{m_1,m_2,\ldots,m_k} z_1^{m_1} z_2^{m_2} \cdots z_k^{m_k} \qquad (2.2.12)$$

is called the sequence multidimensional generating function.

The elements of the sequence $\{\mathbf{A}_{m_1,m_2,\dots,m_k}\}$ are found from the generating function by inversion formulas similar to (2.2.3) and (2.2.4):

$$\mathbf{A}_{m_1,m_2,\dots,m_k} = \frac{1}{m_1!\cdots m_k!}\frac{\partial^N \boldsymbol{\Phi}(z_1,\dots,z_k)}{\partial z_1^{m_1}\cdots\partial z_k^{m_k}}\Big|_{\mathbf{0}} \tag{2.2.13}$$

$$\mathbf{A}_{m_1,m_2,\dots,m_k} = \frac{1}{(2\pi \mathrm{j})^k}\int_\gamma\cdots\int \frac{\boldsymbol{\Phi}(z_1,z_2,\dots,z_k)}{z_1^{m_1+1}\cdots z_k^{m_k+1}}dz_1\cdots dz_k \tag{2.2.14}$$

where $\mathbf{0}$=(0,...,0) is the k-dimensional space origin. The integral representation (2.2.14) is convenient in many calculations related to matrix probabilities. It is particularly true for summing matrix series and for obtaining asymptotic formulas (see Appendices 2 and 5).

2.2.2 Matrix z-Transforms. In some applications it is convenient to use the so-called *z-transform* of a matrix distribution:

$$\boldsymbol{\Psi}_\xi(z) = \boldsymbol{\Phi}_\xi(1/z) = \sum_x \mathbf{P}_\xi(x)z^{-x} \tag{2.2.15}$$

Since the series on the right-hand side of the $\boldsymbol{\Phi}_\xi(z)$ definition (2.2.1) converges inside the unit disk of the complex z-plane, the series on the right-hand side of (2.2.15) converges outside the unit disk ($|z|\geq 1$).

As we see, the z-transform of a distribution is only notationally different from the generating function. To find the z-transform inversion we may use equation (2.2.4) with the substitution $z=1/\zeta$:

$$\mathbf{P}_\xi(x) = \frac{1}{2\pi \mathrm{j}}\oint_{\gamma_1} \boldsymbol{\Psi}_\xi(\zeta)\zeta^{x-1}d\zeta \tag{2.2.16}$$

where γ_1 is a counterclockwise contour that encircles the origin and lies entirely in the region of convergence of $\boldsymbol{\Psi}_\xi(z)$.

The basic reason for using $\boldsymbol{\Psi}_\xi(z)$ instead of $\boldsymbol{\Phi}_\xi(z)$ is that often it simplifies the residue theorem application. In the case of (2.2.4), we must find all the residues of the function $\boldsymbol{\Phi}_\xi(z)$ at the singularities located outside the unit circle, including the point of infinity $z=\infty$. The residue at the infinity is calculated differently than all the other residues. To simplify the residue theorem application, it is convenient to transform ∞ into 0 by the inversion $1/z$ of the complex plane. Thus, the integral (2.2.16) may be computed using all the residues at the singularities of $\boldsymbol{\Psi}_\xi(z)$ located inside the unit circle.

It is also convenient to use z-transforms to find the random variable ξ matrix negative-binomial moments

$$\mathbf{N}_m = \sum_n \binom{m+n-1}{m} \mathbf{P}_\xi(n)$$

Indeed, if we differentiate (2.2.15) m times and then let $z=1$, we obtain

$$\mathbf{N}_m = \frac{(-1)^m}{m!} \mathbf{\Psi}_\xi^{(m)}(1) \tag{2.2.17}$$

In the particular case of $m=1$, this equation gives the matrix mean $\mathbf{N}_1$. The scalar mean is given by

$$\bar{\xi} = -\psi'_\xi(1) = \mathbf{p}\mathbf{N}_1\mathbf{1} = -\mathbf{p}\mathbf{\Psi}'_\xi(1)\mathbf{1}$$

2.2.3 Matrix Fourier Transform. The matrix Fourier transform (or characteristic function) of a matrix distribution $\mathbf{P}_\xi(x)$ is defined as

$$\mathbf{F}(\omega) = \mathbf{\Phi}(e^{-j\omega}) = \sum_{i=0}^{\infty} \mathbf{P}_\xi(x) e^{-j\omega k} \tag{2.2.18}$$

Since $|e^{-j\omega}|=1$, the matrix Fourier transform of the distribution coincides with its generating function or z-transform on the unit circle. Hence, all basic properties of generating functions are applicable to Fourier transforms.

The Fourier transform inversion formula may be obtained from (2.2.4) by replacing z with $e^{-j\omega}$:

$$\mathbf{P}_\xi(x) = \frac{1}{2\pi} \int_0^{2\pi} \mathbf{F}_\xi(\omega) e^{j\omega x} d\omega \tag{2.2.19}$$

Although it is convenient to find binomial moments using the generating functions, the central moments

$$\mathbf{M}_k = \sum_x \mathbf{P}_\xi(x) x^k$$

are usually computed using the Fourier transform derivatives:

$$\mathbf{M}_k = j^k \mathbf{F}_\xi^{(k)}(0) \tag{2.2.20}$$

EXAMPLE 2.2.3: Let us compute the standard deviation of the number of errors in a block according to Gilbert's model.

The error number distribution

$$p_n(m) = \boldsymbol{\pi}\mathbf{P}_n(m)\mathbf{1}$$

has the Fourier transform

$$f(z) = \boldsymbol{\pi}(\mathbf{P}(0) + \mathbf{P}(1)\mathrm{e}^{-\mathrm{j}\omega})^n\mathbf{1}$$

which can be found from the generating function of Example 2.2.2. Using equation (2.2.20), we obtain

$$a_n = m_1 = \mathrm{j}f'(0) = n\boldsymbol{\pi}\mathbf{P}(1)\mathbf{1} = np_e$$

$$m_2 = -f''(0) = (n-1)(n-2)p_e^2 + 2\boldsymbol{\pi}\mathbf{P}(1)\mathbf{U}_n\mathbf{P}(1)\mathbf{1}$$

where

$$\mathbf{U}_n = [(n-1)\mathbf{I}-n\mathbf{Q}+\mathbf{Q}^n](\mathbf{I}-\mathbf{Q})^{-2} \quad \mathbf{Q} = \mathbf{P}-\mathbf{1}\boldsymbol{\pi}$$

The distribution variance is expressed as[39]

$$\sigma_n^2 = m_2 - m_1^2$$

Since the matrix $\mathbf{P}$ is regular, $\mathbf{Q}^n \to 0$ when $n \to \infty$, and we obtain

$$\sigma_n^2 \sim n[p_e - 3p_e^2 + 2\boldsymbol{\pi}\mathbf{P}(1)(\mathbf{I}-\mathbf{Q})^{-1}\mathbf{P}(1)\mathbf{1}]$$ ■

2.2.4 Matrix Discrete Fourier Transform. Consider now the distribution Fourier transform in the discrete points uniformly distributed on the unit circle of the complex z-plane $z_k = \mathrm{e}^{2\pi\mathrm{j}k/p}$:

$$\mathbf{F}_k = \mathbf{\Phi}(z_k) = \mathbf{F}(-2\pi k/p) \tag{2.2.21}$$

This transform is called the distribution discrete Fourier transform. In general, this transform is not invertible. That is, we cannot uniquely determine the matrix distribution $\mathbf{P}_\xi(x)$ from its discrete Fourier transform.

Indeed, because of the exponential function periodicity ($z_k=z_{k+p}$), the number of different equations (2.2.21) cannot exceed p. So we have a system of p linear equations with an infinite number of unknowns that in general does not have a unique solution. However, if the generating function is a polynomial whose degree is less than p, this system always has a unique solution, because its determinant (being Vandermonde's determinant) is not equal to 0.

If we denote $\alpha=z_1=e^{-2\pi j/p}$, the system (2.2.21) takes the form

$$\mathbf{F}_k=\sum_{i=0}^{p-1}\alpha^{ik}\mathbf{P}_\xi(i) \quad k=0,1,\ldots,p-1 \tag{2.2.22}$$

In order to find the unique solution of this system we multiply its k-th equation by α^{-mk} and then sum all the equations:

$$\sum_{k=0}^{p-1}\alpha^{-mk}\mathbf{F}_k=\sum_{i=0}^{p-1}\mathbf{P}_\xi(i)\sum_{k=0}^{p-1}\alpha^{k(i-m)} \tag{2.2.23}$$

If we make use of the easily verifiable identity

$$\sum_{k=0}^{p-1}\alpha^{k(i-m)}=\begin{cases}0 & \text{if } i\neq m+lp\\ p & \text{if } i=m+lp\end{cases} \tag{2.2.24}$$

where l is an arbitrary integer, then (2.2.23) becomes

$$\sum_{k=0}^{p-1}\alpha^{-mk}\mathbf{F}_k=p\mathbf{P}_\xi(m)$$

So the unique solution of system (2.2.22) can be written as

$$\mathbf{P}_\xi(m)=p^{-1}\sum_{k=0}^{p-1}\alpha^{-mk}\mathbf{F}_k \tag{2.2.25}$$

This formula is extremely useful for the matrix distribution calculations because of the utility of the Fast Fourier Transform (FFT).

Consider now the general case when the generating function degree may be greater than p. In this case, replacing the upper limit of summation in (2.2.22) by ∞ and repeating the previous transformations we obtain

$$\sum_{l=0}^{\infty}\mathbf{P}_\xi(m+lp)=p^{-1}\sum_{k=0}^{p-1}\alpha^{-mk}\mathbf{F}_k \tag{2.2.26}$$

This equation is obviously the generalization of (2.2.25), which is obtained when $\mathbf{P}_\xi(m+lp)=0$ for $m=0,1,...,p-1$, and $l>0$. Equation (2.2.26) may be used to calculate sums of the elements whose indices have the form $m+lp$.

EXAMPLE 2.2.4: Let us find the probability of an odd number of errors in a block of length n according to Gilbert's model.

The matrix probability of an odd number of errors is the probability that the number of errors is $2l+1$:

$$\mathbf{P}_{odd} = \sum_l \mathbf{P}_n(2l+1)$$

This sum can be found from equation (2.2.26) when $p=2$ and $m=1$:

$$\mathbf{P}_{odd} = 2^{-1}(\mathbf{F}_0 - \mathbf{F}_1)$$

Using the results of Example 2.2.2 we obtain

$$\mathbf{F}_0 = \mathbf{\Phi}(1) = (\mathbf{P}(0) + \mathbf{P}(1))^n = \mathbf{P}^n$$

$$\mathbf{F}_1 = \mathbf{\Phi}(-1) = (\mathbf{P}(0) - \mathbf{P}(1))^n$$

Thus

$$\mathbf{P}_{odd} = [\mathbf{P}^n + (\mathbf{P}(0) - \mathbf{P}(1))^n]/2$$ ■

In the future, we will need the multidimensional generalization of equation (2.2.25). Let us consider an n-dimensional discrete Fourier transform

$$\mathbf{F_k} = \sum_{\mathbf{i}=0}^{\mathbf{q}} \boldsymbol{\alpha}^{\mathbf{ik}} \mathbf{P}_\xi(\mathbf{i}) \tag{2.2.27}$$

where

$$\mathbf{k}=(k_1,k_2,...,k_n) \quad \mathbf{i}=(i_1,i_2,...,i_n) \quad \mathbf{q}=(p_1-1,p_2-1,...,p_n-1)$$

$$\xi=(\xi_1,\xi_2,...,\xi_n) \quad \boldsymbol{\alpha}^{\mathbf{ik}}=\alpha_1^{i_1k_1}\alpha_2^{i_2k_2}\cdots\alpha_n^{i_nk_n} \quad \text{and} \quad \alpha_i=e^{-2\pi j/p_i}$$

As with the one-dimensional Fourier transform, we derive the transform inversion formula

$$\mathbf{P}_\xi(\mathbf{m}) = V^{-1}\sum_{\mathbf{k}=0}^{\mathbf{q}} \boldsymbol{\alpha}^{-\mathbf{mk}}\mathbf{F}_\mathbf{k} \tag{2.2.28}$$

where $V=p_1p_2\cdots p_n$. The generalized equation (2.2.26) is given by

$$\sum_{\mathbf{l}=0}^{\infty}\mathbf{P}_\xi(\boldsymbol{\mu}(\mathbf{l})) = V^{-1}\sum_{\mathbf{k}=0}^{\mathbf{q}} \boldsymbol{\alpha}^{-\mathbf{km}}\mathbf{F}_\mathbf{k} \tag{2.2.29}$$

where $\boldsymbol{\mu}(\mathbf{l})=(m_1+l_1p_1, m_2+l_2p_2, ..., m_n+l_np_n)$. In the particular case $p_1 = p_2 = \cdots = p_n = p$ and this formula can be simplified:

$$\sum_{\mathbf{l}=0}^{\infty}\mathbf{P}_\xi(\boldsymbol{\mu}(\mathbf{l})) = p^{-n}\sum_{\mathbf{k}=0}^{\mathbf{q}} \alpha^{-\mathbf{k}\cdot\mathbf{m}}\mathbf{F}_\mathbf{k} \tag{2.2.30}$$

where $\mathbf{k}\cdot\mathbf{m}=k_1m_1+k_2m_2+\cdots+k_nm_n$ is a scalar product.

2.2.5 Matrix Transforms and Difference Equations. In many cases it is convenient to obtain generating functions with the help of recurrence relations. For example, formula (2.2.10) was obtained using equations (2.2.9). Consider a difference equation

$$\mathbf{W}_n = \sum_{m=1}^{n}\mathbf{A}_m\mathbf{W}_{n-m} + \mathbf{B}_n \tag{2.2.31}$$

Equations of this type are called convolution equations. When the matrices $\mathbf{W}_0$, $\mathbf{A}_1, \mathbf{A}_2, ...,$ and $\mathbf{B}_1, \mathbf{B}_2, ...,$ are given, the matrices

$$\mathbf{W}_1 = \mathbf{A}_1\mathbf{W}_0 + \mathbf{B}_1 \quad \mathbf{W}_2 = \mathbf{A}_1\mathbf{W}_1 + \mathbf{A}_2\mathbf{W}_0 + \mathbf{B}_2 \cdots$$

can be found by sequential substitution, so that the equation has a unique solution.

Define formally $\mathbf{A}_0=0$. Then equation (2.2.31) can be rewritten in the form

$$\mathbf{W}_n = \sum_{m=0}^{n}\mathbf{A}_m\mathbf{W}_{n-m} + \mathbf{B}_n \tag{2.2.32}$$

The latter equation has the following form in terms of generating functions:

$$\mathbf{W}(z) = \mathbf{A}(z)\mathbf{W}(z) + \mathbf{B}(z)$$

If $\mathbf{A}(z)$ is a square matrix, then we obtain the solution generating function

$$\mathbf{W}(z) = [\mathbf{I} - \mathbf{A}(z)]^{-1}\mathbf{B}(z) \tag{2.2.33}$$

The converse problem often interests us in calculations of matrix distributions. Given generating functions of unknown matrix probabilities, find the recurrence relations for their computation. Usually, these equations are obtained by factoring the generating function and using the convolution theorem.

EXAMPLE 2.2.5: Let us find the recurrence relations for the matrix distribution $\mathbf{P}_n(t)$ of t errors in a block of n bits according to the Gilbert model.

As we saw in example 2.2.2, the generating function of this distribution has the form

$$\mathbf{\Phi}_t^n(z) = [\mathbf{P}(0) + z\mathbf{P}(1)]^n$$

Since $[\mathbf{P}(0) + z\mathbf{P}(1)]^n = [\mathbf{P}(0) + z\mathbf{P}(1)]^{n-m}[\mathbf{P}(0) + z\mathbf{P}(1)]^m$, we obtain the equation

$$\mathbf{P}_n(t) = \sum_{\tau}\mathbf{P}_{n-m}(\tau)\mathbf{P}_m(t-\tau) \tag{2.2.34}$$

This equation is useful for fast computation of the distribution. ■

Let us consider closely the case when the elements of the generating function are rational fractions. As is shown in Appendix 2.1, the generating function can be written in the form of (A.2.2):

$$\mathbf{W}(z) = \mathbf{V}(z)/q(z) \tag{2.2.35}$$

where

$$q(z) = \sum_{i=0}^{k} q_i z^i \qquad \mathbf{V}(z) = \sum_{i=0}^{r} \mathbf{V}_i z^i$$

Bringing the previous relation to the common denominator, we obtain $q(z)\mathbf{W}(z) = \mathbf{V}(z)$. The comparison of the coefficients at the same powers of z leads us to the desired recurrence relations

$$\mathbf{W}_m = (1/q_0)\mathbf{V}_m - \sum_{i=1}^{k}(q_i/q_0)\mathbf{W}_{m-i} \tag{2.2.36}$$

where $\mathbf{W}_i = 0$ when $i < 0$. The corresponding scalar distribution $w_m = \mathbf{p}\mathbf{W}_m\mathbf{1}$, obviously, satisfies the equation

$$w_m = (1/q_0)v_m - \sum_{i=1}^{k}(q_i/q_0)w_{m-i} \tag{2.2.37}$$

where $v_m = \mathbf{pV}_m\mathbf{1}$.

Consider now some matrix distributions that generalize the most often used scalar distributions.

2.3 MATRIX DISTRIBUTIONS

2.3.1 Matrix Multinomial Distribution. Suppose that at every moment t one and only one event from a set $A_1, A_2, ..., A_r$ can occur. Each event depends only on $\mathbf{e}_t$ and has constant matrix probability:

$$\mathbf{P}_i = \mathbf{P}(A_i;t-1,t)$$

Let us find the matrix probability $\mathbf{P}_n(m_1,m_2,...,m_r)$ that in the time interval $(t,t+n]$ the event A_1 occurs m_1 times, A_2 occurs m_2, ..., and A_r occurs m_r times $(m_1+m_2+...+m_r=n)$.

In general, the direct calculation of the distribution $\mathbf{P}_n(m_1,m_2,...,m_r)$ is complicated by the fact that the matrices $\mathbf{P}_i$ and $\mathbf{P}_j$ do not commute. For example, the direct formula becomes complex even for n=5, r=2, m_1=3, m_2=2:

$$\begin{aligned}\mathbf{P}_5(3,2) = {} & \mathbf{P}_1^3\mathbf{P}_2^2+\mathbf{P}_1^2\mathbf{P}_2^2\mathbf{P}_1+\mathbf{P}_1\mathbf{P}_2^2\mathbf{P}_1^2+\mathbf{P}_2^2\mathbf{P}_1^3+\mathbf{P}_2\mathbf{P}_1^3\mathbf{P}_2+\\ & \mathbf{P}_2\mathbf{P}_1^2\mathbf{P}_2\mathbf{P}_1+\mathbf{P}_2\mathbf{P}_1\mathbf{P}_2\mathbf{P}_1^2+\mathbf{P}_1\mathbf{P}_2\mathbf{P}_1\mathbf{P}_2\mathbf{P}_1+\\ & \mathbf{P}_1\mathbf{P}_2\mathbf{P}_1^2\mathbf{P}_2+\mathbf{P}_1^2\mathbf{P}_2\mathbf{P}_1\mathbf{P}_2\end{aligned}$$

One can propose methods to calculate matrix probabilities based on recurrence relations which are obtained using the formula of total probability:

$$\begin{aligned}&\mathbf{P}_n(m_1,m_2,...,m_r) = \\ &\sum_{\mu_1,...,\mu_r}\mathbf{P}_{n-k}(m_1-\mu_1,m_2-\mu_2,...,m_r-\mu_r)\mathbf{P}_k(\mu_1,\mu_2,...,\mu_r)\end{aligned} \tag{2.3.1}$$

For k=1 we obtain the important particular case of this formula

$$\mathbf{P}_n(m_1,m_2,...,m_r) = \sum_{j=1}^{r}\mathbf{P}_{n-1}(m_1,...,m_j-1,...,m_r)\mathbf{P}_j \tag{2.3.2}$$

The recurrence relations (2.3.1) and (2.3.2) permit us to calculate the distribution $\mathbf{P}_n(m_1,m_2,...,m_r)$ for a not very large number of SSM states and n. Therefore, it is important to obtain approximate formulas and asymptotic estimations of this distribution. To solve this problem we use several generating functions of the matrix multinomial distribution.

The simplest generating function is given by the convolution theorem, or by using the recurrence relations (2.3.1):

$$\mathbf{\Phi}_n(z_1,z_2,\ldots,z_r) = (\sum_{i=1}^{r}\mathbf{P}_i z_i)^n \tag{2.3.3}$$

where

$$\mathbf{\Phi}_n(z_1,z_2,\ldots,z_r) = \sum_{m_1,\ldots,m_r} \mathbf{P}_n(m_1,m_2,\ldots,m_r) z_1^{m_1} z_2^{m_2} \cdots z_r^{m_r}$$

The elements of the matrix $\mathbf{\Phi}_n(z_1,z_2,\ldots,z_r)$ are polynomials in $z_1,z_2,\cdots z_r$. The matrix $\mathbf{P}_n(m_1,m_2,\ldots,m_r)$ can be found as a coefficient of $z_1^{m_1} z_2^{m_2} \cdots z_r^{m_r}$ in the expansion of this polynomial. In order to find the asymptotic formula when $n \to \infty$, it is convenient to use the generating function

$$\mathbf{\Psi}(w;z_1,z_2,\ldots,z_r) = \sum_{n=0}^{\infty} \mathbf{\Phi}_n(z_1,z_2,\ldots,z_r) w^n$$

where we assume $\mathbf{\Phi}_0 = \mathbf{I}$ for the convenience of summation. Obviously, the series on the right-hand side of the previous equation converges if $|z_i|<1$, $(i=1,2,\ldots,r)$, $|w|<1$, and by (2.3.3) its sum is

$$\mathbf{\Psi}(w;z_1,z_2,\ldots,z_r) = (\mathbf{I} - w\sum_{i=1}^{r}\mathbf{P}_i z_i)^{-1} \tag{2.3.4}$$

2.3.2 Matrix Binomial Distribution

2.3.2.1 Generating Functions. Let us study more closely a particular case, $r=2$, of the matrix multinomial distribution, the matrix binomial distribution, that plays the most important role in applications.

Denote $\mathbf{P}_1(0)=\mathbf{P}(0)$, $\mathbf{P}_1(1)=\mathbf{P}(1)$, $\mathbf{P}_n(m)=\mathbf{P}_n(n-m,m)$, $\mathbf{\Phi}_n(1,z)=\mathbf{\Phi}_n(z)$, and $\mathbf{\Psi}(w;1,z)=\mathbf{\Psi}(w;z)$. Using these simplified notations we can rewrite the basic equations from the previous section. Formula (2.3.1) now has the form

$$\mathbf{P}_n(m) = \sum_{\mu}\mathbf{P}_{n-\nu}(m-\mu)\mathbf{P}_\nu(\mu) \tag{2.3.5}$$

and formula (2.3.2) can be rewritten as

$$\mathbf{P}_n(m) = \mathbf{P}_{n-1}(m)\mathbf{P}(0) + \mathbf{P}_{n-1}(m-1)\mathbf{P}(1) \tag{2.3.6}$$

Generating functions (2.3.3) and (2.3.4) become

$$\mathbf{\Phi}_n(z) = [\mathbf{P}(0) + \mathbf{P}(1)z]^n \tag{2.3.7}$$

and

$$\mathbf{\Psi}(w;z) = [\mathbf{I} - \mathbf{P}(0)w - \mathbf{P}(1)wz]^{-1} \tag{2.3.8}$$

These equations have exactly the same form as in the case of the simple binomial distribution

$$P_n(m) = \binom{n}{m} p^m q^{n-m}$$

This is obviously a particular case of the matrix distribution when all the matrices are scalars (their sizes are equal to one). However, in the case of matrix probabilities this simple binomial formula is not generally valid.

One can take advantage of the relative simplicity of the generating functions and calculate the matrix distribution using the inverse transforms described in Sec. 2.2. In particular, the inversion formulas (2.2.4) and (2.2.14) yield

$$\mathbf{P}_n(m) = \frac{1}{2\pi \mathrm{j}} \oint_{\gamma} [\mathbf{P}(0)+\mathbf{P}(1)z]^n z^{-m-1} dz \tag{2.3.9}$$

$$\mathbf{P}_n(m) = \frac{1}{(2\pi \mathrm{j})^2} \int\int_{\gamma_1} [\mathbf{I}-\mathbf{P}(0)w-\mathbf{P}(1)wz]^{-1} z^{-m-1} w^{-n-1} dw dz \tag{2.3.10}$$

One more generating function may be obtained if we expand the generating function $\mathbf{\Psi}(w;z)$ into a power series:

$$\mathbf{\Psi}(w;z) = \sum_{m=0}^{\infty} [(\mathbf{I} - \mathbf{P}(0)w)^{-1}\mathbf{P}(1)w]^m (\mathbf{I}-\mathbf{P}(0)w)^{-1} z^m$$

The coefficient at z^m in this expansion is a generating function

$$\mathbf{\Psi}_m(w) = \sum_{n=m}^{\infty} \mathbf{P}_n(m) w^n = [(\mathbf{I} - \mathbf{P}(0)w)^{-1}\mathbf{P}(1)w]^m (\mathbf{I} - \mathbf{P}(0)w)^{-1} \tag{2.3.11}$$

which is convenient to use for asymptotic expansions when $n \to \infty$.

One can obtain recurrence relations that differ from the previous ones by using this generating function. Indeed, rewriting the function $\mathbf{\Psi}_m(w)$ in the form of the product $\mathbf{\Psi}_m(w) = \mathbf{\Psi}_k(w)\mathbf{P}(1)w\mathbf{\Psi}_{m-k-1}(w)$, we obtain from the convolution theorem

$$\mathbf{P}_n(m) = \sum_{\nu} \mathbf{P}_\nu(k)\mathbf{P}(1)\mathbf{P}_{n-\nu-1}(m-k-1) \tag{2.3.12}$$

and in particular

$$\mathbf{P}_n(m) = \sum_{\nu=0}^{n-m} \mathbf{P}^\nu(0)\mathbf{P}(1)\mathbf{P}_{n-\nu-1}(m-1) \tag{2.3.13}$$

This equation may be used to calculate $\mathbf{P}_n(m)$ for small m. For example,

$$\mathbf{P}_n(0) = \mathbf{P}^n(0) \quad \mathbf{P}_n(1) = \sum_{\nu=0}^{n-1} \mathbf{P}^\nu(0)\mathbf{P}(1)\mathbf{P}^{n-\nu-1}(0)$$

2.3.2.2 Fourier Transform. The relative simplicity of the generating functions of the matrix binomial distribution makes the transform method an attractive tool for calculating the distribution. Since generating function (2.3.7) of the distribution is a polynomial, we may use inversion formula (2.2.25) of the discrete Fourier transform:

$$\mathbf{P}_n(m) = p^{-1} \sum_{k=0}^{p-1} \alpha^{-mk}(\mathbf{P}(0) + \mathbf{P}(1)\alpha^k)^n \tag{2.3.14}$$

where $\alpha = e^{-2\pi j/p}$, $p > n$. Taking into account the symmetry of the function α^k, one may decrease the upper limit of summation in the above formula:

$$\mathbf{P}_n(m) = p^{-1}\mathbf{P}^n + \eta_{mp}(\mathbf{P}(0) - \mathbf{P}(1))^n + 2p^{-1}\mathrm{Re}\sum_{k=1}^{\nu}(\mathbf{P}(0) + \mathbf{P}(1)\alpha^k)^n\alpha^{-mk} \tag{2.3.15}$$

where $\nu = \lfloor 0.5p \rfloor$ is the integer part of $0.5p$, $\eta_{mp} = 0.5(-1)^m[1+(-1)^p]p^{-1}$. A dramatic increase in the speed of computation can be achieved by using the FFT.[47] If we select $p = 2^r$, then equation (2.3.14) becomes

$$\mathbf{P}_n(m) = 2^{-r}\sum_{k=0}^{p/2-1}(\mathbf{P}(0) + \mathbf{P}(1)\alpha^k)^n\alpha^{-mk} + 2^{-r}\sum_{k=p/2-1}^{p-1}(\mathbf{P}(0) + \mathbf{P}(1)\alpha^k)^n\alpha^{-mk}$$

which after obvious simplifications takes the form

$$\mathbf{P}_n(m) = 2^{-r} \sum_{k=0}^{p/2-1} [(\mathbf{P}(0)+\mathbf{P}(1)\alpha^k)^n + (-1)^m (\mathbf{P}(0)-\mathbf{P}(1)\alpha^k)^n]\alpha^{-mk}$$

If $m=2r$ is even, the previous equation becomes

$$\mathbf{P}_n(2l) = 2^{-r} \sum_{k=0}^{p/2-1} [(\mathbf{P}(0)+\mathbf{P}(1)\alpha^k)^n + (\mathbf{P}(0)-\mathbf{P}(1)\alpha^k)^n]\beta^{-lk} \tag{2.3.16}$$

where $\beta=\alpha^2$. The right-hand side of this equation is a $p/2$ point Fourier transform, and therefore we may use the $p/2$ point transform (instead of the original p point transform) to calculate the probability $\mathbf{P}_n(m)$ for even m. A similar result is valid for an odd m:

$$\mathbf{P}_n(2l+1) = 2^{-r} \sum_{k=0}^{p/2-1} [(\mathbf{P}(0)+\mathbf{P}(1)\alpha^k)^n - (\mathbf{P}(0)-\mathbf{P}(1)\alpha^k)^n]\alpha^k\beta^{-lk} \tag{2.3.17}$$

Thus, instead of having to compute the p point DFT for the function $2^{-r}(\mathbf{P}(0)+\mathbf{P}(1)\alpha^k)^n$, we may use the $p/2$ point DFTs for the function $2^{-r}[(\mathbf{P}(0)+\mathbf{P}(1)\alpha^k)^n + (\mathbf{P}(0)-\mathbf{P}(1)\alpha^k)^n]$ to calculate $\mathbf{P}_n(m)$ for an even m and $p/2$ point DFT for the function $2^{-r}\alpha^k[(\mathbf{P}(0)+\mathbf{P}(1)\alpha^k)^n - (\mathbf{P}(0)-\mathbf{P}(1)\alpha^k)^n]$ to calculate $\mathbf{P}_n(m)$ for an odd m. Using similar reasoning, we may break each of the two $p/2$ point DFTs into two $p/4$ point DFTs, and so on.

This method of calculating the DFT is called the Fast Fourier Transform (FFT). Generally, when p is large the FFT substantially increases the efficiency of computation. However, it requires a significant amount of computer memory to store p matrices $(\mathbf{P}(0)+\mathbf{P}(1)\alpha^k)^n$ and their transformations. In many applications we need to compute $\mathbf{P}_n(m)$ for only several values of m. In this case, we may prefer to use the original formula (2.3.14), which allows us to calculate $(\mathbf{P}(0)+\mathbf{P}(1)\alpha^k)^n$ on the fly rather than storing it for further use as required by the FFT. This method is especially useful for obtaining the matrix probabilities

$$\mathbf{P}_n(m_1<m\le m_2) = \sum_{m=m_1+1}^{m_2} \mathbf{P}_n(m)$$

that the number of occurrences of the event A_2 is greater than m_1 but not greater than m_2.

If we replace $\mathbf{P}_n(m)$ with the right-hand side of equation (2.3.14) and sum the geometric progressions, then

$$\mathbf{P}_n(m_1<m\le m_2) = \xi_{m_1,m_2}\mathbf{P}^n+\eta_{m_1m_2}(\mathbf{P}(0)-\mathbf{P}(1))^n+ \\ \mathrm{Re}\sum_{k=0}^{p/2-1}(\mathbf{P}(0)+\mathbf{P}(1)\alpha^k)^n\zeta_{m_1m_2k} \tag{2.3.18}$$

where

$$\xi_{m_1m_2} = (m_2 - m_1)/p$$

$$\eta_{m_1m_2} = [(-1)^{m_2}-(-1)^{m_1}][1-(-1)^p]/2p$$

$$\zeta_{m_1m_2k} = 2(\alpha^{(m_1+1)k}-\alpha^{(m_2+1)k})/p(1-\alpha^k)$$

This formula not only allows one to calculate $\mathbf{P}_n(m_1<m\le m_2)$ for large m_2-m_1 faster than with the FFT, but also delivers higher accuracy than does the FFT.

2.3.2.3 Asymptotic Expansion. As we mentioned above, it is convenient to use z–transforms when dealing with asymptotic expansions. If we replace $w=1/z$ in equation (2.3.11), we obtain the z-transform

$$\mathbf{\Gamma}_m(z) = z\,[(\mathbf{I}z - \mathbf{P}(0))^{-1}\mathbf{P}(1)]^m[\mathbf{I}z - \mathbf{P}(0)]^{-1} \tag{2.3.19}$$

Since this is a rational function, we may use the results of Appendix 2 to find the asymptotic relations. To transform this function to the form of (A.2.2), we express the inverse characteristic matrix (see Appendix 5) as

$$[\,\mathbf{I}z - \mathbf{P}(0)\,]^{-1} = \mathbf{B}(z)/\Delta(z) \tag{2.3.20}$$

where $\Delta(z) = \det(\mathbf{I}z - \mathbf{P}(0))$ is the characteristic polynomial of the matrix $\mathbf{P}(0)$ and $\mathbf{B}(z)$ is the characteristic adjoint matrix; i. e., its elements are adjuncts of the corresponding elements of the matrix $(\mathbf{I}z - \mathbf{P}(0)$. Equation (2.3.19) becomes

$$\mathbf{\Gamma}_m(z) = z\,[\mathbf{B}(z)\mathbf{P}(1)]^m\mathbf{B}(z)/\Delta^{m+1}(z) \tag{2.3.21}$$

The right-hand side of this equation has the form of (A.2.2), and therefore the distribution $\mathbf{P}_n(m)$ is given by formula (A.2.6).

Consider the most important particular case when the eigenvalues of the matrix $\mathbf{P}(0)$ are different or, in other words, the roots $z_1,z_2,...,z_k$ of the polynomial $\Delta(z)$ are all different. In this case

$$q(z) = \Delta^{m+1}(z) = (z-z_1)^{m+1}(z-z_2)^{m+1}\cdots(z-z_k)^{m+1} \tag{2.3.22}$$

By formula (A.2.6) we obtain

$$\mathbf{P}_n(m) = \sum_{j=1}^{k}\sum_{\nu=0}^{m}\mathbf{A}_{j\nu}\binom{n}{\nu}z_j^{n-\nu} \tag{2.3.23}$$

where

$$\mathbf{A}_{j\nu} = \mathbf{\Theta}_m^{(m-\nu)}(z_j)/(m-\nu)! \qquad \mathbf{\Theta}_m(z) = (z-z_j)^{m+1}\mathbf{\Gamma}_m(z)z^{-1}$$

It is convenient to use these equations if the dimensions of the matrices and m are not large.

Since the matrix z-transform $\mathbf{\Gamma}_m(z)$ has the form of (A.2.2), we can use formula (A.2.10) for the $\mathbf{P}_n(m)$ asymptotic expansion. Suppose that the matrix $\mathbf{P}(0)$ has a simple positive eigenvalue z_1 with the largest absolute value. In this case, the asymptotic formula (A.2.10) becomes

$$\mathbf{P}_n(m) \sim \mathbf{A}_{1m}\binom{n}{m}z_1^{n-m} \tag{2.3.24}$$

where

$$\mathbf{A}_{1m} = [\mathbf{B}(z_1)\mathbf{P}(1)]^m\mathbf{B}(z_1)/[\Delta'(z_1)]^{m+1}$$

This formula is exact if the matrices $\mathbf{P}(0)=p$ and $\mathbf{P}(1)=q$ are scalars (first-order matrices). Indeed, in this case the characteristic equation $\Delta(z)=z-q$, has only one root, $z_1=q$. Since $\mathbf{B}(z) = 1$ and $\mathbf{A}_{1m} = p^m$, we obtain, as expected, the usual binomial distribution

$$\mathbf{P}_n(m) = \binom{n}{m}p^m q^{n-m}$$

Using Theorem A.2.1, we obtain the asymptotic formula

$$\mathbf{P}_n(m) \sim [\mathbf{B}(z_1)/\Delta'(z_1)][n\mathbf{P}(1)\mathbf{B}(z_1)/\Delta'(z_1)]^m z_1^{n-m}/(r-1)! \tag{2.3.25}$$

which is analogous to the Poisson formula for approximating the binomial distribution.[39] Switching the matrices $\mathbf{P}(1)$ and $\mathbf{P}(0)$, as is usually done in the study of binomial distribution, we can obtain a formula analogous to (2.3.20) if m is close to n.

EXAMPLE 2.3.1: For Gilbert's model (see Sec. 1.1.3), let us compute the matrix probability $\mathbf{P}_n(m)$ that the total length of bursts inside a block of length n is equal to m.

The probability $\mathbf{P}_n(m)$ is equal to the probability that the Markov chain visits the second state m times during the transmission of the n consecutive bits. Therefore, this distribution represents the matrix binomial distribution with matrices

$$\mathbf{P}(0) = \begin{bmatrix} Q & 0 \\ p & 0 \end{bmatrix} \quad \mathbf{P}(1) = \begin{bmatrix} 0 & P \\ 0 & q \end{bmatrix}$$

We find the distribution using the z-transform (2.3.19). The inverse matrix

$$(\mathbf{I}z-\mathbf{P}(0))^{-1} = \begin{bmatrix} z-Q & 0 \\ -p & z \end{bmatrix}^{-1} = \mathbf{B}(z)/\Delta(z)$$

where

$$\mathbf{B}(z) = \begin{bmatrix} z & 0 \\ p & z-Q \end{bmatrix} \quad \Delta(w) = z - Q$$

Then, after matrix multiplication in (2.3.19), we obtain

$$\mathbf{\Gamma}_m(z) = \frac{zr^{m-1}(z)}{(z-Q)^{m+1}} \begin{bmatrix} Ppz & Pz(z-Q) \\ pr(z) & (z-Q)r(z) \end{bmatrix}$$

where $r(z) = qz+(p-Q)$. And, finally, equation (2.3.23) gives

$$\mathbf{P}_n(m) = \sum_{i=1}^{2}\sum_{\nu=0}^{m} \mathbf{A}_{i\nu}\binom{n}{\nu}Q^{n-\nu}$$

where

$$\mathbf{A}_{i\nu} = \mathbf{\Theta}_m^{(m-\nu)}(Q)/(m-\nu)! \qquad \mathbf{\Theta}_m(z) = (z-Q)^{m+1}\mathbf{\Gamma}_m(z)z^{-1}$$

After some elementary transformations the previous equation takes the form

$$\mathbf{A}_{i\nu} = [\mathbf{M}_1\binom{m+1}{m-\nu} + \mathbf{M}_2\binom{m}{m-\nu} + \mathbf{M}_3\binom{m-1}{m-\nu}]q^{m-\nu}(Pp)^{\nu-1}$$

$$\mathbf{M}_1 = \begin{bmatrix} 0 & P^3p^2q^{-2} \\ 0 & P^2p^2q^{-1} \end{bmatrix} \quad \mathbf{M}_2 = \begin{bmatrix} P^2p^2q^{-1} & Pp(Q+Qp-2p) \\ Pp^2 & -P^2p^2q^{-1} \end{bmatrix}$$

$$\mathbf{M}_3 = \begin{bmatrix} Pp(Q-p)q^{-1} & P^2p(p-Q)q^{-2} \\ 0 & 0 \end{bmatrix}$$ ■

2.3.2.4 The Central Limit Theorem. As in the case of the ordinary binomial distribution,

$$p_n(m) = \mathbf{pP}_n(m)\mathbf{1}$$

tends to have a normal distribution when $n \to \infty$ under some mild regularity conditions. More precisely, if a Markov chain with the matrix $\mathbf{P} = \mathbf{P}(0) + \mathbf{P}(1)$ is regular and $\mathbf{P}(0) \neq 0$ and $\mathbf{P}(1) \neq 0$, then $(m-a_n)/\sigma_n$ is asymptotically (0,1) normal.

Indeed, according to (2.3.7), the distribution characteristic function (see also Example 2.2.2) is equal to

$$\phi_n(t) = \mathbf{pQ}^n(t)\mathbf{1} \tag{2.3.26}$$

where

$$\mathbf{Q}(t) = \mathbf{P}(0) + \mathbf{P}(1)\mathrm{e}^{-\mathrm{j}t}$$

Since $\mathbf{Q}(0) = \mathbf{P}$ is a regular matrix, $\mathbf{Q}(t)$ has a simple eigenvalue $\lambda(t)$ for small t such that $\lambda(0) = 1$. Then, asymptotically, when $n \to \infty$ we obtain from the matrix $\mathbf{Q}(t)$ spectral representation (see Appendix 5.1.5)

$$\phi_n(t) \sim q(t)\lambda_1^n(t)$$

This means (see Ref. [48] p. 37) that, asymptotically, $\phi_n(t)$ may be regarded as a characteristic function of a sum of independent equally distributed variables, which, as is well known,[39] is asymptotically normal.

2.3.3 Matrix Poisson Distribution. We can also derive a matrix analog of Poisson distribution as a limiting case of the matrix binomial distribution. Let us denote $n = t/\Delta t$ and assume that

$$\mathbf{P} = \mathbf{I} + (\mathbf{C} + \mathbf{D})\Delta t + o(\Delta t) \quad \mathbf{P}(1) = \mathbf{D}\Delta t + o(\Delta t)$$

when $\Delta t \to 0$. Generating function (2.3.7) has a limit

$$\lim_{n \to \infty} [\mathbf{P}(0) + z\mathbf{P}(1)]^n = \mathrm{e}^{(\mathbf{C}+z\mathbf{D})t}$$

which can be treated as a generating function of matrix Poisson distribution. This generating function [41,123,124,125] has numerous applications in queueing theory. A natural limiting case of the model with different state error probabilities considered in Sec. 1.1.2 is the so-called Markov modulated Poisson process which is a doubly stochastic Poisson process with the rate depending on the state of a continuous-time Markov chain.

The matrix Poisson distribution satisfies relations that are similar to the matrix binomial distribution. In particular, since

$$e^{(\mathbf{C}+z\mathbf{D})t} = e^{(\mathbf{C}+z\mathbf{D})(t-\tau)}e^{(\mathbf{C}+z\mathbf{D})\tau}$$

according to the convolution theorem, the matrix Poisson distribution $\mathbf{P}(m,t)$ satisfies the equation

$$\mathbf{P}(m,t) = \sum_{\mu=0}^{m}\mathbf{P}(m-\mu,t-\tau)\mathbf{P}(\mu,\tau)$$

which is similar to equation (2.3.5). We cannot use this equation the same way as (2.3.5) for calculating matrix probabilities recursively, because the right-hand side and the left-hand side of it contain matrix probabilities of m events at different time intervals. This equation can be used to calculate the probabilities $\mathbf{P}(m,t)$, given the probabilities $\mathbf{P}(0,\tau),\mathbf{P}(1,\tau),...,\mathbf{P}(m,\tau)$ for $\tau = t/2$. In order to have the same time intervals in both sides of the equation we let $\tau \to 0$. Then we obtain the following system of recursive differential equations:[125]

$$\frac{d}{dt}\mathbf{P}(m,t) = \mathbf{P}(m,t)\mathbf{C} + \mathbf{P}(m-1,t)\mathbf{D} \quad m \geq 1 \quad t \geq 0$$

$$\mathbf{P}(0,t) = e^{\mathbf{C}t}$$

This may be considered a generalization of (2.3.6). One can obtain the matrix Poisson distribution by solving this system:

$$\mathbf{P}(m,t) = \int_0^t \mathbf{P}(m-1,\tau)\mathbf{D}e^{\mathbf{C}(t-\tau)}d\tau$$

Several suggestions for solving this system numerically have been given in Ref. [41].

The analog of generating function (2.3.8) is the Laplace transform of the above-defined matrix Poisson generating function:

$$\Psi(p;z) = \int_0^\infty e^{(\mathbf{C}+z\mathbf{D})t} e^{-pt} dt = (\mathbf{I}p - \mathbf{C} - z\mathbf{D})^{-1}$$

The coefficient at z^m in this function's expansion is a Laplace transform

$$\Psi_m(p) = \int_0^\infty \mathbf{P}(m,t) e^{-pt} dt = [\mathbf{I}p - \mathbf{C})^{-1}\mathbf{D}]^m (\mathbf{I}p - \mathbf{C})^{-1}$$

which, according to the convolution theorem, leads to the recurrence equation

$$\mathbf{P}(m,t) = \int_0^t \mathbf{P}(k,\tau)\mathbf{D}\mathbf{P}(m-k-1,t-\tau)d\tau$$

This equation is similar to (2.3.12).

In many applications of matrix Poisson distribution it is convenient to use its integral representations. Using Cauchy's integral we obtain the equation

$$\mathbf{P}(m,t) = \frac{1}{2\pi j} \oint_\gamma e^{(\mathbf{C}+z\mathbf{D})t} z^{-m-1} dz$$

This equation is similar to (2.3.9). The matrix probability $\mathbf{P}(m_1<m\le m_2)$ that the number of occurrences of the event A_2 is greater than m_1 but not greater than m_2 can be expressed as

$$\mathbf{P}(m_1<m\le m_2,t) = \frac{1}{2\pi j} \oint_\gamma e^{(\mathbf{C}+z\mathbf{D})t} \frac{z^{-m_1} - z^{-m_2}}{z^2-z} dz$$

By replacing z with $e^{j\phi}$ we can convert these integrals into Fourier transforms. The inversion of the Laplace transform then yields

$$\mathbf{P}(m,t) = \frac{1}{2\pi j} \int_{a-j\infty}^{a+j\infty} \Psi_m(p) e^{pt} dp$$

Since $\Psi_m(p)$ is a rational function, this integral can be calculated using the residue theorem.

2.3.4 Matrix Pascal Distribution. Along with matrix binomial distribution, we often use matrix Pascal distribution and matrix-geometric distribution in particular.

Assume that two events A_1 and A_2 are mutually exclusive, depend only on the value $\mathbf{e}_t$, and have matrix probabilities $\mathbf{P}(A_1) = \mathbf{P}(0)$ and $\mathbf{P}(A_2) = \mathbf{P}(1)$. Let us find the probability distribution $\mathbf{P}(l,r)$ that event A_2 occurs for the r-th time on the

Markov chain's l-th step.

If $r=1$, then clearly the matrix probability that event A_2 occurs for the first time on the l-th step is equal to

$$\mathbf{P}(l, 1) = \mathbf{P}^{l-1}(A_1)\mathbf{P}(A_2) = \mathbf{P}^{l-1}(0)\mathbf{P}(1) \tag{2.3.27}$$

This distribution, called the matrix-geometric distribution, has the generating function

$$\mathbf{\Theta}_1(z) = \sum_{l=1}^{\infty}\mathbf{P}^{l-1}(0)\mathbf{P}(1)z^l = (\mathbf{I} - \mathbf{P}(0)z)^{-1}\mathbf{P}(1)z \tag{2.3.28}$$

In general, the variable l can be represented as a sum of r conditionally independent variables $l=l_1+l_2+\cdots+l_r$, each having geometric distribution (2.3.27). By the convolution theorem, the generating function $\mathbf{\Theta}_r(z)$ of the matrix distribution $\mathbf{P}(l,r)$ is equal to the product of the generating functions of the addends:

$$\mathbf{\Theta}_r(z) = \mathbf{\Theta}_1^r(z) = [(\mathbf{I} - \mathbf{P}(0)z)^{-1}\mathbf{P}(1)z]^r \tag{2.3.29}$$

The distribution $\mathbf{P}(l,r)$ is called the matrix Pascal distribution. Since

$$\mathbf{\Theta}_r(z) = \mathbf{\Theta}_{r_1}(z)\mathbf{\Theta}_{r-r_1}(z)$$

this distribution satisfies the recurrence relations

$$\mathbf{P}(l,r) = \sum_x \mathbf{P}(x,r_1)\mathbf{P}(l-x,r-r_1) \tag{2.3.30}$$

and in particular

$$\mathbf{P}(l,r) = \sum_x \mathbf{P}^{x-1}(0)\mathbf{P}(1)\mathbf{P}(l-x,r-1) \tag{2.3.31}$$

The generating function $\mathbf{\Theta}_r(z)$ differs from $\mathbf{\Psi}_r(z)$, defined by formula (2.3.11), by only a factor:

$$\mathbf{\Theta}_r(z) = \mathbf{\Psi}_r(z)[\mathbf{I} - \mathbf{P}(0)z\,] \tag{2.3.32}$$

and therefore satisfies the similar asymptotic relations

$$\mathbf{P}(l,r) \sim [\mathbf{B}(z_1)\mathbf{P}(1)/\Delta'(z_1)]^r l^{r-1} z_1^{l-r}/(r-1)! \tag{2.3.33}$$

where z_1 is the greatest eigenvalue of the matrix $\mathbf{P}(0)$.

2.4 MARKOV FUNCTIONS

2.4.1 Block Matrix Probabilities. As was shown in the previous chapter, it is sometimes convenient to describe the SSM model as a Markov function. In this case the set of the Markov chain states can be partitioned into subsets corresponding to the events, and the transition probability matrix can be presented in the block form

$$\mathbf{P} = block[\mathbf{P}_{ij}]_{s,s} = \begin{bmatrix} \mathbf{P}_{11} & \mathbf{P}_{12} & \cdots & \mathbf{P}_{1s} \\ \mathbf{P}_{21} & \mathbf{P}_{22} & \cdots & \mathbf{P}_{2s} \\ \cdots & \cdots & \cdots & \cdots \\ \mathbf{P}_{s1} & \mathbf{P}_{s2} & \cdots & \mathbf{P}_{ss} \end{bmatrix} \tag{2.4.1}$$

The case when events A_i do not depend on the states deterministically can be transformed into a case of the deterministic dependency by increasing the number of the Markov chain states. Indeed, consider a new Markov chain whose states are defined as vectors (i,k), where i is the state of the original chain and k is the event A_k index. If the so-defined chain is in the state (i,k), then the event A_k occurs. The new chain matrix has the form

$$\mathbf{P}_1 = \begin{bmatrix} \mathbf{P}(A_1) & \mathbf{P}(A_2) & \cdots & \mathbf{P}(A_s) \\ \mathbf{P}(A_1) & \mathbf{P}(A_2) & \cdots & \mathbf{P}(A_s) \\ \cdots & \cdots & \cdots & \cdots \\ \mathbf{P}(A_1) & \mathbf{P}(A_2) & \cdots & \mathbf{P}(A_s) \end{bmatrix}$$

The price that we pay for such a description is increase in the number of states by the factor of s. Sometimes, however, by using specifics of the process it is possible to achieve the equivalent description with fewer states. We illustrate this statement in the following example.

EXAMPLE 2.4.1: As we saw in Example 1.2.3, Gilbert's model with two states can be described as a function of the Markov chain with three states and the matrix

$$\mathbf{P}_1 = \begin{bmatrix} Q & Ph & P(1-h) \\ p & qh & q(1-h) \\ p & qh & q(1-h) \end{bmatrix}$$

when errors occur only in the third state. If A_2 denotes an error event and A_1 its complement (e_t=0), then this matrix $\mathbf{P}_1$ can be written in form (2.4.1)

$$\mathbf{P}_1 = \begin{bmatrix} \mathbf{P}_{11} & \mathbf{P}_{12} \\ \mathbf{P}_{21} & \mathbf{P}_{22} \end{bmatrix}$$

where

$$\mathbf{P}_{11} = \begin{bmatrix} Q & Ph \\ p & qh \end{bmatrix} \quad \mathbf{P}_{12} = \begin{bmatrix} P(1-h) \\ q(1-h) \end{bmatrix} \quad \mathbf{P}_{21} = \begin{bmatrix} p & qh \end{bmatrix} \quad \mathbf{P}_{22} = q(1-h)$$ ■

Let us express the principal distributions using the transition matrix blocks. Since the Markov function is a particular case of the SSM (see Sec. 1.2.4), the corresponding formulas are corollaries of the formulas of the previous sections. The matrices $\mathbf{P}(A_i)$ have the form

$$\mathbf{P}(A_i) = \begin{bmatrix} 0 & \dots & 0 & \mathbf{P}_{1i} & 0 & \dots & 0 \\ 0 & \dots & 0 & \mathbf{P}_{2i} & 0 & \dots & 0 \\ \cdot & \cdot & \cdot & \cdot & \cdot & \cdot & \cdot \\ 0 & \dots & 0 & \mathbf{P}_{si} & 0 & \dots & 0 \end{bmatrix} \tag{2.4.2}$$

2.4.2 Matrix Multinomial Distribution. All the formulas from section 2.3 are preserved if the matrices $\mathbf{P}(A_i)$ have the form of (2.4.2). Noting that the matrices have many zero elements, we can rewrite the formulas using block matrices whose size is considerably smaller than the size of the matrix $\mathbf{P}(A_i)$. Let $\mathbf{P}_{ij}(m_1,m_2,\dots,m_r;n)$ be the blocks of the matrix $\mathbf{P}_n(m_1,m_2,\dots,m_r)$. Then it follows from (2.4.1) that

$$\mathbf{P}_{ij}(m_1,m_2,\dots,m_r;n) = \sum_{k,\mu_1,\dots,\mu_r} \mathbf{P}_{ik}(m_1-\mu_1,m_2-\mu_2,\dots,m_r-\mu_r;n-m)\mathbf{P}_{kj}(\mu_1,\mu_2,\dots,\mu_r;m) \tag{2.4.3}$$

and in particular

$$\mathbf{P}_{ij}(m_1,m_2,\dots,m_r;n) = \sum_{k=1}^{s} \mathbf{P}_{ik}(m_1,\dots,m_j-1,\dots,m_r;n-1)\mathbf{P}_{kj} \tag{2.4.4}$$

The corresponding generating functions can also be rewritten in the block form

$$\mathbf{\Phi}_{ij}(z_1,z_2,\dots,z_r;n) = \sum_{k=1}^{s}\mathbf{\Phi}_{ik}(z_1,z_2,\dots,z_r;n-m)\mathbf{\Phi}_{kj}(z_1,z_2,\dots,z_r;m) \qquad (2.4.5)$$

where $\mathbf{\Phi}_{ij}(z_1,z_2,\dots,z_r;n)$ are the blocks of the generating function (2.3.3), which has the form

$$\mathbf{\Phi}_n(z_1,z_2,\dots,z_r) = [\,\mathbf{PT}(z_1,z_2,\dots,z_r)\,]^n \qquad (2.4.6)$$

where

$$\mathbf{T} = block\ diag\,\{\mathbf{I}_i z_i\} = \sum_{k=1}^{s}{}^{(\cdot)}\mathbf{I}_k z_k \qquad (2.4.7)$$

The symbol $\sum^{(\cdot)}$ denotes a direct sum of matrices which is defined by (2.4.7).

In the particular case of matrix binomial distribution, we have

$$\mathbf{P}(A_1) = \mathbf{P}(0) = \begin{bmatrix}\mathbf{P}_{11} & 0\\ \mathbf{P}_{21} & 0\end{bmatrix} \qquad \mathbf{P}(A_2) = \mathbf{P}(1) = \begin{bmatrix}0 & \mathbf{P}_{12}\\ 0 & \mathbf{P}_{22}\end{bmatrix} \qquad (2.4.8)$$

and the recurrence relations (2.4.4) become

$$\begin{aligned}\mathbf{P}_{i1}(m;n) &= \mathbf{P}_{i1}(m;n-1)\mathbf{P}_{11} + \mathbf{P}_{i2}(m;n-1)\mathbf{P}_{21}\\ \mathbf{P}_{i2}(m;n) &= \mathbf{P}_{i1}(m-1;n-1)\mathbf{P}_{12} + \mathbf{P}_{i2}(m-1;n-1)\mathbf{P}_{22}\end{aligned} \qquad (2.4.9)$$

Generating function (2.3.7) is given by

$$\mathbf{\Phi}_n(z) = \begin{bmatrix}\mathbf{P}_{11} & \mathbf{P}_{12}z\\ \mathbf{P}_{21} & \mathbf{P}_{22}z\end{bmatrix}^n \qquad (2.4.10)$$

Finally, by (2.3.19) and (2.4.8) we have

$$\mathbf{\Gamma}_m(z) = z\,[(\mathbf{I}z - \mathbf{P}(0))^{-1}\mathbf{P}(1)]^m(\mathbf{I}z - \mathbf{P}(0))^{-1}$$

where

$$(\mathbf{I}z - \mathbf{P}(0))^{-1} = \begin{bmatrix}(\mathbf{I}z-\mathbf{P}_{11})^{-1} & 0\\ z^{-1}\mathbf{P}_{21}(\mathbf{I}z-\mathbf{P}_{11})^{-1} & z^{-1}\mathbf{I}\end{bmatrix}$$

After some straightforward transformations we obtain

$$\mathbf{\Gamma}_m(z) = z^{-m}\Delta_1^{-m-1}(z)\mathbf{A}(z)\mathbf{C}^{m-1}(z)\mathbf{D}(z)$$

where

$$\mathbf{A}(z) = \begin{bmatrix} (\mathbf{I}z - \mathbf{P}_{11})^{-1}\mathbf{P}_{12}z \\ \mathbf{C}(z) \end{bmatrix}$$

$$\mathbf{C}(z) = \mathbf{P}_{21}\mathbf{B}_1(z)\mathbf{P}_{12}+\Delta_1(z)\mathbf{P}_{22} \quad \mathbf{D}(z) = \begin{bmatrix}\mathbf{P}_{21}\mathbf{B}_1(z) & \mathbf{I}\end{bmatrix}$$

$\Delta_1(z)=\det(\mathbf{I}z-\mathbf{P}_{11})$ is the characteristic polynomial of the matrix $\mathbf{P}_{11}$, and $\mathbf{B}_1(z)=\Delta_1(z)(\mathbf{I}z-\mathbf{P}_{11})^{-1}$ is its adjoint matrix. Since $\mathbf{\Gamma}_m(z)$ is a rational function, the matrices $\mathbf{P}_{ij}(m;n)$ can be expressed in a form similar to that of (2.3.23). However, the sizes of the matrices in these equations are normally significantly smaller than the sizes of matrices in equation (2.3.23).

Suppose that the characteristic polynomial $\Delta_1(z)$ has a simple root z_1 whose absolute value is greater than the absolute values of its remaining roots. For this to be true, by the theorem of Frobenius, it is sufficient that the matrix $\mathbf{P}_{11}$ is irreducible and acyclic (see Appendix 6). (If, in addition, $\mathbf{P}_{12}\mathbf{1} > 0$, then $z_1 < 1$.) Then the asymptotic representations of the matrices $\mathbf{P}_{ij}(m;n)$ have the form of (2.3.24).

Analyzing the results of this section, we conclude that block matrix distributions have a more complex form than in terms of matrices $\mathbf{P}(A_i)$. However, since the orders of the blocks are smaller than the order of the whole matrix, it is worthwhile to use the formulas in the block form in practical calculations.

2.4.3 Interval Distributions of Markov Functions. Let $\mathbf{P}_{ij}(l)$ be the matrix probability that a Markov chain leaving the states of the set A_i for the first time enters the states of the set A_j on the l-th step. Denote $B_j=\overline{A}_j$ the complement of the subset A_j with respect to the whole set of states. Then the distribution $\mathbf{P}_{ij}(l)$ is given by

$$\begin{aligned} &\mathbf{P}_{ij}(1) = \mathbf{P}_{ij} \\ &\mathbf{P}_{ij}(l) = \mathbf{P}(B_j \mid A_i)\mathbf{P}^{l-2}(B_j \mid B_j)\mathbf{P}(A_j \mid B_j) \quad \text{for } l > 1 \end{aligned} \tag{2.4.11}$$

where the matrix $\mathbf{P}(B_j \mid A_i)$ is the matrix obtained from matrix $\mathbf{P}$ by eliminating all the columns corresponding to the states of A_j and all the rows corresponding to the states that do not belong to the set A_i; matrix $\mathbf{P}(B_j \mid B_j)$ is the matrix obtained from matrix $\mathbf{P}$ by eliminating all the rows and columns corresponding to the states of A_j; and matrix $\mathbf{P}(A_j \mid B_j)$ is the matrix obtained from matrix $\mathbf{P}$ by eliminating all the columns corresponding to the states that do not belong to A_j and all the rows corresponding to the states of the set A_j.

For example, if $i=1$, $j=2$, $s=3$, and

$$\mathbf{P} = \begin{bmatrix} \mathbf{P}_{11} & \mathbf{P}_{12} & \mathbf{P}_{13} \\ \mathbf{P}_{21} & \mathbf{P}_{22} & \mathbf{P}_{23} \\ \mathbf{P}_{31} & \mathbf{P}_{32} & \mathbf{P}_{33} \end{bmatrix}$$

then

$$\mathbf{P}(B_2|A_1) = \begin{bmatrix} \mathbf{P}_{11} & \mathbf{P}_{13} \end{bmatrix} \quad \mathbf{P}(B_2|B_2) = \begin{bmatrix} \mathbf{P}_{11} & \mathbf{P}_{13} \\ \mathbf{P}_{31} & \mathbf{P}_{33} \end{bmatrix} \quad \mathbf{P}(A_2|B_2) = \begin{bmatrix} \mathbf{P}_{12} \\ \mathbf{P}_{32} \end{bmatrix}$$

The generating function of the matrices $\mathbf{P}_{ij}(l)$ is given by

$$\boldsymbol{\theta}_{ij}(z) = \sum_{l=1}^{\infty} \mathbf{P}_{ij}(l) z^l = z\mathbf{P}_{ij} + \sum_{l=2}^{\infty} z^2 \mathbf{P}(B_j|A_i)[z\mathbf{P}(B_j|B_j)]^{l-2}\mathbf{P}(A_j|B_j)$$

After summing the matrix-geometric progression (see Appendix 5.1.3) we obtain

$$\boldsymbol{\theta}_{ij}(z) = z\mathbf{P}_{ij} + z^2\mathbf{P}(B_j|A_i)[\mathbf{I} - z\mathbf{P}(B_j|B_j)]^{-1}\mathbf{P}(A_j|B_j) \tag{2.4.12}$$

In the particular case when $i=j$, the previous expression gives the generating function of the distribution of the time of the first return to the subset A_i:

$$\boldsymbol{\theta}_{ii}(z) = z\mathbf{P}_{ii} + z^2\mathbf{P}(B_i|A_i)[\mathbf{I} - z\mathbf{P}(B_j|B_j)]^{-1}\mathbf{P}(A_j|B_j) \tag{2.4.13}$$

Quite analogously, one can find the generating functions of the so-called tabu distributions.[49] Let ${}_h\mathbf{P}_{ij}(l)$ be the matrix probability that the Markov chain leaving the states of the set A_i arrives at a state of the set A_j for the first time in l steps without visiting the states of the set A_h or returning to A_i. If we denote $B_g = \bar{A}_i \cap \bar{A}_j \cap \bar{A}_h$, then this distribution can be written in the form

$$\begin{aligned} {}_h\mathbf{P}_{ij}(1) &= \mathbf{P}_{ij} \\ {}_h\mathbf{P}_{ij}(l) &= \mathbf{P}(B_g|A_i)\mathbf{P}^{l-2}(B_g|B_g)\mathbf{P}(A_j|B_g) \quad \text{for } l>1 \end{aligned} \tag{2.4.14}$$

But then the generating function is given by

$${}_h\boldsymbol{\theta}_{ij}(z) = z\mathbf{P}_{ij} + z\mathbf{P}(B_g|A_i)[\mathbf{I} - z\mathbf{P}(B_g|B_g)]^{-1}z\mathbf{P}(A_j|B_g) \tag{2.4.15}$$

Using these generating functions and applying the convolution theorem, one can find the generating function of the interval sum distributions. For example, the generating function of the matrix probabilities that the Markov chain leaving the states of the set A_i enters a state of the set A_j for the r-th time on the l-th step is equal to the product $\boldsymbol{\theta}_{ij}(z)\boldsymbol{\theta}_{jj}^{r-1}(z)$ of the generating functions defined above.

It is convenient to use the generating functions of this section to find mean values for the intervals. For example, the mean interval of entering the states of A_j can be found using generating function (2.4.12):

$$\mathbf{m}_{ij} = \boldsymbol{\theta}'_{ij}(1)\mathbf{1}$$

After taking the derivative we obtain

$$\boldsymbol{\theta}'_{ij}(1) = \mathbf{P}_{ij}\mathbf{1} + 2\mathbf{P}(B_j|A_i)[\mathbf{I} - \mathbf{P}(B_j|B_j)]^{-1}\mathbf{P}(A_j|B_j)\mathbf{1} +$$
$$\mathbf{P}(B_j|A_i)[\mathbf{I} - \mathbf{P}(B_j|B_j)]^{-2}\mathbf{P}(A_j|B_j)\mathbf{1}$$

Since $\mathbf{P}$ is a stochastic matrix, $\mathbf{P1} = \mathbf{1}$ and, therefore,

$$\mathbf{P}(B_j|B_j)\mathbf{1} + \mathbf{P}(A_j|B_j)\mathbf{1} = \mathbf{1}$$

This can be rewritten as

$$[\mathbf{I} - \mathbf{P}(B_j|B_j)]^{-1}\mathbf{P}(A_j|B_j)\mathbf{1} = \mathbf{1}$$

Using this formula we obtain

$$\mathbf{m}_{ij} = \mathbf{1} + \mathbf{P}(B_j|A_i)[\mathbf{I} - \mathbf{P}(B_j|B_j)]^{-1}\mathbf{1}$$

2.4.4 Signal Flow Graph Applications. When dealing with Markov functions, we find it convenient to compute matrix probabilities and generating functions with the help of signal-flow graphs.[40,50] We assume that the events depend deterministically on Markov chain states and that the transition matrix has the form of (2.4.1). Accordingly, we consider a block stochastic graph (see Appendix 2.3) whose matrix branch transmissions are $\mathbf{T}_{ij}=\mathbf{P}_{ij}$. In this section we consider problems that can be solved with the help of signal-flow graphs.

2.4.4.1 Exclusion of Unobserved States. Suppose that we observe a Markov chain only when it is in a subset A_i. It is obvious that the new process is the Markov chain. If we denote, as in the previous section, $B_i=\overline{A}_i$ the subset A_i complement, then the transition matrix of the new chain is given by[40]

$$\overline{\mathbf{P}}_i = \mathbf{P}_{ii} + \sum_{m=2}^{\infty}\mathbf{P}(B_i|A_i)\mathbf{P}(B_i|B_i)^{m-2}\mathbf{P}(A_i|B_i)$$

or

$$\bar{\mathbf{P}}_i = \mathbf{P}_{ii} + \mathbf{P}(B_i | A_i)[\mathbf{I} - \mathbf{P}(B_i | B_i)]^{-1}\mathbf{P}(A_i | B_i) \tag{2.4.16}$$

(provided, of course, that the set B_i is not absorbing).

This expression has exactly the same form as formula (A.2.23). Therefore according to the results of Appendix 2 we conclude that a stochastic signal-flow graph of the observed process is the *residue*[50] of the original graph after absorption of the nodes that correspond to the unobserved states. The transfer matrix of the residue graph is the transfer matrix of the observed process.

2.4.4.2 The Generating Function of the Distribution of Transition Time. If we compare equation (2.4.15) with expression (A.2.23), we conclude that the matrix generating function ${}_h\boldsymbol{\theta}_{ij}(z)$ can be computed as the transfer matrix from the subgraph corresponding to the states of the set A_i to the subgraph corresponding to the states of the set A_j in the residue graph obtained from the original stochastic graph with the matrix $z\mathbf{P}$, after absorption of the subgraph corresponding to the states of the set B_g.[51]

2.4.4.3 Multinomial Distribution Graph. We have shown in Appendix 2 that the elementary operations of matrix algebra can be performed using transformation of some signal-flow graphs. Since the generating function of the matrix multinomial distribution may be expressed as a matrix product by equation (2.4.6), it can be found as a transfer matrix from subgraph G_1 to subgraph G_n of the graph shown in Fig. 2.1.

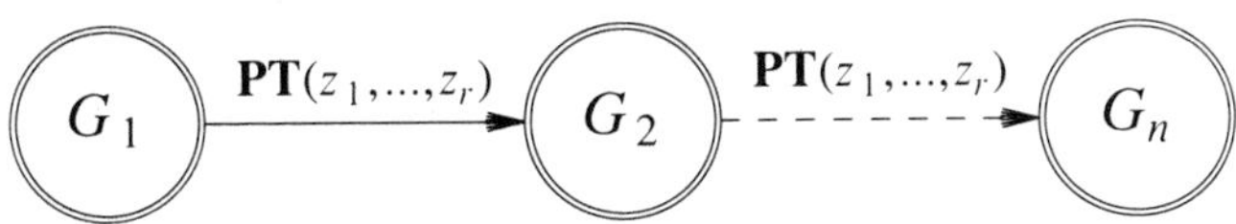

Figure 2.1. Matrix multinomial distribution transition graph.

2.4.4.4 Computing Scalar Probabilities and Generating Functions. As shown above, scalar probabilities and generating functions can be found from their matrix counterparts as matrix products

$$Pr(A) = \mathbf{pP}(A)\mathbf{1} \quad \phi(z) = \mathbf{p\Phi}(z)\mathbf{1}$$

Since matrix products may be interpreted using signal-flow graphs, the above equations can be presented as transfers from node S to node F of the graph

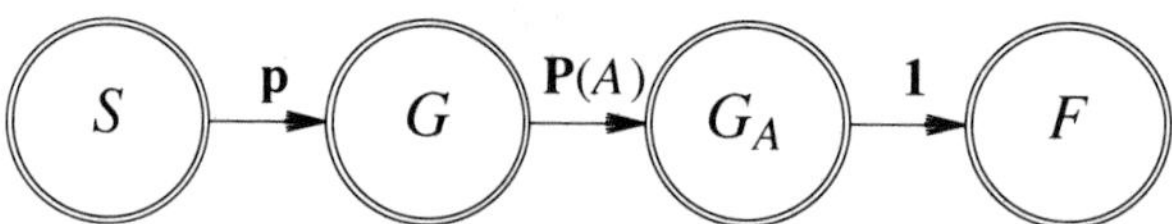

Figure 2.2. Scalar probability calculation.

depicted in Fig. 2.2.

2.4.4.5 Stationary Probabilities. We can find stationary probabilities of a Markov chain using signal-flow graphs,[52] since the stationary distribution satisfies the equation

$$\boldsymbol{\pi}\mathbf{P} = \boldsymbol{\pi} \qquad \boldsymbol{\pi}\mathbf{1} = 1 \tag{2.4.17}$$

However, Mason's formula[50] cannot be applied directly to the equation $\boldsymbol{\pi}\mathbf{P}=\boldsymbol{\pi}$, because of the matrix $\mathbf{I}-\mathbf{P}$ singularity [$\det(\mathbf{I}-\mathbf{P}) = 0$]. To solve the system, the graph with the matrix $\mathbf{P}$ should be replaced by another graph, or subgraphs of the graph are considered.

We can propose other methods of computing the stationary probabilities. One of them consists of computing the matrix $(\mathbf{I}z-\mathbf{P})^{-1}$ when $|z|>1$. This transformation is performed by absorbing the loop in the graph of Fig. 2.3.

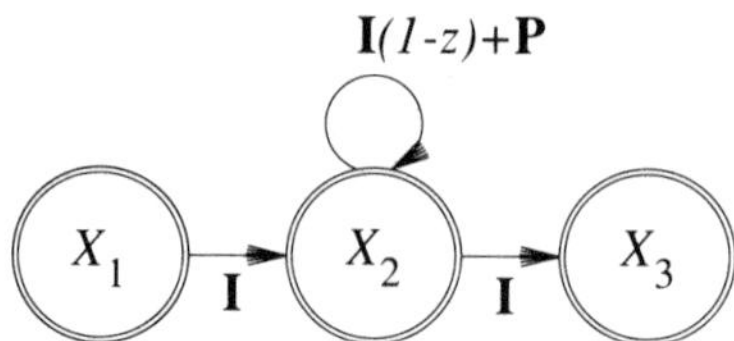

Figure 2.3. Characteristic matrix graph.

In Sec. 1.2 we saw that a stationary distribution can be found as

$$\lim_{k\to\infty} \mathbf{P}^k = \mathbf{1}\boldsymbol{\pi}$$

On the other hand, the same limit may be obtained from the sequence-generating function [39]

$$\mathbf{1P} = \lim_{z\downarrow 1}(z-1)(\mathbf{I}z-\mathbf{P})^{-1} = \mathbf{B}(1)/\Delta'(1)$$

where $\Delta(z)$ is the matrix characteristic polynomial and $\mathbf{B}(z)$ is its adjoint matrix (see Appendix 5).

Since all the rows of the matrix $\mathbf{1\pi}$ are identical, it is sufficient to find only one row of the matrix $(\mathbf{I}z - \mathbf{P})^{-1}$, as is illustrated in the following example.

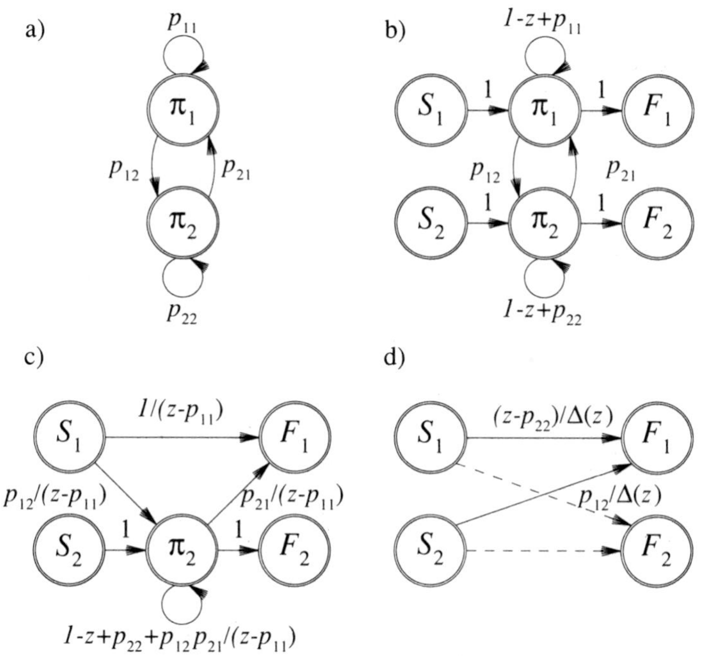

Figure 2.4. Reduction of a characteristic matrix graph.

EXAMPLE 2.4.2: Consider a Markov chain with the matrix

$$\mathbf{P} = \begin{bmatrix} p_{11} & p_{12} \\ p_{21} & p_{22} \end{bmatrix}$$

whose graph is shown in Fig. 2.4a.

In order to invert the matrix $\mathbf{I}z-\mathbf{P}$ we create the block graph shown in Fig. 2.3. This graph is shown in Fig. 2.4b. After the absorption of the nodes π_1 and π_2 using methods described in Ref. [50], we obtain subsequently the graphs shown in Figs. 2.4c and 2.4d.

The first line of the matrix $(\mathbf{I}z - \mathbf{P})^{-1}$ is presented by the transitions from vertices S_1, S_2 to the node F_1 which is

$$\frac{1}{\Delta(z)} \begin{bmatrix} z-p_{22} & 1 \end{bmatrix}$$

where $\Delta(z)=(z-p_{22})(z-p_{11})-p_{12}p_{21}$. After multiplying this row matrix by $z-1$, we obtain, when $z \to \infty$,

$$\boldsymbol{\pi} = \frac{1}{\Delta'(1)} \begin{bmatrix} 1-p_{22} & 1 \end{bmatrix} = \begin{bmatrix} \dfrac{p_{21}}{p_{12}+p_{21}} & \dfrac{p_{12}}{p_{12}+p_{21}} \end{bmatrix}$$ ■

It is also possible to find the stationary probabilities by representing system (2.4.17) in the form of a graph. All the vertices of the stochastic graph with the matrix **P** have to be connected with an additional node that is ascribed the value of 1, and all the transfers to this node should be equal to 1. One has to take into account that in this case Mason's formula is inapplicable and the graph should be solved by absorption of the nodes.[50]

EXAMPLE 2.4.3: Using the second method, let us find the stationary distribution for the Markov chain considered in Example 2.4.2.

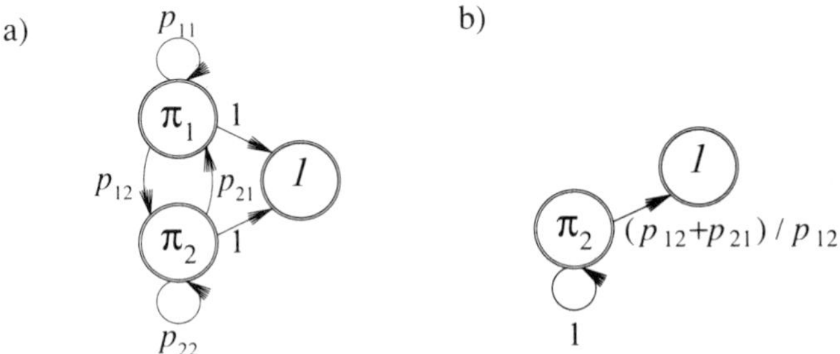

Figure 2.5. Stationary Distribution Calculation.

The graph for the system (2.4.17) is depicted in Fig. 2.5a. After absorbing the node π_1, we obtain the graph shown in Fig. 2.5b. From this graph it follows that $\pi_2(p_{21}+p_{12})/p_{12}=1$. Thus $\pi_2 = p_{12}/(p_{21}+p_{12})$.

■

2.5 MONTE CARLO METHOD

The analytical methods of calculating basic SSM characteristics often lead to complex expressions. Sometimes it is convenient to use computer simulation (the Monte Carlo method) to calculate these characteristics. If the sample size is large, the difference between a characteristic and its estimated value obtained by simulation will be small.

The advantage of the Monte Carlo method is its relative simplicity: in many cases it is easier to program the system operation than to evaluate its performance analytically. We can also use raw data obtained from channel-error statistics measurements to evaluate system performance characteristics.

The main disadvantage of the Monte Carlo method is its low degree of accuracy when it is used to estimate small values and the amount of statistical data is limited. To achieve adequate accuracy, the simulation time may have to be very long.

In this section we compare the Monte Carlo method of error number distribution calculation with the analytical methods described in the previous sections. But first we consider methods of SSM simulation using a random or pseudo-random number generator.

2.5.1 Error Source Simulation. Let us assume that an SSM error source model is presented by its canonical form: its states are governed by the Markov chain with the transition matrix **P**, and conditional probabilities of errors are either 0 or 1 (see Sec. 1.2.3). The error source simulation may be performed using the Markov chain definition: simulate its initial state s_0, according to the initial distribution **p**, then simulate the next state s_1 with the transition probability $p_{s_0 s_1}$, and so on. This method uses one random number to generate the next state and would be very inefficient for real channel models, since errors appear infrequently and therefore the Markov chain does not change its states frequently. One may take advantage of these error distribution specifics by simulating intervals between errors.

According to the results of Sec. 1.4.2, a Markov chain with the matrix P is equivalent to the semi-Markov process with the matrix

$$\mathbf{Q} = [q_{ij}]_{m,m} \quad q_{ii} = 0 \quad q_{ij} = p_{ij}/(1-p_{ii}) \text{ for } i \neq j \tag{2.5.1}$$

and geometric interval distributions

$$w_i(l) = (1-p_{ii})p_{ii}^{\ l-1} \quad i = 1,2,\ldots,m \tag{2.5.2}$$

Using this description, one may perform the simulation in the following way: According to the initial distribution $\boldsymbol{\pi}=(\pi_1,\pi_2,\ldots,\pi_n)$, select an initial state s_0, then, according to distributions (2.5.1) and (2.5.2), select the next state and the transition interval, and so on.

To be more specific, let us consider a sequence $\xi_1,\xi_2,\ldots,\xi_i,\ldots$ of samples of the [0,1] uniformly distributed random numbers. Then the initial state s_0 is selected if

$$\pi_0+\pi_1+\cdots+\pi_{s_0-1} \leq \xi_1 < \pi_1+\cdots+\pi_{s_0} \tag{2.5.3}$$

where $\pi_0 = 0$. Next we simulate the state-holding time, which is a series of $l_1 = \lceil \log(1-\xi_2)/\log p_{s_0 s_0} \rceil$ states s_0. If the probability of the error **e** is equal to 1 in this state, a series of l_1 errors (including the initial error) is generated.

We select s_1 as the next state of the semi-Markov process if

$$(q_{s_0,0}+q_{s_0,1}+\cdots+q_{s_0,s_1-1}) \leq \xi_3 < (p_{s_0,1}+\cdots+p_{s_0,s_1}) \tag{2.5.4}$$

where $p_{s_0,0} = 0$, $q_{ij} = p_{ij}/(1-p_{ii})$, and holding time $l_2 = \lceil \log(1-\xi_4)/\log p_{s_1,s_1} \rceil$ of the state s_1, and so on.

If a Markov chain is semi-Markov lumpable by a partition $\{A_1, A_2, ..., A_s\}$ and we have a generator that produces random numbers whose distribution has the form

$$w_{ij}(l) = \mathbf{q}_i \mathbf{P}_{ii}^{l-1} \mathbf{P}_{ij} \mathbf{1} / \mathbf{q}_i \mathbf{Q}_{ij} \mathbf{1} \tag{2.5.5}$$

then it is possible to speed up the simulation process by considering the lumped process, which has fewer states than the original chain. However, often we need to solve the opposite problem: creating a random-number generator for a distribution of the form

$$p(l) = \mathbf{a}\mathbf{B}^{l-1}\mathbf{c} \tag{2.5.6}$$

The standard method of solving this problem requires the inversion of the cumulative distribution function

$$F(x) = \sum_{l=1}^{x} p(l) = \mathbf{a}(\mathbf{I} - \mathbf{B}^x)(\mathbf{I}-\mathbf{B})^{-1}\mathbf{c} \tag{2.5.7}$$

since $\zeta = \lceil F^{-1}(\xi) \rceil$ has a cumulative distribution $F(x)$ if ξ is $[0,1]$ uniformly distributed. In general, inverting the function (2.5.7) is a complex problem and cannot be solved analytically. However, it is possible to solve the problem indirectly. We know that the similarity transformation

$$\mathbf{a}_1 = \mathbf{a}\mathbf{T} \quad \mathbf{T}\mathbf{B}_1 = \mathbf{B}\mathbf{T} \quad \mathbf{T}\mathbf{c}_1 = \mathbf{c} \tag{2.5.8}$$

does not change the distribution (see Theorem 1.4.1). If we can find non-negative matrices $\mathbf{a}_1$, $\mathbf{B}_1$, and $\mathbf{c}_1$ that satisfy the conditions

$$\mathbf{c}_1 = (\mathbf{I} - \mathbf{B}_1)\mathbf{1} \quad \text{and} \quad \mathbf{a}_1 \mathbf{1} = 1 \tag{2.5.9}$$

then the random variable l can be modeled as an interval between two consecutive returns into the last state of the Markov chain with the matrix

$$\begin{bmatrix} \mathbf{B}_1 & \mathbf{c}_1 \\ \mathbf{a}_1 & 0 \end{bmatrix} \tag{2.5.10}$$

As an example, consider a variable l with the polygeometric probability distribution

$$p(l) = \sum_{i=1}^{m} a_i(1-q_i)q_i^{l-1} \tag{2.5.11}$$

with $a_i>0$ and $q_i>0$. In this case we satisfy the above conditions by selecting $\mathbf{a}_1 = (a_1, a_2, \dots, a_m)$ and

$$\mathbf{B}_1 = diag\{q_i\} = \begin{bmatrix} q_1 & 0 & \dots & 0 \\ 0 & q_2 & \dots & 0 \\ \dots & \dots & \dots & \dots \\ 0 & 0 & \dots & q_m \end{bmatrix} \tag{2.5.12}$$

Therefore, the polygeometrically distributed variable can be simulated as an interval between two consecutive returns into the last state of the chain with the matrix

$$\mathbf{P} = \begin{bmatrix} q_1 & 0 & \cdots & 0 & 1-q_1 \\ 0 & q_2 & \cdots & 0 & 1-q_2 \\ \dots & \dots & \dots & \dots & \dots \\ 0 & 0 & \cdots & q_m & 1-q_m \\ a_1 & a_2 & \cdots & a_m & 0 \end{bmatrix} \tag{2.5.13}$$

Using the above-described method of simulating the Markov chain, we select a transition state i with the probability a_i and then select the state-holding time according to geometric distribution

$$w_i(l) = (1-q_i)q_i^{l-1} \tag{2.5.14}$$

2.5.2 Performance Characteristic Calculation. As we mentioned before, sometimes it is simpler to find system performance characteristics using the Monte Carlo method. For example, the error number distribution

$$p_n(m) = \mathbf{p}\mathbf{P}_n(m)\mathbf{1} \tag{2.5.15}$$

where $\mathbf{P}_n(m)$ is the matrix binomial distribution defined in Sec. 2.3.2 may be obtained by direct simulation of the Markov chain with the matrix

$$\begin{bmatrix} \mathbf{P}(0) & \mathbf{P}(1) \\ \mathbf{P}(0) & \mathbf{P}(1) \end{bmatrix} \tag{2.5.16}$$

The probability that the number of errors in a block of length n is equal to m is estimated by

$$p_n(m) \approx \frac{\nu_m}{N} \tag{2.5.17}$$

where ν_m is the number of length n blocks that contain exactly m errors and Nn is the total number of simulated symbols. We give more examples of applying this method in the following chapters.

Since any model is an approximation of the real error statistics and random generator quality may be poor, it is important to verify the agreement between experimental and simulated data. This problem is a particular case of statistical hypothesis testing, which we consider in the next chapter.

It can be shown that the number of times a regular Markov chain enters its state is asymptotically normal. Since an error number is a sum of the numbers of times the chain enters "bad" states, it is also asymptotically normal. Therefore, assuming that the regularity conditions are satisfied, we can use the standard χ^2 or information criteria of closeness between the experimental data and the result of simulation.

Let $p_i = p_n(m_{i-1} \le m < m_i)$ be the probability that the number of errors in a block satisfies the condition $m_{i-1} \le m < m_i$:

$$p_i = \sum_{m=m_{i-1}}^{m_i-1} p_n(m)$$

Denote

$$\hat{p}_{i,exp} = \nu_{i,exp} / N_{exp}$$

the corresponding experimental frequencies and

$$\hat{p}_{i,sim} = \nu_{i,sim} / N_{sim}$$

the estimates obtained by simulation.

If the estimates $\hat{p}_{i,exp}$ and $\hat{p}_{i,sim}$ are asymptotically normal, then[53]

$$\chi^2 = N_{sim} N_{exp} \sum_{i=1}^{r} (\hat{p}_{i,exp} - \hat{p}_{i,sim})^2 / (\hat{p}_{i,exp} + \hat{p}_{i,sim})$$

and

$$2I = 2 N_{exp} N_{sim} \sum_{i=1}^{r} (\hat{p}_{i,exp} - \hat{p}_{i,sim}) \ln(\hat{p}_{i,exp} / \hat{p}_{i,sim}) / (N_{exp} + N_{sim})$$

have asymptotically the χ^2 distribution with $r-1$ degrees of freedom.

These criteria values allow us to decide whether the experimental data agree with the simulation results. In the next chapter we discuss how the values should be selected.

2.6 CALCULATING ERROR NUMBER DISTRIBUTION

In this section we calculate error number distribution for the particular models described in Sec. 1.5. If we are interested in probabilities of small numbers of errors in a block, then we may use recursive equations (2.3.5). These equations can be applied similarly to fast computations of the real number powers. Express the block size as a binary number

$$n = 2^k + a_1 2^{k-1} + \cdots + a_{k-1} 2 + a_k$$

and then find matrix distributions of the number of errors for the blocks whose length is a power of two. For example, to calculate the error number distribution in a block of 100 bits we use the binary representation $100 = 2^{64} + 2^{32} + 2^4$. Then we calculate $\mathbf{P}_2(m)$, $\mathbf{P}_4(m)$, $\mathbf{P}_8(m)$, $\mathbf{P}_{16}(m)$, and $\mathbf{P}_{32}(m)$, using formula (2.3.5) with $n=v$. Next we calculate

$$\mathbf{P}_{36}(m) = \sum_{\mu} \mathbf{P}_{32}(m-\mu)\mathbf{P}_4(\mu)$$

$$\mathbf{P}_{64}(m) = \sum_{\mu} \mathbf{P}_{32}(m-\mu)\mathbf{P}_{32}(\mu)$$

And finally,

$$\mathbf{P}_{100}(m) = \sum_{\mu} \mathbf{P}_{64}(m-\mu)\mathbf{P}_{36}(\mu)$$

The results of calculations using this method for the models described in Sec. 1.5 are presented in Table 2.1.

Table 2.1

Computed distribution of the number of errors in a block of 100 bits.

Number of errors	Telephone channel	Radio channel	Troposcatter channel
0	0.7200	0.9092	0.8121
1	0.0457	0.0311	0.0612
2	0.0391	0.0205	0.0451
3	0.0333	0.0135	0.0316
4	0.0283	0.0089	0.0206
5	0.0239	0.0058	0.0123
6	0.0201	0.0038	0.0067
7	0.0168	0.0025	0.0033
>7	0.0728	0.0047	0.0071

Table 2.2

Simulated distribution of the number of errors in a block of 100 bits.

Number of errors	Telephone channel	Radio channel	Troposcatter channel
0	0.7213	0.9089	0.8170
1	0.0458	0.0311	0.0615
2	0.0371	0.0208	0.0452
3	0.0336	0.0136	0.0312
4	0.0267	0.0090	0.0204
5	0.0236	0.0058	0.0123
6	0.0214	0.0037	0.0067
7	0.0172	0.0025	0.0032
>7	0.0733	0.0046	0.0025

In this table, the telephone channel parameters are given by matrix (1.5.5), the radio channel parameters by (1.5.6), and the troposcatter channel parameters by (1.5.7). The results of the Monte Carlo simulation are shown in Table 2.2.

For the T1 channel error source model the results of calculations are compared with the experimental data in Tables 2.3 and 2.4.

Table 2.3

Simulated, computed, and measured distributions of the number of errors in a block of 100 bits for the T1 channel.

Number of errors	Simulated	Computed	Experimental data
0	0.999999759885	0.999999770083	0.999999769597
1	0.000000190491	0.000000183577	0.000000186894
2	0.000000038302	0.000000037183	0.000000033933
3	0.000000007581	0.000000007388	0.000000006293
>3	0.000000002859	0.000000001767	0.000000003283

In evaluating the performance of communication protocols it is often necessary to know the distribution of the number of sub-blocks with errors in some larger block. A typical example is the distribution of error-free seconds. This distribution can be found using formula (2.3.5) in which $\mathbf{P}_1(0) = \mathbf{P}^k(0)$, $\mathbf{P}_1(1) = \mathbf{P}^k - \mathbf{P}^k(0)$, and k is the sub-block size.

For the Gilbert model and $k=100$ these matrices are given by

$$\mathbf{P}^{100}(0) = \begin{bmatrix} 0.7731 & 0.0000 & 0.0000 \\ 0.0000 & 0.0000 & 0.0000 \\ 0.1206 & 0.0000 & 0.0000 \end{bmatrix} \quad \mathbf{P}^{100}-\mathbf{P}^{100}(0) = \begin{bmatrix} 0.1587 & 0.0534 & 0.0148 \\ 0.8903 & 0.0871 & 0.0226 \\ 0.7761 & 0.0819 & 0.0214 \end{bmatrix}$$

Table 2.4

Simulated, computed, and measured distributions of the number of errors in a block of 10^8 bits for the T1 channel.

Number of errors	Simulated	Computed	Experimental data
0	0.79313	0.80164	0.80651
1	0.14115	0.13626	0.12972
2	0.04424	0.04309	0.04433
3	0.01361	0.01330	0.01177
4	0.00429	0.00402	0.00547
>4	0.00358	0.00169	0.00219

The results of calculations using formula (2.3.5) and Monte Carlo simulation of the distribution of the number of errored sub-blocks of 100 bits in a block of 1000 are presented in Table 2.5.

Table 2.5

Simulated and computed distributions of the number of sub-blocks of 100 bits with errors in a block of 1000 bits.

Number of sub-blocks with errors	Simulated	Computed
0	0.0793	0.0711
1	0.1633	0.1600
2	0.2054	0.2265
3	0.2162	0.2221
4	0.1847	0.1650
>4	0.1511	0.1553

In evaluating the performance of concatenated codes described in Chapter 4 it is important to know the distribution of the sub-blocks with undetected errors, sub-blocks with errors detected by some error-detecting code, and the error-free sub-blocks. This distribution, which is an example of multinomial distribution can be computed using equation (2.3.1). For instance, if a simple parity check is used for error detection, then the matrix probability of undetected errors is the probability that the number of errors in a sub-block is an even number greater than zero:

$$\mathbf{P}_u = \sum_{m=2,4,\ldots} \mathbf{P}_k(m) = 0.5\,[\,\mathbf{P}^k + (\,\mathbf{P}(0) - \mathbf{P}(1)\,)^k\,] - \mathbf{P}^k(0)$$

The matrix probability of an error-free sub-block is $\mathbf{P}_c = \mathbf{P}^k(0)$, and the matrix probability of detected errors is the probability that their number is odd:

$$\mathbf{P}_d = \sum_{m=1,3,\ldots} \mathbf{P}_k(m) = 0.5\,[\,\mathbf{P}^k - (\,\mathbf{P}(0) - \mathbf{P}(1)\,)^k\,]$$

We can now apply equation (2.3.1) with the initial values $\mathbf{P}_1(A_0)=\mathbf{P}_c$, $\mathbf{P}_1(A_1)=\mathbf{P}_u$, and $\mathbf{P}_1(A_2)=\mathbf{P}_d$ to calculate the multinomial distribution of different sub-blocks.

2.7 CONCLUSION

The matrix-train form of error sequence probabilities enables us to introduce matrix probabilities by stripping off the initial state probability matrix and the final matrix **1**. The matrix probability of a sequence of conditionally independent events can be expressed similarly to the probability of independent events. However, since matrix products are noncommutative, matrix probabilities of combinations of events have a much more complex form than their scalar counterparts.

In this chapter we have generalized methods of classical probability theory to matrix probabilities. We analyzed the most commonly used matrix distributions and developed numerous methods of their calculation. In particular, the transform methods proved to be the most efficient in many cases. We showed that signal-flow graphs with block transitions can be used to solve various problems related to matrix probabilities. In particular, by using matrix probabilities as transitions on a graph describing a finite-state machine we were able to calculate probabilities and distributions that characterize the machine's performance. These methods are applied to performance analysis of communication protocols in Chapter 5. Because of the generality of the finite-state machine model, these methods can be successfully used in applications different from those considered in this book (such as queueing theory, voice and image recognition).

We illustrated the methods by calculating the probability distribution of the numbers of errors in a block. The results of calculations agree with experimental data and computer simulation.

Some additional material like the derivation of asymptotic and approximate formulae for matrix probabilities is given in Appendix 2. We have also analyzed basic operations with block graphs in this appendix, to make the book self-contained.

CHAPTER 3

MODEL PARAMETER ESTIMATION

A model of an error source is developed and its parameters are estimated by experimental data processing. Usually, measurements are performed with the help of special equipment and then are processed by a computer. The essential factors in collecting experimental data are a test sequence and sample size selection to obtain good estimates of the model parameters. As a rule, a model is developed in several stages; at each stage, as new data become available, the model is refined. For instance, at the first stage we can collect experimental results about the distribution of errors, assuming that the channel is symmetric; at the second stage we can, in addition, analyze the nature of asymmetry; and at the third stage we can take into account the influence of other channels on the channel under consideration.

The fundamental problem we solve in this chapter can be formulated as follows: Given a sequence $\{e_i\}$ of zeros and ones (the zeros representing correct symbols, the ones representing errors), it is necessary to build a model that approximates the process $\{e_i\}$. We develop methods of the process approximation as an output of some SSM. Because an SSM model can be described in many different ways, we consider several approaches to model building. After a brief statistical introduction, we consider a matrix-geometric distribution parameter estimation which plays a fundamental role in the SSM model building. Next we consider the model approximation based on multiple Markov chains and semi-Markov processes. Then we examine the question of the model building on the basis of matrix processes. And finally we indicate the modifications in experimental data processing methodology for the development of a model of the source of errors in several channels.

3.1 STATISTICAL INFERENCE

An error source model development is a problem of mathematical statistics. Our goal is to show that the major steps in model building can be described as particular cases of a known statistical methods. We describe here some of the methods that are used at various stages of the statistical analysis.

3.1.1 Distribution Parameter Estimation. Let $Pr(\xi = x) = f(x, \boldsymbol{\tau})$ be a random variable distribution with the unknown parameter $\boldsymbol{\tau} = (\tau_1, \tau_2, ..., \tau_k)$ which should be estimated using a sequence of N independent samples $x_1, x_2, ..., x_N$ of this variable. The parameter estimate $\hat{\boldsymbol{\tau}}$ is defined as some function of these samples:

$$\hat{\boldsymbol{\tau}} = \phi(x_1, x_2, ..., x_N) \tag{3.1.1}$$

such that its value $\hat{\boldsymbol{\tau}}$ is in some sense close to the actual parameter value.

An estimate is said to be *efficient* if its mean square deviation from the real parameter $\boldsymbol{\tau}$ is the smallest:

$$\mathbf{E}\{[\phi(\xi_1, \xi_2, ..., \xi_N) - \boldsymbol{\tau}]^2\} = \min \tag{3.1.2}$$

It is said to be *unbiased* if its mean is equal to $\boldsymbol{\tau}$:

$$\mathbf{E}\{\phi(\xi_1, \xi_2, ..., \xi_N)\} = \boldsymbol{\tau} \tag{3.1.3}$$

It is also important that an estimate $\hat{\boldsymbol{\tau}}$ converge to $\boldsymbol{\tau}$ in probability:

$$P \lim_{N \to \infty} \phi(\xi_1, \xi_2, ..., \xi_N) = \boldsymbol{\tau} \tag{3.1.4}$$

In this case the estimate is said to be *consistent*.

EXAMPLE 3.1.1: Consider a channel with independent errors and bit error probability p. For this model, intervals between consecutive errors are geometrically distributed

$$p(l) = p(1-p)^{l-1} \tag{3.1.5}$$

This distribution has only one parameter ($\tau = p$). If we have a sequence $x_1, x_2, ..., x_N$ of measured intervals between errors, then their mean is inversely proportional to the bit error rate, which can be used as an estimate of the bit error probability:

$$\hat{p} = N \,/\, \sum_{i=1}^{N} x_i \tag{3.1.6}$$

This estimate is unbiased, consistent, and efficient. ■

One of the most powerful methods of parameter estimation, which is called the *maximum likelihood* (ML) method, is based on the assumption that for the best estimate an experiment's result has the largest probability. Suppose that $L(\mathbf{x},\boldsymbol{\tau})$ is the probability that N samples of a variable ξ are represented by $\mathbf{x} = (x_1, x_2, \ldots, x_N)$. This function is called the *likelihood function*, and the value of the parameter $\hat{\boldsymbol{\tau}}$ that delivers its global maximum is called the ML estimate:

$$\max_{\boldsymbol{\tau}} L(\mathbf{x},\boldsymbol{\tau}) = L(\mathbf{x},\hat{\boldsymbol{\tau}}) \tag{3.1.7}$$

If the set Ω of all possible values of the random variable ξ is partitioned into r disjoint subsets

$$\Omega = \bigcup_i A_i \quad A_i \cap A_j = \varnothing \quad \text{for } i \neq j$$

then the likelihood function can be expressed as

$$L(\mathbf{x},\boldsymbol{\tau}) = \prod_{i=1}^{r} P_i^{\nu_i}(\boldsymbol{\tau}) \tag{3.1.8}$$

where $P_i(\boldsymbol{\tau})$ is the probability that the random variable ξ belongs to the subset A_i, $N = \nu_1 + \nu_2 + \cdots + \nu_r$. The maximization of this function is equivalent to the minimization of the so-called *information* criterion [53]

$$I = \sum_{i=1}^{r} \nu_i \ln(\nu_i / N) - \ln L(\mathbf{x},\boldsymbol{\tau}) \tag{3.1.9}$$

or

$$I = \sum_{i=1}^{r} \nu_i \ln[\nu_i / N P_i(\tau)] \tag{3.1.10}$$

Another measure of the estimate quality, which is very popular in statistical applications, is the χ^2 criterion

$$\chi^2 = \sum_{i=1}^{r}(\nu_i - NP_i)^2 / NP_i \tag{3.1.11}$$

Using the approximation $\ln x \approx (x - x^{-1})/2$ for $x \approx 1$, it is easy to prove that

$$2\sum_i a_i \ln(a_i/b_i) \approx \sum_i (a_i - b_i)^2/b_i \quad \text{if} \quad \sum_i a_i = \sum_i b_i \tag{3.1.12}$$

It can be shown, applying this formula to (3.1.9), that the ML estimation method is asymptotically equivalent to the minimum of I or χ^2 criterion,[53] and that under general assumptions [54] it delivers a consistent estimate.

The defined above χ^2 in which $\boldsymbol{\tau}$ is replaced with its ML estimate $\hat{\boldsymbol{\tau}}$ and $N \to \infty$ has a so-called χ^2 distribution with $r-k-1$ degrees of freedom † and the probability density

$$K_{r-k-1}(x) = A_{r-k-1}x^{(r-k-3)}e^{-0.5x} \tag{3.1.13}$$

where

$$A_m = \begin{cases} [2^{i+1}i\,!]^{-1} & \text{for } m=2i+2 \\ [1 \cdot 3 \cdots (2i-1)2^{-i}\sqrt{2\pi}\,]^{-1} & \text{for } m=2i+1 \end{cases}$$

EXAMPLE 3.1.2: If $x_1, x_2, \ldots, x_N$ is a sequence of observed intervals between errors that occur independently with the probability p, then the likelihood function is equal to

$$L(\mathbf{x}, p) = \prod_{i=1}^{N} p(1-p)^{x_i - 1}$$

The ML estimate can be found from the necessary condition of this function maximum

$$\frac{\partial L(\mathbf{x}, p)}{\partial p} = 0 \tag{3.1.14}$$

This equation has a unique solution, which is given by equation (3.1.6). ■

† The distribution parameter $r-k-1$ is called a number of degrees of freedom because it is equal to the total number r of independent sets minus the number of connections between them: k equations to estimate parameters $(\tau_1, \tau_2, \ldots, \tau_k)$ and one identity that states that the total sum of probabilities $P_i(\boldsymbol{\tau})$ is equal to 1.

Even though the ML method delivers good parameter estimates, its practical application is rather limited by the complexity of the computations required. In many cases, significantly simpler equations can be obtained using the *method of moments*. According to this method, the distribution characteristics are calculated theoretically and then equated to their experimental estimates, thus creating a system of equations for the unknown parameters.

EXAMPLE 3.1.3: If $x_1, x_2, \ldots, x_N$ is a sequence of observed intervals between errors that occur independently with probability p, then the mean interval between errors is

$$\bar{l} = \sum_{l=1}^{\infty} l\, p(1-p)^{l-1} = 1/p \tag{3.1.15}$$

By equating $\bar{l}$ to the average observed interval between errors, we obtain the following equation of the method of moments, that is equivalent to (3.1.6)

$$\frac{1}{N}\sum_{i=1}^{N} x_i = 1/\hat{p}$$

■

3.1.2 Hypothesis Testing. Statistical hypothesis testing is one of the basic methods of model building. We make assumptions about the model's structure and parameters and then, on the basis of some experimental data $\mathbf{x} = (x_1, x_2, \ldots, x_n)$, judge whether this model describes the data accurately. In most cases, any experimental data are compatible with a model. So our inference should be based not on compatibility with the model but rather on the model's likelihood.

To test a hypothesis H_0, we need some criterion (or measure) $w(\mathbf{x})$ of closeness between the experimental data and a theoretical prediction based on the hypothesis. Usually, the criterion grows when the difference between experimental data and theoretical prediction grows. If the criterion value is greater than some critical value w_{cr}, the hypothesis H_0 is rejected and some alternative hypothesis H_1 is therefore accepted. If we define the hypothesis H_0 rejection region as $D_0 = \{\mathbf{x}: w(\mathbf{x}) > w_{cr}\}$, then the hypothesis H_0 is rejected when $\mathbf{x} \in D_0$. The complement $\bar{D}_0$ of the rejection region is usually called the acceptance region.

There is a possibility of committing the Type I error by rejecting H_0 when it is true with the probability

$$Pr(\mathbf{x} \in D_0 \,|\, H_0) = \alpha \tag{3.1.16}$$

or the Type II error of accepting H_0 when it is false with the probability

$$Pr(\mathbf{x} \notin D_0 \,|\, H_1) = \beta \tag{3.1.17}$$

The Type I error probability α is called the criterion *significance level* and the probability $1-\beta$ is called the criterion *power*. Naturally, we want to select a critical region to minimize the probabilities of errors of both types. However, it is not possible to make these probabilities arbitrarily small simultaneously. Usually, the critical region D_0 is selected to minimize the Type II error probability for some small fixed Type I error probability. The region D_0 that maximizes the criterion power subject to $Pr(\mathbf{x} \in D_0 | H_0) \le \alpha$ is called the most powerful test of size α.

EXAMPLE 3.1.4: For the geometric distribution (3.1.5), consider the problem of testing the hypothesis

$$H_0\colon p = p_0 \quad \text{against} \quad H_1\colon p \neq p_0 \tag{3.1.18}$$

The distribution parameter estimate (3.1.6) can be rewritten as

$$\hat{p} = N/T \quad \text{where } T = \sum_{i=1}^{N} x_i \tag{3.1.19}$$

is the total number of bits in all the intervals. Let

$$A(p_0) = (p_0 - \delta_\alpha, p_0 + \delta_\alpha) \tag{3.1.20}$$

be an acceptance region for the hypothesis H_0. If the hypothesis H_0 is true, then $\hat{p}$ has a binomial distribution

$$Pr(\hat{p} = m/T) = \binom{T}{m} p_0^m q_0^{T-m} \tag{3.1.21}$$

so that δ_α can be found as a solution of the following equation:

$$Pr(p_0 - \delta_\alpha \le \hat{p} \le p_0 + \delta_\alpha) = \sum_{i=a}^{b} \binom{T}{m} p_0^m q_0^{T-m} = 1 - \alpha \tag{3.1.22}$$

where $a = \max[0, (p_0 - \delta_\alpha)T]$, $b = \min[T, (p_0 + \delta_\alpha)T]$. If this equation does not have a solution, we may find a smaller acceptance region and introduce an uncertainty region, where a decision is made using a random number generator.

For large T the binomial distribution can be approximated by the Gaussian distribution, and equation (3.1.22) becomes

$$Pr(p_0 - \delta_\alpha \le \hat{p} \le p_0 + \delta_\alpha) = erf(\delta_\alpha \sqrt{T/2p_0 q_0}) = 1 - \alpha \tag{3.1.23}$$

when $\delta_\alpha < p_0 < T - \delta_\alpha$. Let Z_α be the root of the equation $erfc(Z_\alpha) = \alpha$. Then the acceptance interval may be expressed as

$$A(p_0) = (p_0 - Z_\alpha\sqrt{2p_0(1-p_0)/T},\ p_0 + Z_\alpha\sqrt{2p_0(1-p_0)/T}) \qquad (3.1.24)$$

3.1.3 Goodness of Fit. The following problem plays a fundamental role in model building. Suppose that the set Ω of all possible values of the random variable ξ is partitioned into r disjoint subsets:

$$\Omega = \bigcup_i A_i \quad A_i \cap A_j = \varnothing \ \text{ for } i \neq j$$

Let $Pr(\xi \in A_i) = P_i(\boldsymbol{\tau})$ be the probability that the random variable belongs to the subset A_i and $\hat{P}_i = \nu_i / N$ is the corresponding empirical frequency. Our goal is to estimate the distribution parameters $\boldsymbol{\tau} = (\tau_1, \tau_2, ..., \tau_k)$ and decide whether or not the distribution $\{P_i(\hat{\tau})\}$ agrees with the experimental data $\hat{P}_i$. We use the χ^2 or the information criterion I to solve this problem.

Denote $\hat{\boldsymbol{\tau}} = (\hat{\tau}_1, \hat{\tau}_2, ..., \hat{\tau}_k)$ an estimate of the parameter obtained by the ML (minimum I or χ^2) method. The hypothesis that $\{P_i(\hat{\boldsymbol{\tau}})\}$ is the probability distribution of outcomes can be expressed as

$$H_0:\ Pr(\xi \in A_i) = P_i(\hat{\boldsymbol{\tau}}) \quad (i=1,2,...,r) \quad \text{against} \quad H_1 = \overline{H}_0$$

To test this hypothesis, we can use the criterion χ^2 or $2I$ defined by equations (3.1.10), (3.1.11), in which τ is replaced by $\hat{\boldsymbol{\tau}}$. Both criteria have the χ^2 distribution (3.1.13) with $r-k-1$ degrees of freedom. Let $\chi^2_{\gamma,m}$ be a root of the equation

$$Pr(\chi^2 < \chi^2_{\gamma,m}) = \int_0^{\chi^2_{\gamma,m}} K_m(x)dx = \gamma$$

Then the significance α rejection region may be defined as

$$D_0 = \{\chi^2:\ \chi^2 > \chi^2_{1-\alpha,r-k-1}\}$$

Thus we say that the distribution agrees with the experimental data if the calculated value χ^2 does not exceed $\chi^2_{1-\alpha,r-k-1}$. The probability of rejecting H_0 when it is true is less than α. Since $\chi^2 \approx 2I$, the same rejection region may be used for the information criterion.

EXAMPLE 3.1.5: Suppose that random vectors $\mathbf{v}=(\xi,\eta)$ are independent. We want to test the null hypothesis that vector components ξ and η are independent.

Let p_{ij} be a vector **v** probability distribution and $p_{i.}, p_{.j}$ be the distributions of its components

$$p_{i.} = \sum_{j}^{s} p_{ij} \quad p_{.j} = \sum_{i}^{r} p_{ij}$$

$i = 1,2,...,r$, $j = 1,2,...,s$. Then the null hypothesis is equivalent to the hypothesis that the vector probability distribution has the form

$$p_{ij} = p_{i.}p_{.j} \tag{3.1.25}$$

In other words, we need to test the goodness of fit of the distribution that depends on $r+s-2$ independent parameters.

Suppose that in a sequence of independent trials a vector (i,j) occurs ν_{ij} times. If the hypothesis H_0 is true, then the likelihood function is given by

$$L = \prod_{i=1}^{r}\prod_{j=1}^{s}(p_{i.}p_{.j})^{\nu_{ij}} = \prod_{i=1}^{r} p_{i.}^{\nu_{i.}} \prod_{j=1}^{s} p_{.j}^{\nu_{.j}}$$

where

$$\nu_{i.} = \sum_{j=1}^{s} \nu_{ij} \quad \nu_{.j} = \sum_{i=1}^{r} \nu_{ij}$$

This function reaches its maximal value when

$$p_{i.} = \hat{p}_{i.} = \nu_{i.}/n \quad p_{.j} = \hat{p}_{.j} = \nu_{.j}/n \quad n = \sum_{i=1}^{r} \nu_{i.}$$

The test information criterion (3.1.10) takes the form

$$I = \sum_{i=1}^{r}\sum_{j=1}^{s} \nu_{ij} \ln(n\nu_{ij}/\nu_{i.}\nu_{.j}) \tag{3.1.26}$$

It is distributed as $0.5\chi^2$ with $rs-(r+s-2)-1 = (r-1)(s-1)$ degrees of freedom. The expression (3.1.11) reduces to

$$\chi^2 = \sum_{i=1}^{r}\sum_{j=1}^{s} (n\nu_{ij} - \nu_{i.}\nu_{.j})^2/n\nu_{i.}\nu_{.j} \tag{3.1.27}$$

This also has a limiting χ^2 distribution with $(r-1)(s-1)$ degrees of freedom. The null hypothesis is rejected if

$$\chi^2 > \chi^2_{(1-\alpha),(r-1)(s-1)} \qquad \blacksquare$$

3.1.4 Confidence Limits. The parameter estimates considered above are called point estimates in mathematical statistics. They do not give us an idea of the estimate variations. It is clear that even if the estimate is good, its reliability may be poor for a small number of samples. The quantitative characterization of the estimate reliability is given by the theory of confidence limits. The confidence region $\Theta(\mathbf{x})$ for the parameter τ with the confidence level $1-\alpha$ is defined as a region that contains the parameter with probability $1-\alpha$:

$$Pr(\tau \in \Theta(\mathbf{x})) = 1-\alpha$$

Thus, if α and the region are small, then with high confidence we can say that we have found a reliable estimate. In the case of single-parameter estimation, the confidence region is usually an interval, so that the previous equation becomes

$$Pr(\tau_L < \tau \leq \tau_H) = 1-\alpha$$

The interval limits are called the confidence limits.

To find a confidence region, let us consider a hypothesis

$$H_0: \ \boldsymbol{\tau} = \boldsymbol{\tau}_0 \quad \text{against} \quad H_1: \ \boldsymbol{\tau} \neq \boldsymbol{\tau}_0$$

and its acceptance region $A(\boldsymbol{\tau}_0)$ of significance α

$$Pr(\mathbf{x} \in A(\boldsymbol{\tau})) = 1-\alpha$$

A family of acceptance regions $\{A(\boldsymbol{\tau})\}$ for all possible $\boldsymbol{\tau}$ defines a region $U(\alpha)$ in the $(\mathbf{x},\boldsymbol{\tau})$ domain. The confidence region $\Theta(\mathbf{x})$ contains all $\boldsymbol{\tau}$ for which $(\mathbf{x},\boldsymbol{\tau}) \in U(\alpha)$, since

$$Pr(\mathbf{x} \in A(\boldsymbol{\tau})) = Pr\{(\mathbf{x},\boldsymbol{\tau}) \in U(\alpha)\} = 1-\alpha \qquad (3.1.28)$$

EXAMPLE 3.1.6: Let us find the confidence limits for the parameter p of the geometric distribution (3.1.5). Assuming that the number of samples is large and $\delta_\alpha < p_0 < T-\delta_\alpha$, we obtain the family of acceptance regions (3.1.24)

$$A(p) = (p - Z_\alpha \sqrt{2p(1-p)/T},\ p + Z_\alpha \sqrt{2p(1-p)/T}) \tag{3.1.29}$$

The region $U(\alpha)$ in this case lies inside the ellipse $\hat{p} = p \pm Z_\alpha \sqrt{2p(1-p)/T}$ and square $0 \le p \le 1$, $0 \le \hat{p} \le 1$. Solving these equations with respect to p, we obtain the confidence interval

$$p_{H,L} = \frac{T}{T+2Z_\alpha^2}\left[\hat{p} \pm \frac{Z_\alpha}{T} + Z_\alpha \sqrt{2\hat{p}(1-\hat{p})/T + (Z_\alpha^2/T)^2}\right]$$

Since T is assumed to be large, these equations may be simplified:

$$p_L = \hat{p} - Z_\alpha \sqrt{2\hat{p}(1-\hat{p})/T}$$

$$p_H = \hat{p} + Z_\alpha \sqrt{2\hat{p}(1-\hat{p})/T}$$

■

3.1.5 Steps in Model Building. The initial step in model building is model selection. For the error statistics description we select a matrix process or SSM model. The next step is the model parameter estimation; we need to estimate the sizes and elements of matrices describing the model. Finally, we need to test the model's agreement with the experimental data, using some criteria $W_1, W_2, \ldots, W_m$. For example, we may test a hypothesis that the matrix process rank is equal to r against the alternative hypothesis that it is greater than r, a hypothesis that the transition matrix is equal to $\mathbf{M}$ versus the alternative hypothesis that it is not equal to $\mathbf{M}$, and so on. Some distributions' goodness of fit may be tested.

If one of the tests fails, it does not necessarily mean that the experimental data contradict the model. However, since the test significance level α is small, we have enough background to reject the model. On the other hand, if all the criteria do not reject the model, it does not necessarily mean that the model agrees with the experimental data. If sample size is small, many models show good agreement with the experimental data. If a model is not rejected by several criteria for the large sample size, then we may infer, with high probability, that it agrees with the experimental data.

3.2 MARKOV CHAIN MODEL BUILDING

3.2.1 Statistical Tests for Simple Markov Chains. There are many publications devoted to statistical inference concerning Markov chains.[55—60] We consider here only the methods that are used for SSM model building.

Let $x_1, x_2, \ldots, x_n$ be a sequence of samples of some s-ary stochastic process ξ_t ($\xi_t = 1, 2, \ldots, s$). We want to test a hypothesis H_0 that this process is a simple (first order) regular Markov chain with the transition probability matrix $\mathbf{P}(\boldsymbol{\tau}) = [\, p_{ij}(\boldsymbol{\tau}) \,]_{s,s}$, with k independent parameters $\boldsymbol{\tau} = (\tau_1, \tau_2, \ldots, \tau_k)$. Within this hypothesis, the likelihood function is given by

$$L(\boldsymbol{\tau}) = p_{x_1} \prod_{i=1}^{n-1} p_{x_i x_{i+1}}(\boldsymbol{\tau}) = p_{x_1} \prod_{ij} p_{ij}^{n_{ij}}(\boldsymbol{\tau}) \tag{3.2.1}$$

where n_{ij} is the observed frequency of transitions from i to j in the sequence $x_, x_2,...,x_n$. The logarithmic likelihood function is equal to

$$\ln L = J_0 + J_1 \tag{3.2.2}$$

where

$$J_0 = \ln p_{x_1} \quad J_1(\boldsymbol{\tau}) = \sum_{ij} n_{ij} \ln p_{ij}(\boldsymbol{\tau}) \tag{3.2.3}$$

The ML parameter estimate $\hat{\boldsymbol{\tau}} = (\hat{\tau}_1, \hat{\tau}_2, ..., \hat{\tau}_k)$ asymptotically maximizes J_1, since $J_0 / J_1 \to 0$ when $n \to \infty$, and can be found as a solution of the system

$$\frac{\partial J_1(\boldsymbol{\tau})}{\partial \tau_m} = \sum_{ij} \frac{n_{ij}}{p_{ij}} \frac{\partial p_{ij}}{\partial \tau_m} = 0 \quad m=1,2,...,k \tag{3.2.4}$$

assuming that all the derivatives exist.

It is shown [55,58] that if H_0 is true, the hypothesis information criterion

$$I = \sum_{ij} n_{ij} \ln \left[n_{ij} / n_{i.} p_{ij}(\hat{\boldsymbol{\tau}}) \right] \tag{3.2.5}$$

is distributed asymptotically as $0.5\chi^2$ for a family of functions $p_{ij}(\boldsymbol{\tau})$ that satisfy the following conditions.[60] They have third-order continuous partial derivatives in some open subset Θ of the Euclidean space E^k. And the Jacobian matrix

$$\left[\frac{\partial p_u(\boldsymbol{\tau})}{\partial \tau_v} \right]_{d,k} \qquad u = (i,j) \in D$$

has rank k throughout Θ. Here D is a set of $u = (i,j)$ such that for each $u \in D$ $p_u(\boldsymbol{\tau}) > 0$ throughout Θ, d is the size of D, and the chain is assumed to be regular. The summation in (3.2.5) is performed over all $(i,j) \in D$, and the number of degrees of freedom in the limiting distribution is

$$f = N_{nz} - N_{st} - N_{par} \tag{3.2.6}$$

where $N_{nz} = d$ is the total number of $p_{ij}(\hat{\boldsymbol{\tau}})$ which are not equal to zero, $N_{st} = s$ is the total number of states, and $N_{par} = k$ is the total number of independent parameters that have been estimated.

The hypothesis χ^2 criterion

$$\chi^2 = \sum_{ij} \frac{[n_{ij} - n_{i.} p_{ij}(\hat{\boldsymbol{\tau}})]^2}{n_{i.} p_{ij}(\hat{\boldsymbol{\tau}})} \tag{3.2.7}$$

is asymptotically equivalent to $2I$. [58] For the functions that we use to create an SSM model these conditions are satisfied.

EXAMPLE 3.2.1: Consider a particular case of independent and identically distributed variables. The transition probabilities do not depend on i and therefore can be expressed as $p_{ij} = \tau_j$ for all i. The number of independent parameters is equal to $N_{par} = s-1$, since $\tau_1 + \tau_2 + \cdots + \tau_s = 1$. The likelihood function J_1 reaches its maximal value when

$$\tau_j = \hat{\tau}_j = n_{.j} / n$$

and the information criterion (3.2.5) coincides with (3.1.26). If all the probabilities $\tau_j > 0$, then, according to (3.2.6), the number of degrees of freedom is equal to $s^2 - s - (s-1) = (s-1)^2$, which coincides with the results obtained before.

■

3.2.2 Multiple Markov Chains. As we have seen in Sec. 1.1.5, the SSM error source model can be built on the basis of order k Markov chains. Different sequences of errors $e_1, e_2, \ldots, e_k$ constitute states of the SSM. The state transition probabilities are given by

$$p_{(d_1 \cdots d_k)(e_1 \cdots e_k)} = \begin{cases} p_{e_1 e_2 \cdots e_k e_{k+1}} & \text{if } e_1 = d_2, \ldots, e_{k-1} = d_k \\ 0 & \text{otherwise} \end{cases} \tag{3.2.8}$$

where $p_{e_1 e_2 \cdots e_k e_{k+1}}$ is defined as the probability of an error e_{k+1} at the moment t, given that errors e_1, e_2, ..., e_k occurred at the moments $t-k$, $t-k+1$, ..., $t-1$, respectively:

$$p_{e_1 e_2 \cdots e_{k+1}} = Pr(e_{k+1} \mid e_1, e_2, \ldots, e_k)$$

These transition probabilities, along with the chain order (k), are the model parameters, and they must be estimated from the experimental data.

To estimate the chain order we test the null hypothesis that the chain has order $k-1$ against the alternative hypothesis that its order is k. The hypothesis to be tested may be expressed as

$$H_0\colon\ p_{e_1 e_2 \cdots e_{k+1}} = p_{e_2 \cdots e_{k+1}} \quad \text{for all} \quad e_1, e_2, \ldots, e_{k+1} \tag{3.2.9}$$

This is a particular case of the hypothesis considered in the previous section, which can be tested using criterion (3.2.5) or (3.2.7). The values on the right-hand side of equations (3.2.9) are the transition probability parameters that must be estimated.

Suppose that $e_1, e_2, \ldots, e_n$ are the outcomes of some experiment. If H_0 is true, the likelihood function is given by

$$L = p(e_1 e_2 \cdots e_{k-1}) \prod_{m=k}^{n-k+1} p_{e_m e_{m+1} \cdots e_{m+k-1}} =$$
$$p(e_1 e_2 \cdots e_{k-1}) \prod_{x_1,\ldots,x_k} p_{x_1 x_2 \cdots x_k}^{n(x_1,\ldots,x_k)}$$

where $n(x_1, x_2, \ldots, x_k)$ is the observed frequency of k-tuples $x_1, x_2, \ldots, x_k$ in the sequence $e_1, e_2, \ldots, e_n$. The logarithmic likelihood function is equal to

$$\ln L = J_0 + J_1$$

where

$$J_0 = \ln p(e_1 e_2 \cdots e_{k-1})$$

$$J_1 = \sum_{x_1,\ldots,x_k} n(x_1, \ldots, x_k) \ln p_{x_1 x_2 \cdots x_k}$$

The maximum likelihood parameter estimates $\hat{p}_{x_1 x_2 \cdots x_k}$ for large n are those that maximize J_1 and can be found as a solution to the system of equations (3.2.4). It is easy to see that

$$\hat{p}_{x_1 x_2 \cdots x_k} = n(x_1, x_2, \ldots, x_k) / n(x_1, x_2, \ldots, x_{k-1}) \tag{3.2.10}$$

The test information criterion (3.2.5) takes the following form:

$$I = \sum_{x_1,\ldots,x_{k+1}} n(x_1, x_2, \ldots, x_{k+1}) \ln(\hat{p}_{x_1 x_2 \cdots x_{k+1}} / \hat{p}_{x_2 x_3 \cdots x_{k+1}}) \tag{3.2.11}$$

and is distributed asymptotically as $0.5\chi^2$. To calculate the number of degrees of freedom, let us assume that all the probabilities (3.2.9) are different from zero. This is also sufficient for the chain regularity. Then, according to (3.2.8), the total number of chain's nonzero transition probabilities is $N_{nz} = 2^{k+1}$. The number of states is $N_{st} = 2^k$. The total number of independent parameters is $N_{par} = 2^{k-1}$, so that the total number of degrees of freedom is equal to $2^{k+1} - 2^k - 2^{k-1} = 2^{k-1}$. The hypothesis criterion (3.2.7) takes the form

$$\chi^2 = \sum_{x_1,\ldots,x_{k+1}} n^*(x_1,x_2,\ldots,x_k)(\hat{p}_{x_1x_2\cdots x_{k+1}} - \hat{p}_{x_2\cdots x_{k+1}})^2 / \hat{p}_{x_2x_3\cdots x_{k+1}}$$

$$n^*(x_1,x_2,\ldots,x_k) = \sum_l n(x_1,x_2,\ldots,x_k,l)$$

This criterion has a limiting χ^2 distribution with 2^{k-1} degrees of freedom if H_0 is true. Note that $n^*(x_1,x_2,\ldots,x_k)$ may differ from $n(x_1,x_2,\ldots,x_k)$ by 1, so that for a large sample sequence we may replace one by the other.

We can start building an error source model, assuming that it is a classical binary symmetrical channel without memory. In other words, we make an assumption that errors are independent and occur with probability p_1. This model may be considered as a zero-order Markov chain. The equation (3.2.10) gives us the ML probability of error estimate $\hat{p}_1 = n(1) / n$, where $n(1)$ is the total number of errors in the sequence of n bits. Also $\hat{p}_0 = 1-\hat{p}_1 = n(0) / n$, where $n(0)$ is the number of correct bits. Information criterion (3.2.11) takes the form

$$I = \sum_{x_1x_2} n(x_1,x_2) \ln (\hat{p}_{x_1x_2} / \hat{p}_{x_2})$$

where

$$\hat{p}_{x_1x_2} = n(x_1,x_2) / n(x_1)$$

This equation obviously coincides with (3.1.26). We reject the hypothesis that the channel is a binary symmetrical channel without memory if $I > 0.5\chi^2_{(1-\alpha),1}$.

If the hypothesis that errors are independent is rejected, we try to test the hypothesis that errors form a first-order Markov chain against the alternative hypothesis that they form a second-order Markov chain. The hypothesis information criterion (3.2.11) is given by

$$I = \sum_{x_1x_2x_3} n(x_1,x_2,x_3) \ln (\hat{p}_{x_1x_2x_3} / \hat{p}_{x_2x_3}) \qquad (3.2.12)$$

where

$$\hat{p}_{x_1x_2x_3} = n(x_1,x_2,x_3) / n(x_1,x_2)$$

If this hypothesis is rejected, we test the hypothesis that the error source is described by a second-order Markov chain and so on, until an r-th-order Markov chain fits the experimental data reasonably well.

The advantage of this method of model building is that states of the SSM are well defined (they are different sequences of errors). The disadvantage of this method is the usually tremendous size of matrices that describe the model. If an error source is described by an order r chain, then $\mathbf{P}(0)$ and $\mathbf{P}(1)$ are order 2^r square matrices (see Sec. 1.1.5 and Example 1.1.1). Normally, the size of a channel memory r is measured by hundreds of bits so that it becomes practically impossible to use this model. However, as we will see later, this approach can be successfully used to describe error statistics inside bursts or clusters of short bursts.

3.3 SEMI-MARKOV PROCESS HYPOTHESIS TESTING

3.3.1 ML Parameter Estimation. Let $x_1, x_2, \ldots, x_n$ be a sequence of samples of some s-ary stochastic process. We want to test the hypothesis H_0 that this process is a semi-Markov process with the interval transition distributions $f_{ij}(l;\boldsymbol{\tau})$ that depends on κ independent parameters $\boldsymbol{\tau} = (\tau_1, \tau_2, \ldots, \tau_\kappa)$. Within this hypothesis, according to the results of Sec. 1.4, the likelihood function is given by

$$L(\boldsymbol{\tau}) = p_{y_0} f^{(0)}_{y_0 y_1}(l_0) \prod_{i=1}^{n} f_{y_i y_{i+1}}(l_i;\boldsymbol{\tau}) \tag{3.3.1}$$

or

$$L(\boldsymbol{\tau}) = p_{y_0} f^{(0)}_{y_0 y_1}(l_0) \prod_{\xi,\eta,l} [f_{\xi\eta}(l;\boldsymbol{\tau})]^{n(\xi,\eta,l)} \tag{3.3.2}$$

where $n(\xi,\eta,l)$ is the frequency of transitions from state ξ to state η that took l steps. The logarithmic likelihood function is equal to

$$\ln L(\boldsymbol{\tau}) = J_0 + J_1(\boldsymbol{\tau}) \tag{3.3.3}$$

where

$$J_0 = \ln p_{y_0} f^{(0)}_{y_0 y_1}(l_0) \tag{3.3.4}$$

$$J_1(\boldsymbol{\tau}) = \sum_{\xi,\eta,l} n(\xi,\eta,l) \ln f_{\xi\eta}(l;\boldsymbol{\tau}) \tag{3.3.5}$$

Since $J_0/J_1 \to 0$ when $n \to \infty$, we can ignore the term J_0 in the expression of the logarithmic likelihood function. The ML parameter estimates are found by minimization of J_1 and represent a solution of the system

$$\frac{\partial J_1(\boldsymbol{\tau})}{\partial \tau_i} = 0 \quad i = 1, 2, \ldots, \kappa$$

If we make no assumptions about the form of the interval transition distributions, this equation has an obvious solution:

$$\hat{f}_{\xi\eta}(l) = n(\xi,\eta,l)\,/\,N(\xi) \tag{3.3.6}$$

where

$$N(\xi) = \sum_{\eta,l} n(\xi,\eta,l)$$

3.3.2 Grouped Data. Usually, the interval $[1,\infty)$ is partitioned into a finite number of subintervals:

$$[1,\infty) = \bigcup_{m=1}^{u(\xi,\eta)} \Big[x_m(\xi,\eta), x_{m+1}(\xi,\eta)\Big)$$

for each type of transition $\xi \to \eta$. Let

$$F_{\xi\eta}(m;\boldsymbol{\tau}) = \sum_{l=x_m}^{x_{m+1}-1} f_{\xi\eta}(l;\boldsymbol{\tau}) = Pr(x_m(\xi,\eta) \le l < x_{m+1}(\xi,\eta),\ \eta \,|\, \xi)$$

be the probability that $l \in [x_m(\xi,\eta), x_{m+1}(\xi,\eta))$, and the process transfers from state ξ to η. Denote $\mathbf{z}_i = (y_{i+1}, m_i)$, where m_i is the index of the subinterval that contains l_i:

$$x_{m_i}(y_i, y_{i+1}) \le l_i < x_{m_i+1}(y_i, y_{i+1})$$

If H_0 is true, the process $\{\mathbf{z}_i\}$ is a Markov chain, because the conditional probability

$$Pr(\mathbf{z}_i \,|\, \mathbf{z}_0, \mathbf{z}_1, \ldots, \mathbf{z}_{i-1}) = F_{y_i y_{i+1}}(m_i;\boldsymbol{\tau}) = Pr(\mathbf{z}_i \,|\, \mathbf{z}_{i-1})$$

depends only on $\mathbf{z}_{i-1}$. This means that we may apply the methods developed for Markov chain hypothesis testing to semi-Markov process hypothesis testing.

If H_0 is true, then

$$Pr(\zeta,k \,|\, \xi,\eta,m) = F_{\eta\zeta}(k;\boldsymbol{\tau}) \quad \text{for all } \xi,\eta,\zeta,m,k \tag{3.3.7}$$

The ML parameter estimate $\hat{\boldsymbol{\tau}}$ for the grouped data can be found from equation

$$\sum_{\xi,\eta,k} \frac{N(\xi,\eta,k)}{F_{\xi\eta}(k;\boldsymbol{\tau})} \frac{\partial F_{\xi\eta}(k;\boldsymbol{\tau})}{\partial \tau_m} = 0 \tag{3.3.8}$$

where

$$N(\xi,\eta,k) = \sum_{l=x_m}^{x_{m+1}-1} n(\xi,\eta,l)$$

Hypothesis information criterion (3.3.5) may be written as

$$I = \sum_{\xi,\eta,\zeta,m,k} N(\xi,\eta,\zeta,m,k) \ln \left[N(\xi,\eta,\zeta,m,k) \,/\, N(\xi,\eta,m) F_{\eta\zeta}(k;\hat{\boldsymbol{\tau}}) \right] \tag{3.3.9}$$

where $N(\xi,\eta,\zeta,m,k)$ is the frequency of transitions $\xi \to \eta \to \zeta$ and where adjacent transition interval lengths x and y satisfy conditions.

$$x_m(\xi,\eta) \le x < x_{m+1}(\xi,\eta) \qquad x_k(\eta,\zeta) \le y < x_{k+1}(\eta,\zeta)$$

This criterion is asymptotically distributed as $0.5\chi^2$ with f degrees of freedom defined by equation (3.3.5). The total number of the chain $\{\mathbf{z}_i\}$ states is given by

$$N_{st} = \sum_{\xi,\eta} u(\xi,\eta)$$

The total number of the nonzero transition probabilities is equal to

$$N_{nz} = \sum_{\xi,\eta,\zeta} u(\xi,\eta) u(\eta,\zeta)$$

Therefore, the number of degrees of freedom of the χ^2 distribution is equal to

$$f = \sum_{\xi,\eta,\zeta} u(\xi,\eta) u(\eta,\zeta) - \sum_{\xi,\eta} u(\xi,\eta) - \kappa \tag{3.3.10}$$

Criterion (3.3.6) becomes

$$\chi^2 = \sum_{\xi,\eta,\zeta,m,k} \frac{[N(\xi,\eta,\zeta,m,k) - N(\xi,\eta,m) F_{\eta\zeta}(k;\hat{\boldsymbol{\tau}})]^2}{N(\xi,\eta,m) F_{\eta\zeta}(k;\hat{\boldsymbol{\tau}})} \tag{3.3.11}$$

It is asymptotically equivalent to $2I$.

Consider now a semi-Markov process without making any assumptions about the interval distribution form. This case may be treated as an intermediate stage of SSM model building when we just want to test whether the model can be described on the basis of semi-Markov processes. In this case, all the $F_{\xi\eta}(m)$ become parameters and the ML estimates can be found from system (3.3.8), which has the following solution:

$$\hat{F}_{\xi\eta}(m) = N(\xi,\eta,m) / N(\xi) \tag{3.3.12}$$

The total number of independent parameters is equal to

$$\kappa = \sum_{\xi,\eta} u(\xi,\eta) - s$$

since

$$\sum_{\eta,m} F_{\xi\eta}(m) = 1 \qquad \xi = 1,2,\ldots,s$$

Hypothesis information criterion (3.3.9) is given by

$$I = \sum_{\xi,\eta,\zeta,m,k} N(\xi,\eta,\zeta,m,k) \ln \left(\frac{N(\xi,\eta,\zeta,m,k) N(\eta)}{N(\xi,\eta,m) N(\eta,\zeta,k)} \right) \tag{3.3.13}$$

and criterion (3.3.11) becomes

$$\chi^2 = \sum_{\xi,\eta,\zeta,m,k} \frac{[N(\xi,\eta,\zeta,m,k) - N(\xi,\eta,m)\hat{F}_{\eta\zeta}(k)]^2}{N(\xi,\eta,m)\hat{F}_{\eta\zeta}(k)} \tag{3.3.14}$$

Criteria $2I$ and χ^2 are equivalent and have a limiting χ^2 distribution with

$$f = \sum_{\xi,\eta,\zeta} u(\xi,\eta)u(\eta,\zeta) - \sum_{\xi,\eta} u(\xi,\eta) - \sum_{\xi,\eta} u(\xi,\eta) - s \tag{3.3.15}$$

degrees of freedom.

In the particular case in which $x_1(\xi,\eta) = 1$ and $x_2(\xi,\eta) = \infty$ for all ξ and η, the previous criterion coincides with criterion (3.3.11) of the first-order Markov chain hypothesis testing. Another particular case, $s=2$, can be used to test the hypothesis that $x_1, x_2, \ldots, x_n$ are the samples from an alternating renewal process. In this case, the above criteria become the well-known criteria of random variable independence testing.[53]

3.3.3 Underlying Markov Chain Parameter Estimation. As we pointed out in Sec. 1.4, there is an alternative description of semi-Markov processes, which uses the notion of the underlying Markov chain whose transition probabilities are defined by the equation

$$p_{ij}(\boldsymbol{\tau}) = \sum_{l=1}^{\infty} f_{ij}(l;\boldsymbol{\tau})$$

and state-holding time distributions

$$w_{ij}(l;\boldsymbol{\tau}) = f_{ij}(l;\boldsymbol{\tau}) \,/\, p_{ij}(\boldsymbol{\tau})$$

Let us assume that all the transition probabilities and holding time distributions depend on different sets of parameters. Then the interval transition distribution takes the form

$$f_{ij}(l;\boldsymbol{\tau}) = p_{ij}(\boldsymbol{\tau}_1) w_{ij}(l;\boldsymbol{\tau}_2) \tag{3.3.16}$$

This assumption, as we will see, may simplify SSM model building. Likelihood function (3.3.5) can be rewritten as

$$J_1(\boldsymbol{\tau}) = J_{11}(\boldsymbol{\tau}_1) + J_{12}(\boldsymbol{\tau}_2) \tag{3.3.17}$$

where

$$J_{11}(\boldsymbol{\tau}_1) = \sum_{\xi,\eta} n(\xi,\eta) \ln p_{\xi\eta}(\boldsymbol{\tau}_1) \quad n(\xi,\eta) = \sum_{l} n(\xi,\eta,l) \tag{3.3.18}$$

$$J_{12}(\boldsymbol{\tau}_2) = \sum_{\xi,\eta,l} n(\xi,\eta,l) \ln w_{\xi\eta}(l;\boldsymbol{\tau}_2) \tag{3.3.19}$$

It is clear from equation (3.3.17) that the ML estimate of the parameter $\boldsymbol{\tau}_1$ of the underlying Markov process can be found independently of the parameter $\boldsymbol{\tau}_2$ of the state-holding time distributions. If the individual distributions have independent parameters, the likelihood function $J_{12}(\boldsymbol{\tau}_2)$ can also be decomposed:

$$J_{12}(\boldsymbol{\tau}_2) = \sum_{\xi\eta} G_{\xi\eta}(\boldsymbol{\tau}_{\xi\eta})$$

where

$$G_{\xi\eta}(\boldsymbol{\tau}_{\xi\eta}) = \sum_l n(\xi,\eta,l)\ln w_{\xi\eta}(l;\boldsymbol{\tau}_{\xi\eta})$$

The ML estimates $\hat{\boldsymbol{\tau}}_{\xi\eta}$ can be found independently by maximization of $G_{\xi\eta}(\boldsymbol{\tau}_{\xi\eta})$. In the case in which no assumptions about the distributions' dependency on the parameters have been made, the ML estimates are given by

$$\hat{p}_{\xi\eta} = n(\xi,\eta)/n(\xi) \quad n(\xi) = \sum_\eta n(\xi,\eta)$$

$$\hat{w}_{\xi\eta}(l) = n(\xi,\eta,l)/n(\xi,\eta)$$

The null hypothesis information criterion can be decomposed into a sum of information criteria of elementary hypotheses: the hypothesis that the transition process is a Markov chain with the transition probabilities $p_{ij}(\boldsymbol{\tau}_1)$, the hypothesis that the state-holding time distribution is $w_{\xi\eta}(l;\boldsymbol{\tau}_{\xi\eta})$, and the hypothesis that the process is semi-Markov with no assumptions about its interval transition distribution form. For the grouped data we denote

$$W_{\xi\eta}(l;\boldsymbol{\tau}_{\xi\eta}) = \sum_{l=x_m}^{x_{m+1}-1} w_{\xi\eta}(l;\boldsymbol{\tau}_{\xi\eta}) = Pr(x_m(\xi,\eta) \le l < x_{m+1}(\xi,\eta) \mid \xi,\ \eta)$$

Hypothesis information criterion (3.3.9), after some elementary transformations, can be rewritten in the form

$$I = I_1 + \sum_{\xi\eta} I_{\xi\eta} + I_3 \tag{3.3.20}$$

where

$$I_1 = \sum_{\eta,\zeta} N(\eta,\zeta)\ln\left[N(\eta,\zeta)/N(\eta)p_{\eta\zeta}(\hat{\boldsymbol{\tau}}_1)\right] \tag{3.3.21}$$

is information criterion (3.3.5) for the hypothesis that the underlying transition process is a Markov chain, and

$$I_{\xi\eta} = \sum_{\eta,\zeta,k} N(\eta,\zeta,k)\ln\left[N(\eta,\zeta,k)/N(\eta,\zeta)W_{\eta\zeta}(k;\hat{\boldsymbol{\tau}}_{\xi\eta})\right] \tag{3.3.22}$$

is the information criterion for the goodness of fit of the state-holding time distribution, and, finally,

$$I_3 = \sum_{\xi,\eta,\zeta,m,k} N(\xi,\eta,\zeta,m,k) \ln \left\{ \frac{N(\xi,\eta,\zeta,m,k)N(\eta)}{N(\xi,\eta,m)N(\eta,\zeta,k)} \right\} \tag{3.3.23}$$

is information criterion (3.3.13) for the hypothesis that the process can be described by a semi-Markov process when no assumptions have been made about its transition distributions structure.

Each of the above criteria can be used for null hypothesis rejection. First, we can use the criterion I_3 to test whether it is possible to describe the process with a semi-Markov process without making any other assumptions. If the hypothesis passes the first test, then we can test the hypothesis that the transition process is a Markov chain. After that we can test the goodness of fit of all the state-holding time distributions. And, finally, if all these tests do not reject the null hypothesis, we can apply total information criterion (3.3.20).

3.3.4 Autonomous Semi-Markov Processes. If a semi-Markov process is autonomous, then its state-holding time distributions depend only on the transition initial state:

$$w_{\xi\eta}(l;\boldsymbol{\tau}_{\xi\eta}) = w_{\xi}(l;\boldsymbol{\theta}_{\xi}) \tag{3.3.24}$$

In this case the likelihood function $J_{12}(\boldsymbol{\tau}_2)$ can be written as

$$J_{12}(\boldsymbol{\tau}_2) = \sum_{\xi} G_{\xi}(\boldsymbol{\theta}_{\xi}) \tag{3.3.25}$$

where

$$G_{\xi}(\boldsymbol{\theta}_{\xi}) = \sum_{l} n(\xi,l) \ln w_{\xi}(l;\boldsymbol{\theta}_{\xi}) \qquad n(\xi,l) = \sum_{\eta} n(\xi,\eta,l) \tag{3.3.26}$$

The ML estimates are

$$\hat{w}_{\xi}(l) = n(\xi,l) / n(\xi)$$

if no assumptions about the distribution form have been made.

The null hypothesis can be tested using information criterion (3.3.20). For the autonomous semi-Markov processes, this criterion might be rewritten as

$$I = I_1 + \sum_{\eta} I_{a,\eta} + I'_3 \tag{3.3.27}$$

where I_1 is defined by equation (3.3.21); the remaining terms are given by

$$I_{a,\eta} = \sum_k N(\eta,k) \ln \left[N(\eta,k) / N(\eta) W_\eta(k;\hat{\boldsymbol{\theta}}_\eta) \right] \tag{3.3.28}$$

and

$$I'_3 = \sum_{\xi,\eta,\zeta,m,k} N(\xi,\eta,\zeta,m,k) \ln \left[\frac{N(\xi,\eta,\zeta,m,k) N(\eta) N(\eta)}{N(\xi,\eta,m) N(\eta,\zeta) N(\eta,k)} \right] \tag{3.3.29}$$

The autonomous semi-Markov processes are very popular in applications. The number of the state-holding time distributions in this case is equal to the number of states, whereas in general it is equal to the squared number of states.

3.4 MATRIX-GEOMETRIC DISTRIBUTION PARAMETER ESTIMATION

3.4.1 Distribution Simplification. As we have seen in the previous chapters, interval distributions of SSM and matrix processes have a matrix-geometric form

$$p(x) = \mathbf{S}\mathbf{M}^{x-1}\mathbf{F} \tag{3.4.1}$$

where $\mathbf{S}$ is order n matrix row, $\mathbf{F}$ is order n matrix column, and $\mathbf{M}$ is order n square matrix. All the elements of these matrices are assumed to be complex numbers. The absolute values of all the matrix $\mathbf{M}$ eigenvalues are assumed to be less than 1, so that the series

$$\sum_{x=1}^{\infty} p(x) = \mathbf{S}(\mathbf{I} - \mathbf{M})^{-1}\mathbf{F} = 1$$

converges for any $\mathbf{S}$ and $\mathbf{F}$. This distribution parameter estimation plays a fundamental role in SSM model building.

The distribution parameters are the elements of the matrices $\mathbf{S}$, $\mathbf{M}$, and $\mathbf{F}$. We must estimate n elements of the matrix $\mathbf{S}$, n^2 elements of the matrix $\mathbf{M}$, and n elements of the matrix $\mathbf{F}$. This makes n^2+2n-1 parameters to be estimated, since one parameter can be found from the condition that the total sum of the probabilities is equal to 1. It is possible to decrease significantly the number of parameters by using the similarity transformation

$$\mathbf{M}_1 = \mathbf{T}^{-1}\mathbf{M}\mathbf{T}$$

since the distribution has the same form for all similar matrices:

$$\mathbf{S}\mathbf{M}^{x-1}\mathbf{F} = \mathbf{S}_1\mathbf{M}_1^{x-1}\mathbf{F}_1$$

where $\mathbf{S}_1 = \mathbf{ST}$, $\mathbf{F}_1 = \mathbf{T}^{-1}\mathbf{F}$ and any nonsingular $\mathbf{T}$. The Jordan normal form of the matrix (see Ref. [24] and Appendix 5.1.5)

$$\mathbf{J} = block\ diag\,\{\mathbf{J}_i\} \tag{3.4.2}$$

where

$$\mathbf{J}_i = \mu_i \mathbf{I} + \mathbf{B}_i \qquad \mathbf{B}_i = \begin{bmatrix} 0 & 1 & 0 \dots 0 \\ 0 & 0 & 1 \dots 0 \\ \dots & \dots & \dots \\ 0 & 0 & 0 \dots 1 \\ 0 & 0 & 0 \dots 0 \end{bmatrix}_{s_i, s_i} \qquad i = 1,2,\dots,m$$

is similar to $\mathbf{M}$ and has the minimal number of parameters. Its parameters μ_i are the eigenvalues of the matrix $\mathbf{M}$.

Let us assume that the matrix $\mathbf{M}$ in representation (3.4.1) already has the Jordan form. Then equation (3.4.1) can be written as

$$p(x) = \sum_{i=1}^{m} \mathbf{S}_i \mathbf{J}_i^{x-1} \mathbf{F}_i$$

where $\mathbf{S}_i$ and $\mathbf{F}_i$ are the corresponding sub-blocks of the matrices $\mathbf{S}$ and $\mathbf{F}$. Since

$$\mathbf{J}_i^r = (\mu_i \mathbf{I} + \mathbf{B}_i)^r = \sum_{k=0}^{s_i-1} \binom{r}{k} \mu_i^{r-k} \mathbf{S}_i \mathbf{B}_i^k \mathbf{F}_i$$

We assume here that $\binom{0}{0} = 1$ and $\binom{r}{k} = r(r-1) \cdots (r-k+1)/k!$, so that $\binom{r}{k} = 0$ for $0 \le r < k$ (see Appendix 4.1). Distribution (3.4.1) takes the form

$$p(x) = \sum_{i=1}^{m} \sum_{k=0}^{s_i-1} \binom{x-1}{k} \mu_i^{x-k-1} \mathbf{S}_i \mathbf{B}_i^k \mathbf{F}_i \tag{3.4.3}$$

Using this equation, it is easy to prove that the matrix $\mathbf{F}$ can always be replaced with $\mathbf{1}$. Thus, the simplified representation of the distribution is given by the following equation:

$$p(x) = \sum_{i=1}^{m}\sum_{k=0}^{s_i-1}\binom{x-1}{k}\mu_i^{x-k-1}c_{ik} \tag{3.4.4}$$

where

$$c_{ik} = \mathbf{S}_i\mathbf{B}_i^k\mathbf{1} \tag{3.4.5}$$

If some of the eigenvalues that belong to different Jordan blocks are equal ($\mu_i = \mu_j$ for $i \neq j$), then the corresponding terms in the distribution can be factored out. We assume that all the μ_i in equation (3.4.4) are different. It is important to note that, if the coefficients c_{ik} have been found, the elements of the matrix $\mathbf{S}_i$ can be determined from linear equations (3.4.5) that always have a unique solution. This solution can be expressed in the matrix form:

$$\mathbf{S}_i = \mathbf{C}_i\mathbf{R}(\mathbf{I} - \mathbf{B}_i) \tag{3.4.6}$$

where $\mathbf{C}_i = row\{c_{ik}\}$

$$\mathbf{R} = \begin{bmatrix} 0 & 0 & \dots & 0 & 1 \\ 0 & 0 & \dots & 1 & 0 \\ \dots & \dots & \dots & \dots & \dots \\ 1 & 0 & \dots & 0 & 0 \end{bmatrix}$$

Let $\mu_i \neq 0$ for $i=1,2,\dots,m-1$ and $\mu_m = 0$; then, using the complex number trigonometric form

$$\mu_i = q_i(\cos\omega_i + \mathrm{j}\sin\omega_i) \quad \text{for } i=1,2,\dots,m-1$$

we can rewrite the distribution $p(x)$ explicitly in its real form:

$$p(x) = \mathbf{S}\mathbf{J}^{x-1}\mathbf{1} = \sum_{i=1}^{m-1}\sum_{k=0}^{s_i-1}\binom{x-1}{k}[a_{ik}\cos\omega_i(x-k-1) - b_{ik}\sin\omega_i(x-k-1)]q_i^{x-k-1} + a_{m,x-1} \tag{3.4.7}$$

where $a_{ik} = \mathrm{Re}\,c_{ik}$ and $b_{ik} = \mathrm{Im}\,c_{ik}$. If the matrix $\mathbf{M}$ does not have zero eigenvalues, then $a_{m,x-1} = 0$.

In the most important (for applications) case, all the nonzero eigenvalues μ_i are simple (s_i=1 for $i=1,2,\dots,m-1$), and the distribution is given by

$$p(x) = \sum_{i=1}^{m-1} a_{i0} q_i^{x-1} + a_{m,x-1} \tag{3.4.8}$$

This distribution is polygeometric except for the term $a_{m,x-1}$, which can be used to reach better agreement with the experimental data for small x.

Let us consider some methods of estimating distribution (3.4.7) parameters.

3.4.2 ML Parameter Estimation. Estimating the distribution parameters can be carried out using the methods described in Sec. 3.1.1. The logarithmic likelihood function, according to (3.1.8), takes the form

$$L(\mathbf{S},\boldsymbol{\mu},\mathbf{z}) = \sum_{i=1}^{r} \nu_i \ln P_i(\mathbf{S},\boldsymbol{\mu},\mathbf{z})$$

where $\boldsymbol{\mu} = (\mu_1,\mu_2,\ldots,\mu_m)$, $\mathbf{z} = (s_1,s_2,\ldots,s_m)$, and $P_i(\mathbf{S},\boldsymbol{\mu},\mathbf{z}) = Pr(x_{i-1} \le x < x_i)$ is the probability that x satisfies the condition $x_{i-1} \le x < x_i$:

$$P_i(\mathbf{S},\boldsymbol{\mu},\mathbf{z}) = \mathbf{S}\mathbf{b}_i(\boldsymbol{\mu},\mathbf{z}) \quad \mathbf{b}_i(\boldsymbol{\mu},\mathbf{z}) = (\mathbf{J}^{x_{i-1}-1} - \mathbf{J}^{x_i})(\mathbf{I} - \mathbf{J})^{-1}\mathbf{1}$$

ν_i is the corresponding empirical frequency.

The ML parameter estimates are found by maximization of $L(\mathbf{S},\boldsymbol{\mu},\mathbf{z})$, subject to the restriction that the total sum of the probabilities is equal to 1. The matrix order n and the sizes s_i of its Jordan blocks are discrete parameters. Normally, the maximization with respect to these variables is carried out by an exhaustive search: for a selected set of the discrete variables the function is maximized with respect to the rest of the parameters. Then a new set of the variables is selected and the maximum with respect to the continuous parameters is found, and so on. The discrete variables that deliver the global maximum over all possible sets give the ML estimates. In the future we will assume that the discrete variables are fixed and will consider the problem of estimating the rest of the parameters.

The ML estimates can be found from the necessary conditions of the function $L(\mathbf{S},\boldsymbol{\mu})$ maximum:

$$\frac{\partial}{\partial S_j}\Big[\sum_{i=1}^{r} \nu_i \ln P_i(\mathbf{S},\boldsymbol{\mu}) + \lambda \sum_{i=1}^{r} P_i(\mathbf{S},\boldsymbol{\mu}) - \lambda\Big] = 0 \quad j=1,2,\ldots,n$$

$$\frac{\partial}{\partial \mu_k}\Big[\sum_{i=1}^{r} \nu_i \ln P_i(\mathbf{S},\boldsymbol{\mu}) + \lambda \sum_{i=1}^{r} P_i(\mathbf{S},\boldsymbol{\mu}) - \lambda\Big] = 0 \quad k=1,2,\ldots,m$$

$$\sum_{i=1}^{r} P_i(\mathbf{S},\boldsymbol{\mu}) - 1 = 0$$

where λ is a Lagrange's multiplier. After taking the derivatives, we obtain

$$\begin{aligned} &\sum_{i=1}^{r}\left[\frac{\nu_i}{\mathbf{S}\mathbf{b}_i(\boldsymbol{\mu})} + \lambda\right]b_{ij} = 0 \qquad j=1,2,\ldots,n \\ &\sum_{i=1}^{r}\left[\frac{\nu_i}{\mathbf{S}\mathbf{b}_i(\boldsymbol{\mu})} + \lambda\right]d_{ik} = 0 \qquad k=1,2,\ldots,m \\ &\mathbf{S}(\mathbf{I} - \mathbf{J})^{-1}\mathbf{1} - 1 = 0 \end{aligned} \tag{3.4.9}$$

where b_{ij} is the j-th element of the matrix-column $\mathbf{b}_i(\boldsymbol{\mu})$

$$d_{ik} = \mathbf{S}_k[\mathbf{G}_k(x_{i-1}-1) - \mathbf{G}_k(x_i)](\mathbf{I}-\mathbf{J}_k)^{-2}\mathbf{1}$$

$$\mathbf{G}_k(x) = [x(\mathbf{I}-\mathbf{J}_k) + \mathbf{J}_k]\mathbf{J}_k^{x-1}$$

We see that the ML parameter estimates satisfy a system of nonlinear equations. This system can be solved using one of the standard numerical methods. However, these methods converge very slowly when estimating parameters of the distribution of intervals between errors. Some of the eigenvalues μ_i are very close to 1, and small variations of these parameters significantly change the direction of the next iteration. We consider several alternative techniques that use the specifics of the distribution and converge faster than direct methods of solving system (3.4.9).

3.4.3 Sequential Least Mean Square Method. The estimation of parameters can be carried out using the minimum χ^2 method, which is asymptotically equivalent to the ML method considered in the previous section. The χ^2 criterion (3.1.11) takes the form

$$\chi^2 = \sum_i [\nu_i - NP_i(\mathbf{S},\boldsymbol{\mu})]^2 / NP_i(\mathbf{S},\boldsymbol{\mu})$$

Normally, the χ^2 or I minimization is carried out using the gradient methods. However, the standard gradient methods applied directly to χ^2 are unstable for the majority of practical cases. The optimal solution may be found by sequential application of the least squares (LS) techniques.

Suppose that $\boldsymbol{\theta}^{(s)}=(\mathbf{S}^{(s)},\boldsymbol{\mu}^{(s)})$ is the initial value of the parameter vector. Then the next approximation can be found by minimization of

$$\chi^2_{c+1} = \sum_i [\, \nu_i - NP_i(\boldsymbol{\theta})\,]^2 \,/\, NP_i(\boldsymbol{\theta}^{(s)})$$

Thus, starting with some initial value $\boldsymbol{\theta}^{(0)}$, we find recursively $\boldsymbol{\theta}^{(1)}$, $\boldsymbol{\theta}^{(2)}$ and so on.

Further simplification may be achieved if we use some specifics of the distribution (3.4.7). If $\boldsymbol{\mu}$ is fixed, then the minimization of χ^2_{c+1} is a linear LS problem and the elements of the matrix **S** can be found by solving a system of linear equations or a linear matrix equation.[61] Suppose that $\boldsymbol{\theta}^{(c)}$ is the parameter approximation. The next approximation of the parameter $\mathbf{S}^{(c+1)}$ can be found in two steps. First we find the next approximation of matrix **S** by minimizing

$$\chi^2_{c+1} = \sum_i [\, \nu_i - N\mathbf{S}b_i(\boldsymbol{\mu}^{(c)})\,]^2 \,/\, NP_i(\boldsymbol{\theta}^{(c)})$$

The necessary condition of minimum

$$\frac{\partial \chi^2_{c+1}}{\partial S_i} = 0$$

gives the following linear equation

$$\mathbf{SA} = \mathbf{c} \tag{3.4.10}$$

where

$$\mathbf{A} = \left[\sum_k b_{ki}(\boldsymbol{\mu}^{(c)}) b_{kj}(\boldsymbol{\mu}^{(c)}) \,/\, NP_k(\boldsymbol{\theta}^{(c)}) \right] \quad \mathbf{c} = \left[\sum_k \nu_k b_{kj}(\boldsymbol{\mu}^{(c)}) \,/\, NP_k(\boldsymbol{\theta}^{(c)}) \right]$$

This system is usually called the *normal equation*.[61] Thus, the optimal **S** satisfies some normal equation.

Second, we find the optimal $\boldsymbol{\mu}$ minimizing

$$\chi^2_{c+1} = \sum_i [\, \nu_i - N\mathbf{S}^{(c+1)}b_i(\boldsymbol{\mu})\,]^2 \,/\, NP_i(\boldsymbol{\theta}^{(c)})$$

with respect to $\boldsymbol{\mu}$. Then the associated value **S** is found from the new normal equation and so on until a minimum for χ^2 is found. This method can be used effectively when the number of parameters is small. However, if the number of parameters is large or unknown, the computations become quite complex. Since direct χ^2 minimization is complex, it is very important to find some quasi-optimal methods of parameter estimation. These methods are based on various linearization techniques of the matrix-geometric function of type (3.4.7).

3.4.4 Piecewise Logarithmic Transformation. The first method is heuristic. It can be used to obtain the initial values of parameters for the method described in the previous section.

If we assume that the distribution $p(x)$ has form (3.4.8)

$$p(x) = \sum_{i=1}^{m-1} a_i(1-q_i)q_i^{x-1} + a_{m,x-1}$$

then $p(x) \approx a_1(1-q_1)q_1^{x-1}$ for $q_1 = \max(q_i)$ and large x. This means that

$$\ln p(x) \approx \ln [a_1(1-q_1)]+(x-1)\ln q_1$$

can be approximated with a straight line. After we approximate $\ln p(x)$ for large x with a linear function $k(x-1)+b$, the distribution parameters are found from obvious equations: $\ln q_1 = k$ and $\ln[a_1(1-q_1)] = b$. Therefore, $a_1 = e^b / (1-e^k)$ and $q_1 = e^k$.

Consider now the function $p^{(1)}(x) = p(x)-a_1(1-q_1)q_1^{x-1}$. Using a similar approach, we can extract the next term, $a_2(1-q_2)q_2^{x-1}$ (if $q_2 = \max(q_i)$, $i \neq 1$), and so on. Finally, we estimate $a_{m,x-1}$ by the differences between the approximation and the empirical frequencies for small x.

This method may be used in conjunction with the method of moments which gives additional equations for the parameters

$$\sum_x p(x) = \sum_{i=1}^{m-1} a_i + \sum_x a_{m,x-1} = 1$$

$$\bar{x} = \sum_x xp(x) = \sum_{i=1}^{m-1} a_i / (1 - q_i) + \sum_x xa_{m,x-1}$$

The most basic deficiency of this method is the arbitrariness in selecting the distribution form: we can approximate by polygeometric distributions, by polynomials, by arbitrary sequences $a_{m,x-1}$, or by any combination of the above. Naturally, in this situation one should choose the approximation which contains the least number of parameters. We also note an instability of the χ^2 minimization in estimating distribution (3.4.7) parameters. This especially concerns the parameter q_1 (small variation of the parameter significantly changes the magnitude of χ^2). We will show a different way to estimate distribution (3.4.7) parameters which in a certain sense is devoid of these deficiencies.

3.4.5 Prony's Method. The other method is based on the fact that the distribution $p(x)$ satisfies a *linear* difference equation (see Sec. 1.3.3). Therefore, if we can estimate the coefficients of this equation, then the distribution $p(x)$ may be found as a solution of the equation.

In order to find the correspondence between parameters of the distribution and its linear equation, we use the z-transform

$$\rho(z) = \sum_{x=0}^{\infty} p(x) z^{-x} \tag{3.4.11}$$

which, according to (3.4.1), is equal to

$$\rho(z) = \mathbf{S}(\mathbf{I}z - \mathbf{M})^{-1}\mathbf{F}$$

If we replace the inverse matrix in this equation with $(\mathbf{I}z - \mathbf{M})^{-1} = \mathbf{B}(z) / \Delta(z)$, where $\mathbf{B}(z)$ is the adjoint matrix and $\Delta(z)$ is the characteristic polynomial of the matrix $\mathbf{M}$ (see Appendix 5), then $\rho(z)$ may be represented by the rational function

$$\rho(z) = u(z) / \Delta(z) \tag{3.4.12}$$

where

$$u(z) = \mathbf{SB}(z)\mathbf{F} = \sum_{k=1}^{c} u_k z^k \qquad c < r$$

According to (3.4.2), the characteristic polynomial $\Delta(z) = \det(\mathbf{I}z - \mathbf{M})$ has the form

$$\Delta(z) = \prod_{i=1}^{m} (z - \mu_i)^{s_i} = z^r + \sum_{i=1}^{r} \alpha_i z^{r-i} \tag{3.4.13}$$

Now the problem of estimating the distribution $p(x)$ parameters can be stated in terms of its z-transform: Given the estimates $\hat{p}(x)$, find an approximation of the distribution z-transform with rational function (3.4.12). The estimates of μ_i are found from characteristic equation (3.4.13), and the elements of the matrix $\mathbf{S}$ can be found by decomposition of the rational function into partial fractions. Indeed, let

$$\rho(z) = \sum_{i=1}^{m} \sum_{k=0}^{s_i - 1} c_{ik} (z - \mu_i)^{-k-1}$$

be the decomposition of the rational function $\rho(z)$ into partial fractions. Using an easily verifiable formula (see Appendix 5)

$$(z-\mu)^{-k-1} = \sum_{x=1}^{\infty} \binom{x-1}{k} \mu^{x-k-1} z^{-x} \tag{3.4.14}$$

and equation (3.4.11), we obtain the equation

$$p(x) = \sum_{i=1}^{m} \sum_{k=0}^{s_i-1} \binom{x-1}{k} \mu_i^{x-k-1} c_{ik}$$

which coincides with (3.4.4). This result proves that equation (3.4.4) always has a unique solution which can be found by the distribution z-transform decomposition into partial fractions. (The direct proof can be obtained by calculating the system determinant, which is the generalized Vandermonde's determinant.) The matrix **S** is found from equation (3.4.6), where $\mathbf{F} = \mathbf{1}$, so that we can construct distribution (3.4.1) from its z-transform, which is a rational function.

The problem of finding rational function approximations for a given analytical function has been intensively studied in the mathematical literature. [62] This problem is also related to the theory of Padé approximations. [63—66]

To obtain the difference equation for $p(x)$, we rewrite equation (3.4.12) in the following form:

$$\Delta(z)\rho(z) = U(z) \quad deg U(z) < r \tag{3.4.15}$$

Comparing the coefficients of the same powers of z on both sides of this equation, we obtain

$$p(x) + \sum_{i=1}^{r} \alpha_i p(x-i) = u_x \tag{3.4.16}$$

where $u_x = 0$ for $x > r$ and $p(x) = 0$ for $x < 1$.

From the mathematical point of view, equation (3.4.4) represents the solution of the linear equation (3.4.16), which satisfies the zero initial conditions. We can also treat the parameters of the system (3.4.16) as an alternative set of parameters that describe equation (3.4.1). As is well known in the theory of linear systems, we can treat (3.4.1) as a solution of the homogeneous system (u_x=0), which satisfies nonzero initial conditions. Therefore, the initial conditions $p(1), p(2), \ldots, p(r)$ can replace the parameters $u_1, u_2, \ldots, u_r$ in describing distribution (3.4.1). The initial r equations of the system (3.4.16) represent the relationship between these two sets of parameters.

If one can estimate the coefficients of this equation and the initial conditions, then distribution (3.4.4) can be found as the unique solution of this equation. The estimates of μ_i are found as the roots of the equation characteristic polynomial (3.4.13). The coefficients c_{ik} of the polynomials can be found from the first r equations of the system of linear equations (3.4.4), in which μ_i and $p(x)$ have been replaced by their estimates. In the particular case where the distribution is polygeometric this method was introduced by Prony[67] in 1795.

The relationship between the system (3.4.16) and the rational function (3.4.13) is well known in the theory of linear systems. The rational function (3.4.12) is interpreted as a linear filter transfer function, the system (3.4.16) describes the filter operation in the time domain, and the sequence $p(x)$ represents the signal on the filter output. The problem of estimating the distribution parameters is equivalent to the linear system identification.

There are many methods of estimating equation (3.4.16) parameters. [68,69,70] The simplest method, suggested by Prony,[67] is to replace $p(x)$ with their estimates $\hat{p}(x) = \nu_x / N$ and then treat equations (3.4.16) as a system of linear equations with $r+c$ unknowns $(\alpha_1, \alpha_2, ..., \alpha_r, u_1, ..., u_c)$. The unknowns α_i can be found independently of u_i if we consider the part of the system (3.4.16) for $x > c$:

$$\hat{p}(x) + \sum_{i=1}^{r} \alpha_i \hat{p}(x-i) = 0 \qquad (3.4.17)$$

If this system has a solution, then the unknowns u_i can be found directly from the part of the system (3.4.16) for $x \le c$. System (3.4.17) is called Toeplitz system.

There are many efficient methods of solving Toeplitz systems. These methods have been developed independently in the following theories: Padé approximations,[63—66] orthogonal polynomials,[71] algebraic codes,[75—78] and linear prediction and spectrum analysis.[69,70,72,73,74] Normally, the solution of a set of r equations would require $O(r^3)$ operations. However, the special structure of the system (3.4.17) allows us to reduce the number of operations. Levinson,[72] Durbin,[79] and Berlekamp[75] developed methods that require $O(r^2)$ operations. The method based on Euclid's algorithm[80] requires only $O(r \log^2 r)$ operations. The main feature of these algorithms is that they are *order recursive*. This means that if we find the best fit for the linear system (3.4.16) of an order r and an approximation criterion is not satisfied, we can increase the system's order and calculate new parameters by a simple modification of the previous parameters without necessity of solving the system from scratch.

EXAMPLE 3.4.1: To illustrate Prony's method we first assume that the equation order is $r=1$. Equation (3.4.16) parameter estimate is found from the first equation of (3.4.17)

$$\hat{\alpha}_1 = -\hat{p}(2) / \hat{p}(1)$$

The characteristic equation

$$\Delta(z) = z + \hat{\alpha}_1 = 0$$

root is estimated by $\hat{\mu}_1 = -\hat{\alpha}_1 = \hat{p}(2) / \hat{p}(1)$. The matrix $\mathbf{S} = S_1$ is found from equation (3.4.4)

$$\hat{p}(1) = S_1 \hat{\mu}_1^0$$

so that $\hat{S}_1 = \hat{p}(1)$. If the null hypothesis that the equation has the order $r=1$ is true, then the remaining probabilities are approximated by

$$\hat{q}_i = \hat{S}_1 \hat{\mu}_1^{i-1} = \hat{p}(1)(\hat{p}(2) / \hat{p}(1))^{i-1}$$

Note that, in this case, we are estimating parameters of the geometric distribution (3.1.5). The estimate that we found is different from the ML estimate given by equation (3.1.6). For the grouped data, the χ^2 criterion is given by

$$\chi^2 = N \sum_{i=1}^{t} [\hat{P}(i) - \hat{Q}(i)]^2 / \hat{Q}(i) \tag{3.4.18}$$

where

$$\hat{P}(i) = \sum_{j=x_{i-1}}^{x_i - 1} \hat{p}(j) \quad \hat{Q}(i) = \sum_{j=x_{i-1}}^{x_i - 1} \hat{q}(j)$$

If the null hypothesis is rejected by this criterion, we try to test the hypothesis that the difference equation has the order $r=2$.

The equation parameters are estimated from the first two equations of system (3.4.17):

$$\hat{p}(3) = \alpha_1 \hat{p}(2) + \alpha_2 \hat{p}(1)$$

$$\hat{p}(4) = \alpha_1 \hat{p}(3) + \alpha_2 \hat{p}(2)$$

Solving this equation, we obtain

$$\hat{\alpha}_1 = \frac{\hat{p}(3)\hat{p}(2) - \hat{p}(4)\hat{p}(1)}{\hat{p}(2)\hat{p}(2) - \hat{p}(3)\hat{p}(1)} \quad \hat{\alpha}_2 = \frac{\hat{p}(4)\hat{p}(2) - \hat{p}(3)\hat{p}(3)}{\hat{p}(2)\hat{p}(2) - \hat{p}(3)\hat{p}(1)}$$

The characteristic polynomial is estimated by

$$\Delta(z) = z^2 + \hat{\alpha}_1 z + \hat{\alpha}_2 .$$

Suppose that this equation has different roots $\hat{\mu}_1$ and $\hat{\mu}_2$. Then we can select $\mathbf{F} = \mathbf{1}$, and the elements of the matrix $\mathbf{S} = [S_1 \ \ S_2]$ can be estimated from the first equations of (3.4.4),

$$\hat{p}(1) = S_1 + S_2$$

$$\hat{p}(2) = S_1\hat{\mu}_1 + S_2\hat{\mu}_2$$

which have the following solution:

$$\hat{S}_1 = \frac{\hat{p}(1)\hat{\mu}_2 - \hat{p}(2)}{\hat{\mu}_2 - \hat{\mu}_1} \qquad \hat{S}_2 = \frac{\hat{p}(2) - \hat{p}(1)\hat{\mu}_1}{\hat{\mu}_2 - \hat{\mu}_1}$$

Next we calculate the approximation of the remaining probabilities using equation (3.4.4):

$$q_i = \hat{S}_1\hat{\mu}_1^{i-1} + \hat{S}_2\hat{\mu}_2^{i-1}$$

If criterion (3.4.18) rejects the hypothesis, we try to fit an order $r=3$ equation, and so on. Finally, if an order r equation agrees with the experimental data, we refine the approximation, using the minimum χ^2 method.

■

Prony's method of estimating is analogous to decoding Reed-Solomon code.[81] The eigenvalues $\boldsymbol{\mu}$ serve as error locators, while $\mathbf{S}$ gives the errors values. This method approximates well the initial part of the distribution (which is good in the case of error correction, since a small number of errors is most probable), but it performs poorly in approximating the distribution tail. To improve the approximation agreement with the experimental data, it is better to estimate the matrix $\mathbf{S}$ using the normal equation (3.4.10) rather than (3.4.4). Originally, we replace $NP_i(\boldsymbol{\theta}^{(0)})$ in equation (3.4.10) with ν_i.

3.4.6 Matrix Echelon Parametrization. The parameters μ_i of distribution (3.4.4) are found as the roots of the characteristic polynomial. There is no need in finding these roots if we use the alternative matrix-echelon parametrization described in Sec. 1.3.4. In this parametrization, the distribution (3.4.1) can be expressed explicitly through the coefficients $\boldsymbol{\alpha} = (\alpha_1, \alpha_2, \ldots, \alpha_r)$ of equation (3.4.16) and the initial conditions $\mathbf{S} = [\ \hat{p}(1)\ \ \hat{p}(2)\ \ \ldots\ \ \hat{p}(r)\ ]$:

$$p(x) = \mathbf{S}\mathbf{M}^{x-1}\mathbf{F}$$

where $\mathbf{F} = [\ 1\ \ 0\ \ 0\ \ \cdots\ \ 0\ ]'$ and

$$\mathbf{M} = \begin{bmatrix} 0 & 0 & \dots & 0 & -\alpha_r \\ 1 & 0 & \dots & 0 & -\alpha_{r-1} \\ \dots & \dots & \dots & \dots & \dots \\ 0 & 0 & \dots & 1 & -\alpha_1 \end{bmatrix}$$

Indeed, it is obvious that $\mathbf{SF} = p(1)$. Since $\mathbf{MF} = [\ 0\ \ 1\ \ 0\ \ \cdots\ \ 0\ ]'$, we obtain $\mathbf{SMF} = p(2)$. Analogously, we prove that equation (3.4.1) is valid for all $x \le r$.

For $x = r+1$, the product in the right-hand side of equation (3.4.1) can be expressed as

$$\mathbf{SM}^r\mathbf{F} = \mathbf{SMM}^{r-1}\mathbf{F} = \mathbf{SM}\ [\ 0\ \ 0\ \ 0\ \ \cdots\ \ 0\ \ 1]'$$

Since

$$\mathbf{SM} = [\ p(2)\ \ p(3)\ \ \cdots\ \ p(r)\ \ -\sum_{i=1}^{r} \alpha_i p(r+1-i)\ ]$$

we obtain

$$\mathbf{SM}^r\mathbf{F} = \mathbf{SM}\ [\ 0\ \ 0\ \ 0\ \ \cdots\ \ 0\ \ 1]' = -\sum_{i=1}^{r} \alpha_i p(r+1-i) = p(r+1)$$

so that (3.4.1) is satisfied for $x = r+1$. Repeating these transformations for $x=r+2,\ r+3,\dots$, we prove that the above-defined matrices satisfy (3.4.1).

The ML estimates of the alternative set of parameters can be found from equation (3.4.9), where $\boldsymbol{\mu}$ is replaced with $\boldsymbol{\alpha}$. This system structure is more complex than (3.4.9), because it is difficult to express powers of matrices using the matrix-echelon form. Similar problems arise when we use the minimum χ^2 method. However, if we do not group the experimental data and use the sequential LS method, nonlinear problem can be reduced to a sequence of linear ones.

Indeed, in this case $\boldsymbol{\alpha}^{(k+1)}$ and $\mathbf{S}^{(k+1)}$ approximations can be found by minimizing the following function with respect to $\alpha_1, \alpha_2, \dots, \alpha_r, u_1, \dots, u_c$:

$$\chi^2_{k+1} = \sum_x [\,\hat{p}(x) + \sum_{i=1}^{r} \alpha_i \hat{p}(x-i) - u_x\,]^2 w_x(k)$$

where $w_x(k) = 1 / Np(x; \mathbf{S}^{(k)}, \boldsymbol{\alpha}^{(k)})$ and $p(x; \mathbf{S}^{(k)}, \boldsymbol{\alpha}^{(k)})$ is the probability approximation found after k-th iteration. Since χ^2_{k+1} is a quadratic function of α_i and u_x, the problem of finding $\boldsymbol{\alpha}^{(k+1)}$ is an LS problem and the parameter value satisfies the following normal equation:

$$\boldsymbol{\alpha}\mathbf{B} = \mathbf{c} \tag{3.4.19}$$

where

$$\mathbf{B} = \left[\sum_x w_x \hat{p}(x-i)\hat{p}(x-j)\right]_{r,r} \qquad \mathbf{c} = -\left[\sum_x w_x \hat{p}(x)\hat{p}(x-j)\right]_{1,r}$$

The parameter $\mathbf{u}^{(k+1)}$ satisfies equation (3.4.16), which becomes

$$\hat{u}_x^{(k+1)} = \hat{p}(x) + \sum_{i=1}^{r} \alpha_i^{(k+1)} \hat{p}(x-i) \qquad x=1,2,...,r$$

It follows from these equations that the initial conditions $\hat{p}(1),\hat{p}(2),...,\hat{p}(r)$ represent the LS estimate of the parameter $\mathbf{S}^{(k+1)}$.

Generally, $\mathbf{B}$ is not a Toeplitz matrix, however, if the weights w_x are constant, this matrix becomes a symmetrical Toeplitz matrix. This symmetry property is important in solving system (3.4.19), because in this case we can apply the classical Levinson-Durbin [72,79] algorithm.

When using the linear system (3.4.16) for the distribution estimation, it is necessary to have the experimental data $\hat{p}(x)$ at the adjacent points ($x = 1,2,3,...$). This is not normally available. At the same time, Prony's method is very sensitive to the initial part of the distribution, which leads to overestimating the equation order. We discuss here several methods of data compression to overcome this deficiency.

3.4.7 Utilization of the Cumulative Distribution Function. If $p(x)$ satisfies a difference equation the gap function $p(mx+1)$ with fixed m also satisfies a difference equation. To find this equation, we first consider the gap function z-transform:

$$\rho^{(m)}(z) = \sum_{x=0}^{\infty} p(mx+1) z^{-x-1}$$

which, according to (3.4.7), is equal to

$$\rho^{(m)}(z) = \mathbf{S}(\mathbf{I}z - \mathbf{M}^m)^{-1}\mathbf{F}$$

It follows from this equation that the gap sequence $p(mx+1)$ characteristic polynomial is det $(\mathbf{I}z-\mathbf{M}^m)$ and the difference equation for the gap sequence can be obtained by the characteristic polynomial decomposition similar to (3.4.13). The roots of this characteristic polynomial are equal to μ_i^m, and, since

$$\lim_{m \to \infty} \mu_i^m = 0$$

for large m, only the terms of the distribution that correspond to the largest $q_i = |\mu_i|$ should be taken into account. Therefore, for large m the gap sequence can be approximated by a function of form (3.4.7) with a smaller number of parameters than the original distribution $p(x)$. Then we can extract the component that corresponds to these parameters from the original distribution, thereby reducing the number of parameters to be estimated.

One of the problems with this approach is that the there may not be enough data for estimating the gap function, especially for large m. More reliable estimates can be found if we make use of the fact that the cumulative interval distribution and $p(x)$ satisfy the same difference equation. Indeed, the cumulative distribution $P(l) = Pr(x \geq l)$, according to (3.4.7), has the form

$$P(l) = \sum_{i=l}^{\infty} p(x) = \mathbf{S}\mathbf{M}^{l-1}(\mathbf{I}-\mathbf{M})^{-1}\mathbf{F}$$

It has the same characteristic polynomial $\Delta(z) = \det(\mathbf{I}z-\mathbf{M})$ as the sequence $\{p(x)\}$, which means that they satisfy the same difference equation. If we find parameters of the distribution tails $P(l)$, then the parameters of the distribution $p(x)$ are found from the identity $p(x)=P(x)-P(x+1)$.

Now we can outline a method of estimating the distribution parameters, which is similar to the piecewise logarithmic transformation. First we find the approximation of the gap sequence of the cumulative distribution $P(ml+1)$ for large m, using Prony's method or piecewise logarithmic transformation by a function of type (3.4.7):

$$Q_1(ml+1) = \mathbf{B}_1\mathbf{C}_1^l\mathbf{D}_1$$

The natural interpolation of this function is given by

$$Q_1(l) = \mathbf{B}_1\mathbf{A}_1^{l-1}\mathbf{D}_1$$

where $\mathbf{A}_1^m = \mathbf{C}_1$.[24] The eigenvalues of the matrix $\mathbf{A}_1$ are equal to the order m roots of the eigenvalues of the matrix $\mathbf{C}_1$. Next we subtract this component from the original cumulative distribution to obtain the residual function $P^{(1)}(l) = P(l)-Q_1(l)$. Repeating similar transformations with respect to this residual function using a smaller gap we extract from the original distribution the components that correspond to intermediate-length intervals, and so on. In the end, we obtain the cumulative function approximation in the form

$$P(l) \approx \sum_i \mathbf{B}_i \mathbf{A}_i^{l-1} \mathbf{D}_i$$

This may be rewritten as

$$P(l) \approx \hat{\mathbf{S}} \hat{\mathbf{M}}^{l-1} \hat{\mathbf{F}}_i$$

where

$$\hat{\mathbf{S}} = block\ row\{\mathbf{B}_i\} \quad \hat{\mathbf{M}} = block\ diag\{\mathbf{A}_i\} \quad \hat{\mathbf{F}} = block\ col\ \{(\mathbf{I}-\mathbf{A}_i)^{-1}\mathbf{D}_i\}$$

If the criterion χ^2 for this approximation is not small enough, we may use the estimated parameter vector as an initial vector for a more powerful but complex method of χ^2 minimization.

In practice it is convenient to select a sequence of gaps that grow exponentially. For example, from 1 to 10 use gap $m=1$ (i.e., consider $P(1)$, $P(2)$,...,$P(10)$), from 10 to 100 use gap $m=10$ (i.e., consider $P(10)$, $P(20)$,...,$P(100)$), from 100 to 1000 use gap $m=100$, and so on. Then start the distribution parameter estimation from the interval with the largest gap. Often the estimates from the adjacent intervals with different gaps do not interfere with each other.

3.4.8 Method of Moments. Let us consider another method of data compression, which also leads to a Toeplitz system for the distribution parameter estimation. Because of distribution (3.4.7) matrix-geometric structure, its moments also have such a structure and therefore satisfy some Toeplitz difference equations. Since in moment calculation we utilize all the available data, this lets us overcome certain ambiguity of gap selection in the previous method. The other advantage of using moments is the possibility of calculating them sequentially so that we can use the efficient methods of recursive estimation discussed above.

Consider first the negative-binomial moments

$$\nu(m) = \sum_x \binom{m+x-1}{m} p(x)$$

which, according to (2.2.17), can be determined by differentiation of the distribution z-transform:

$$\nu(m) = \frac{(-1)^m}{m!} \rho^{(m)}(1) = \mathbf{S}(\mathbf{I}-\mathbf{M})^{-m-1}\mathbf{F}$$

It follows from this equation that the negative-binomial moments of the interval distribution satisfy a linear difference equation. Indeed, the z-transform of the sequence $\{\nu(m)\}$

$$\zeta(z) = \sum_{m=0}^{\infty} \nu(m) z^{-m} = \mathbf{S}(\mathbf{I}(1-z^{-1}) - \mathbf{M})^{-1}\mathbf{F} \tag{3.4.20}$$

is a rational function and therefore the sequence $\{\nu(x)\}$ satisfies the difference equation

$$\nu(x) + \sum_{i=1}^{r} \beta_i \nu(x-i) = 0 \quad \text{for } x > r$$

whose coefficients are determined by the z-transform $\zeta(z)$ poles.

If we compare equation (3.4.20) with (3.4.11), we can observe that $\zeta(z) = \rho(1-z^{-1})$, so that the the function $\zeta(z)$ denominator is equal to $z^r \Delta(1-z^{-1})$ where $\Delta(z)$ is defined by equation (3.4.13). Because of this relationship, the coefficients of the difference equation for $\{\nu(x)\}$ may be expressed as linear combinations of the coefficients of equation (3.4.17).

To find the relations between the coefficients, let us decompose the denominator $z^r \Delta(1-z^{-1})$ of the z-transform $\zeta(z)$ into a power sum. After simple algebraic transformations, we obtain

$$z^r \Delta(1-z^{-1}) = \sum_{m=0}^{r} z^m (-1)^{r-m} \sum_{i=0}^{m} \binom{r-i}{r-m} \alpha_i$$

so that

$$\beta_m = (-1)^m \sum_{i=0}^{r-m} \binom{r-i}{m} \alpha_i$$

where $\alpha_0 = 1$.

The inverse equations

$$\alpha_i = (-1)^{r-i} \sum_{m=r-i}^{r} \binom{m}{r-i} \beta_m \quad (i=0,1,...,m)$$

can be found quite analogously or by direct inversion of the previous equations.

Thus, if we can estimate coefficients of the difference equation for the sequence of negative-binomial moments $\{\nu(x)\}$, then we can find the coefficients of the difference equation of the distribution $\{p(x)\}$, and vice versa.

The roots of the moment sequence characteristic polynomial

$$z^r \Delta(1-z^{-1}) = \det\left[\mathbf{I}(z-1) - z\mathbf{M}\right]$$

have the form $1/(1-\mu_i)$, so that the model parameters μ_i can be calculated directly from the negative-binomial moment characteristic equation.

Similar results may be obtained by using binomial moments

$$\mathbf{S}(\mathbf{I}-\mathbf{M})^{-n-1}\mathbf{M}^n\mathbf{F} = \sum_x \binom{x-n}{n} p(x)$$

whose characteristic equation roots have the form $(2-\mu_i)/(1-\mu_i)$.

3.4.9 Transform Domain Approximation. Another method with similar merits for estimating the parameters of distributions (3.4.7) is the method of the distribution z-transform approximation. Since interval distributions take the form of (3.4.7) if and only if their z-transforms are rational fractions (with real coefficients), to estimate their parameters one can approximate the z-transforms of the empirical distribution $p(x)$ with rational fractions.

An important role is played here by the choice of the approximation error measure. Because the χ^2 criterion is more sensitive to the divergence of the tails of distributions, one should choose a measure of proximity of z-transforms which would have this property. In this sense, the arithmetic mean of deviations of z-transforms on the circle of radius $r \ll 1$ has the necessary property. Indeed, let $\rho(z)$ and $\hat{\rho}(z)$ be the z-transforms of the distribution $p(x)$ and its empirical $\hat{p}(x)$ approximation respectively. If γ is a circle of radius r such that all the poles of the z-transforms are inside the circle. Then Cauchy's formula gives

$$p(x)-\hat{p}(x) = \frac{1}{2\pi j}\oint_\gamma [\rho(z)-\hat{\rho}(z)]z^{x-1}dz$$

This leads to the inequality

$$|\,p(x)-\hat{p}(x)\,| \le \varepsilon r^x$$

where

$$\varepsilon = \frac{1}{2\pi r}\oint_\gamma |\,\rho(z)-\hat{\rho}(z)\,|\;|\,dz\,|$$

This ensures the necessary property of estimations.

Questions of approximating functions with rational fractions in a complex plane are studied intensively in mathematical literature.[62] However, there are several problems in constructing the empirical z-transform $\hat{\rho}(z)$, because the interval distribution $\hat{p}(x)$ is not available for all possible values of its argument. Therefore, one must produce one or another interpolation and extrapolation, since empirical frequencies usually come in a grouped form. It is possible to use geometric interpolation $\hat{p}(x) = a_i q_i^{x-1}$ where $a_i = a_{i-1} q_{i-1}^{x_{i-1}}$ for $x_{i-1} < x \le x_i$ and q_i is found from the condition

$$\hat{p}(x_{i-1} < x \le x_i) = \frac{a_i}{1-q_i}(q_i^{x_{i-1}} - q_i^{x_i})$$

The other alternative is to produce smoothing with the help of geometric progression on longer intervals, taking approximately equal parameters of geometric progression for several adjacent intervals, which is especially important in extrapolation of distribution tails. It is also possible to use a different method of interpolation, specifically the spline method, which in a certain sense generalizes the previous method.

3.5 MATRIX PROCESS PARAMETER ESTIMATION

3.5.1 ML Parameter Estimation. Let $e_1, e_2, ..., e_n$ be a sequence of errors in a binary channel. We want to test a hypothesis H_0 that the source of errors in this channel can be described as a matrix process $\{\mathbf{Y}(\boldsymbol{\tau}), \mathbf{M}(0,\boldsymbol{\tau}), \mathbf{M}(1,\boldsymbol{\tau}), \mathbf{Z}(\boldsymbol{\tau})\}$ whose matrices depend on k independent parameters $\boldsymbol{\tau} = (\tau_1, \tau_2, ..., \tau_k)$. Within this hypothesis the likelihood function is given by

$$L = \mathbf{Y}(\boldsymbol{\tau})\prod_{i=1}^{n}\mathbf{M}(e_i,\boldsymbol{\tau})\mathbf{Z}(\boldsymbol{\tau}) \tag{3.5.1}$$

The ML parameter estimates can be found as a solution of the system

$$\frac{\partial L}{\partial \tau_i} = 0 \quad i=1,2,...,k$$

However, because of the matrix noncommutativity, the derivative in the left-hand side of this equation has a very complex form:

$$\frac{\partial L}{\partial \tau_i} = \frac{\partial \mathbf{Y}(\boldsymbol{\tau})}{\partial \tau_i}\prod_{j=1}^{n}\mathbf{M}(e_j,\boldsymbol{\tau})\mathbf{Z}(\boldsymbol{\tau}) +$$

$$\sum_{m=1}^{n}\prod_{j=1}^{m-1}\mathbf{M}(e_j,\boldsymbol{\tau})\frac{\partial \mathbf{M}(e_m,\boldsymbol{\tau})}{\partial \tau_i}\prod_{j=m+1}^{n}\mathbf{M}(e_j,\boldsymbol{\tau})\mathbf{Z}(\boldsymbol{\tau}) +$$

$$\mathbf{Y}(\boldsymbol{\tau})\prod_{i=1}^{n}\mathbf{M}(e_i,\boldsymbol{\tau})\frac{\partial \mathbf{Z}(\boldsymbol{\tau})}{\partial \tau_i}$$

It is difficult to find the ML parameter estimates using these equations when n is large. Similar difficulties arise if we use the information criterion to test the null hypothesis. Therefore, we must find alternative ways of estimating parameters and testing H_0.

One of the methods consists in using the process basis sequences (see Appendix 1). If the process has finite rank, there exist so-called basis sequences of errors. The probability of any sequence of errors can be uniquely expressed as a linear combination of the probabilities of these basis sequences (see Appendix 1). The description of the method of estimating the process parameters is given in Appendix 3.

3.5.2 Interval Distribution Estimation. In practice, it is convenient to use methods of estimation of parameters of the matrix process by interval distributions, because it is possible to use known methods of mathematical statistics. We will see that the two-dimensional interval distributions enable us to estimate all the model parameters.

Consider first the distribution $p_0(\lambda)$ of intervals between errors and the distribution $p_1(l)$ of lengths of series of consecutive errors. It follows from (1.3.11) and (1.3.12) that these distributions have the following form:

$$p_0(\lambda) = \mathbf{Y}_1\mathbf{M}_{10}\mathbf{M}_{00}^{\lambda-1}\mathbf{M}_{01}\mathbf{Z}_1 \;/\; \mathbf{Y}_1\mathbf{M}_{10}\mathbf{Z}_0 \tag{3.5.2}$$

$$p_1(l) = \mathbf{Y}_0\mathbf{M}_{01}\mathbf{M}_{11}^{l-1}\mathbf{M}_{10}\mathbf{Z}_0 \;/\; \mathbf{Y}_0\mathbf{M}_{01}\mathbf{Z}_1 \tag{3.5.3}$$

It is convenient to use the general form for both previous equations

$$p_e(x) = \mathbf{S}_e\mathbf{M}_{ee}^{x-1}\mathbf{F}_e$$

where

$$\mathbf{S}_e = \mathbf{Y}_{1-e}\mathbf{M}_{1-e,e} \;/\; \mathbf{Y}_{1-e}\mathbf{M}_{1-e,e}\mathbf{Z}_e \quad \mathbf{F}_e = \mathbf{M}_{e,\,1-e}\mathbf{Z}_{1-e}$$

This equation has the matrix-geometric form considered in Sec. 3.5.1. Using the methods described in this section, we can estimate parameters $\mathbf{S}_e$, $\mathbf{M}_{ee}$, and $\mathbf{F}_e$. If these distributions do not contradict experimental data, then the hypothesis H_0 is not rejected, since for two given distributions in form (3.5.2) and (3.5.3) it is always possible to construct a matrix process which has the same distributions of lengths of intervals between errors and lengths of series of errors. Using interval distributions (3.5.2) and (3.5.3), matrices $\mathbf{M}_{00}$ and $\mathbf{M}_{11}$ can be determined in their canonical Jordan form (3.4.2). The estimated sizes of these matrices $\hat{r}_e$ can serve as estimates of ranks r_e of the matrix process. Thus, estimating parameters of distributions (3.5.2) and (3.5.3) using experimental data, we estimate ranks r_e of the process, matrices $\mathbf{M}_{ee}$, $\mathbf{S}_e$, and $\mathbf{F}_e$.

If intervals between consecutive errors and correct symbols are independent, then the matrix process is equivalent to an alternating renewal process and can be constructed using formula (1.4.26), in which all matrices are replaced by their estimates. If the hypothesis of the independence of intervals is rejected, then to estimate the remaining parameters it is necessary to examine some other distributions related to the matrix process.

3.5.3 Utilization of Two-Dimensional Distributions. The parameters that we need to estimate are elements of the matrices $\mathbf{Y}_e$, $\mathbf{Z}_e$, $\mathbf{M}_{01}$, and $\mathbf{M}_{10}$. We can estimate these parameters by solving a system of the method of moments for the probabilities that are independent of the interval distributions used before. When constructing a system of equations one must, for the sake of simplicity of calculation, try to choose such basis sequences whose probabilities depend on unknown parameters in the most simple way (see Appendix 1). Specifically, choosing sequences in the form 000...011...11, we obtain linear equations for the elements of the matrix $\mathbf{M}_{01}$; choosing sequences in the form 11...100...0, we obtain linear equations for the elements of the matrix $\mathbf{M}_{10}$. The two-dimensional distributions of these sequences are given by

$$p_{01}(\lambda,l) = \mathbf{S}_0\mathbf{M}_{00}^{\lambda-1}\mathbf{M}_{01}\mathbf{M}_{11}^{l-1}\mathbf{F}_1 \tag{3.5.4}$$

$$p_{10}(l,\lambda) = \mathbf{S}_1\mathbf{M}_{11}^{l-1}\mathbf{M}_{10}\mathbf{M}_{00}^{\lambda-1}\mathbf{F}_0 \tag{3.5.5}$$

These two equations can be rewritten in a common general form:

$$p_{e,\,1-e}(x,y) = \mathbf{S}_e\mathbf{M}_{ee}^{x-1}\mathbf{M}_{e,\,1-e}\mathbf{M}_{1-e,\,1-e}^{y-1}\mathbf{F}_{1-e} \tag{3.5.6}$$

If we replace in these equations the probabilities $p_{e,\,1-e}(x,y)$ with their estimates for several different values of x and y, we obtain a system of linear equations with unknown elements of the matrix $\mathbf{M}_{e,\,1-e}$. By solving this system we obtain the matrix estimate. The matrix $\mathbf{M}_{e,\,1-e}$ has $r_0 \times r_1$ elements, so that we need at least $r_0 \times r_1$ different estimates $\hat{p}_{e,\,1-e}(x,y)$ to be able to solve this system. If the matrices $\mathbf{M}_{ee}$ in the interval distributions (3.4.7) have the minimal order (or, in other words, r_e are ranks of the matrix process), then the initial $r_0 \times r_1$ equations let us find the unique solution.

It is easy to verify that these initial equations may be written in the following matrix form:

$$\mathbf{G}_e\mathbf{M}_{e,\,1-e}\mathbf{H}_{1-e} = \mathbf{W}_{e,\,1-e} \tag{3.5.7}$$

where

$$\mathbf{W}_{e,\,1-e} = \left[p_{e,\,1-e}(i,j)\right]_{r_e, r_{1-e}}$$

$$\mathbf{G}_e = block\ col\ \{\mathbf{S}_e\mathbf{M}_{ee}^{i-1}\} = \begin{bmatrix} \mathbf{S}_e \\ \mathbf{S}_e\mathbf{M}_{ee} \\ \cdots\cdots \\ \mathbf{S}_e\mathbf{M}_{ee}^{r_e-1} \end{bmatrix}$$

and

$$\mathbf{H}_e = block\ row\{\mathbf{M}_{ee}^{j-1}\mathbf{F}_e\} = \left[\mathbf{F}_e \quad \mathbf{M}_{ee}\mathbf{F}_e \ \ldots \ \mathbf{M}_{ee}^{r_e-1}\mathbf{F}_e\right]$$

This system has a unique solution, because matrices $\mathbf{G}_e$ and $\mathbf{H}_e$ are nonsingular. We will prove that these matrices are nonsingular by showing that their rows (or columns) are linearly independent.

Indeed, let us assume that vectors $\mathbf{S}_e$, $\mathbf{S}_e\mathbf{M}_{ee},\ldots,\mathbf{S}_e\mathbf{M}_{ee}^{r_e-1}$ are linearly dependent. Then there exist numbers $\sigma_1, \sigma_2, \ldots, \sigma_{r_e}$ such that not all of them are equal to zero and

$$\sum_{i=1}^{r_e} \sigma_i \mathbf{S}_e\mathbf{M}_{ee}^{i-1} = 0$$

so that we can express $\mathbf{S}_e\mathbf{M}_{ee}^{k}$ with $k < r_e$ as a linear combination of the vectors corresponding to powers less than k:

$$\mathbf{S}_e\mathbf{M}_{ee}^{k} = \sum_{i=1}^{k} \beta_i \mathbf{S}_e\mathbf{M}_{ee}^{i-1}$$

Multiplying this identity by $\mathbf{M}_{ee}^{x-k-1}\mathbf{F}_e$ we obtain a difference equation for distribution (3.4.7)

$$p_e(x) - \sum_{i=1}^{k} \beta_{k-i+1} p_e(x-i) = 0$$

whose order k is less than r_e. This contradicts the assumption that r_e is the minimal order of this equation (see Sec. 1.3.3).

Similarly, we can prove that vectors $\mathbf{Z}_e$, $\mathbf{M}_{ee}\mathbf{Z}$, ..., $\mathbf{M}_{ee}^{r_e-1}$ are also linearly independent. This completes the proof of matrix $\mathbf{G}_e$ and $\mathbf{H}_e$ nonsingularity.

The unique solution of system (3.5.7) can be expressed in the following matrix form:

$$\mathbf{M}_{e,\,1-e} = \mathbf{G}_e^{-1}\mathbf{W}_{e,\,1-e}\mathbf{H}_e^{-1}$$

We would like to point out that for estimating the matrix $\mathbf{M}_{e,\,1-e}$ it is possible to use estimates of two-dimensional cumulative distributions. According to (3.5.6), the probability

$$p_{e,\,1-e}(x_{i-1} < x \le x_i, y_{j-1} < y \le y_j) = \sum_{x=x_{i-1}+1}^{x_i} \sum_{y=x_{j-1}+1}^{y_j} p_{e,\,1-e}(x,y)$$

takes the form

$$p_{e,\,1-e}(x_{i-1} < x \le x_i, y_{j-1} < y \le y_j) = \mathbf{L}_e(x_{i-1},x_i)\mathbf{M}_{e,\,1-e}\mathbf{R}_e(y_{j-1,}y_j)$$

where

$$\mathbf{L}_e(x_{i-1},x_i) = \mathbf{S}_e(\mathbf{I}-\mathbf{M}_{ee})^{-1}(\mathbf{M}_{ee}^{x_{i-1}}-\mathbf{M}_{ee}^{x_i}) \quad \mathbf{R}_e(y_{j-1,}y_j) = (\mathbf{I}-\mathbf{M}_{ee})^{-1}(\mathbf{M}_{ee}^{y_{j-1}}-\mathbf{M}_{ee}^{y_j})\mathbf{F}_e$$

This equation is similar to (3.5.6) and therefore $\mathbf{M}_{e,\,1-e}$ satisfies a system of linear equations similar to (3.5.7),

$$\mathbf{G}_e^{(1)}\mathbf{M}_{e,\,1-e}\mathbf{H}_{1-e}^{(1)} = \mathbf{W}_{e,\,1-e}^{(1)} \tag{3.5.8}$$

where

$$\mathbf{W}_{e,\,1-e}^{(1)} = \left[p_{e,\,1-e}(x_{i-1} < x \le x_i, y_{i-1} < y \le y_i)\right]_{r_e,r_{1-e}}$$

$$\mathbf{G}_e^{(1)} = block\ col\ \{\mathbf{L}_e(x_{i-1},x_i)\} \quad \text{and} \quad \mathbf{H}_e^{(1)} = block\ row\{\mathbf{R}_e(y_{j-1},y_j)\}$$

We can also use the distribution tails by replacing upper bounds of the intervals with ∞.

3.5.4 Polygeometric Distributions. Let us examine the important special case where distributions (3.4.7) are polygeometric:

$$p_e(x) = \sum_{i=1}^{r_e} a_{ei}(1-q_{ei})q_{ei}^{x-1} \tag{3.5.9}$$

In this case, matrices $\mathbf{M}_{ee}$ in formulation (3.4.2) are diagonal:

$$\mathbf{M}_{ee} = diag\{q_{ei}\} \tag{3.5.10}$$

Matrices $\mathbf{S}_e$ and $\mathbf{F}_e$ may be expressed as

$$\mathbf{S}_e = row\{a_{ei}(1-q_{ei})\} \quad \mathbf{F}_e = \mathbf{1}$$

Matrices $\mathbf{G}_e$ and $\mathbf{H}_e$ take the form

$$\mathbf{G}_e = \begin{bmatrix} a_{e1}(1-q_{e1}) & a_{e2}(1-q_{e2}) & \cdots & a_{er_e}(1-q_{er_e}) \\ a_{e1}(1-q_{e1})q_{e1} & a_{e2}(1-q_{e2})q_{e2} & \cdots & a_{er_e}(1-q_{er_e})q_{er_e} \\ \cdots & \cdots & \cdots & \cdots \\ a_{e1}(1-q_{e1})q_{e1}^{r_e-1} & a_{e2}(1-q_{e2})q_{e2}^{r_e-1} & \cdots & a_{er_e}(1-q_{er_e})q_{er_e}^{r_e-1} \end{bmatrix}$$

$$\mathbf{H}_e = \left[q_{ei}^{j-1} \right]_{r_e, r_e} = \begin{bmatrix} 1 & q_{e1} & q_{e1}^2 & \cdots & q_{e1}^{r_e-1} \\ 1 & q_{e2} & q_{e2}^2 & \cdots & q_{e2}^{r_e-1} \\ \cdots & \cdots & \cdots & \cdots & \cdots \\ 1 & q_{er_e} & q_{er_e}^2 & \cdots & q_{er_e}^{r_e-1} \end{bmatrix}$$

Thus, $\mathbf{H}_e$ becomes Vandermonde's matrix and $\mathbf{G}_e = [diag\{a_{ei}(1-q_{ei})\}\mathbf{H}_e]'$.

Matrices $\mathbf{G}_e^{(1)}$ and $\mathbf{H}_e^{(1)}$ can be found quite analogously. In the case of the distributions tails $[p_{e,\,1-e}(x_{i-1} < x, y_{j-1} < y)]$, they are given by

$$\mathbf{G}_e^{(1)} = \left[a_{ej} q_{ej}^{x_{i-1}} \right]_{r_e, r_e} \quad \mathbf{H}_e^{(1)} = \left[q_{ei}^{y_{j-1}} \right]_{r_e, r_e}$$

3.5.5 Error Bursts. In practice, the following situation is typical: Very long intervals between groups in which error density is relatively high are encountered, but series of consecutive errors are rare. Therefore, the procedure described above does not provide good estimate for the matrix $\mathbf{M}_{11}$. To overcome this deficiency it is convenient to combine the groups with high error density into a single formation—a burst of errors. Single errors (or errors of small multiplicity) belong to the intervals between bursts. Let us denote the resultant process as $\{g_t\}$ (g_t=1, if the symbol at the moment t belongs to the burst, and g_t=0 if the symbol at the moment t belongs to the interval between bursts). We assume that errors within the intervals and bursts are independent and occur with the constant conditional probabilities ε_0 and ε_1, respectively. In addition, let us assume that the process $\{g_t\}$ is a matrix process $\{\mathbf{Y}^{(1)}, \mathbf{M}_0^{(1)}, \mathbf{M}_1^{(1)}, \mathbf{Z}^{(1)}\}$ where $\mathbf{Y}^{(1)} = [\mathbf{Y}_0^{(1)} \;\; \mathbf{Y}_1^{(1)}]$

$$\mathbf{M}_0^{(1)} = \begin{bmatrix} \mathbf{M}_{00}^{(1)} & 0 \\ \mathbf{M}_{10}^{(1)} & 0 \end{bmatrix} \quad \mathbf{M}_1^{(1)} = \begin{bmatrix} 0 & \mathbf{M}_{01}^{(1)} \\ 0 & \mathbf{M}_{11}^{(1)} \end{bmatrix}$$

In this case, the original process $\{e_t\}$ is a matrix process and can be presented as $\{\mathbf{Y},\mathbf{M}_0,\mathbf{M}_1,\mathbf{Z}\}$ with

$$\mathbf{Y} = \begin{bmatrix} \mathbf{Y}_0^{(1)}(1-\varepsilon_0) & \mathbf{Y}_1^{(1)}(1-\varepsilon_1) & \mathbf{Y}_0^{(1)}\varepsilon_0 & \mathbf{Y}_1^{(1)}\varepsilon_1 \end{bmatrix}$$

$$\mathbf{Z} = \begin{bmatrix} \mathbf{Z}^{(1)} \\ \mathbf{Z}^{(1)} \end{bmatrix} \quad \mathbf{M}_0 = \begin{bmatrix} \mathbf{M}_{00} & 0 \\ \mathbf{M}_{11} & 0 \end{bmatrix} \quad \mathbf{M}_1 = \begin{bmatrix} 0 & \mathbf{M}_{11} \\ 0 & \mathbf{M}_{11} \end{bmatrix}$$

$$\mathbf{M}_{00} = \begin{bmatrix} \mathbf{M}_{00}^{(1)}(1-\varepsilon_0) & \mathbf{M}_{01}^{(1)}(1-\varepsilon_1) \\ \mathbf{M}_{10}^{(1)}(1-\varepsilon_0) & \mathbf{M}_{11}^{(1)}(1-\varepsilon_1) \end{bmatrix} \quad \mathbf{M}_{11} = \begin{bmatrix} \mathbf{M}_{00}^{(1)}\varepsilon_0 & \mathbf{M}_{01}^{(1)}\varepsilon_1 \\ \mathbf{M}_{10}^{(1)}\varepsilon_0 & \mathbf{M}_{11}^{(1)}\varepsilon_1 \end{bmatrix}$$

Indeed, the probability of the sequence of errors $e_1,e_2,...,e_n$ is given by

$$Pr(e_1,e_2,...,e_n) = \sum_{g_1,...,g_n} Pr(g_1,g_2,...,g_n)\prod_{i=1}^{n} Pr(e_i \mid g_i)$$

or, since $\{g_t\}$ is a matrix process,

$$Pr(e_1,e_2,...,e_n) = \sum_{g_1,...,g_n} \mathbf{Y}_{g_1} Pr(e_1 \mid g_1)\prod_{i=1}^{n-1}\{\mathbf{M}_{g_i,g_{i+1}}^{(1)} Pr(e_{i+1} \mid g_{i+1})\}\mathbf{Z}_{g_n}^{(1)}$$

where $Pr(e \mid g) = \varepsilon_g^e(1-\varepsilon_g)^{1-e}$. The process $\{e_t\}$ is a matrix process because the previous formula can be written in the following matrix form:

$$Pr(e_1,e_2,...,e_n) = \mathbf{Y}_{e_1}\prod_{i=1}^{n-1}\mathbf{M}_{e_i e_{i+1}}\mathbf{Z}_{e_n}$$

if we use the above defined matrices $\mathbf{Y}_e$, $\mathbf{Z}_e$, and $\mathbf{M}_{e_i e_{i+1}}$.

Thus, having introduced the concept of bursts of errors and the intervals between them, it is possible to produce some smoothing of the experimental data, which simplifies model building.

3.6 SSM MODEL PARAMETER ESTIMATION

3.6.1 Multilayered Error Clusters. As noted in Chapter 1, describing the SSM model based on matrix processes is easier from the point of view of minimizing the dimensions of matrices used in this description and minimizing the number of parameters. However, by so doing we lose the clarity in the model description that is characteristic of the interpretation of Markov functions.

Let us consider the problem of Markov function parameter estimation. The states of Markov chain which take part in the definition of a Markov function are not observed in the experiment; therefore, standard procedures of Markov chain parameters estimating cannot be used.

Since the SSM model is a particular case of a matrix process, it is possible to use the method presented in the previous section to estimate its parameters. If after processing the experimental data we obtain the model of the matrix process $\{\mathbf{Y},\mathbf{M}_0,\mathbf{M}_1,\mathbf{Z}\}$ and the matrices that describe it satisfy the conditions

$$\mathbf{Y} \geq 0 \quad \mathbf{Z} = \mathbf{1} \quad \mathbf{M}_0 \geq 0 \quad \mathbf{M}_1 \geq 0 \quad \mathbf{M}_0\mathbf{1}+\mathbf{M}_1\mathbf{1} = \mathbf{1} \tag{3.6.1}$$

then the matrix process represents an SSM model. If, on the other hand, conditions (3.6.1) are not satisfied, then, possibly, the matrix process is equivalent to a regular Markov function[32,33] whose transition matrices $\mathbf{P}_{e_1e_2}$ have dimensions equal to the process ranks.[28,33,36] In these cases there exist non-singular matrices $\mathbf{T}_e$ such that

$$\mathbf{P}_{e_1e_2} = \mathbf{T}_{e_1}^{-1}\mathbf{M}_{e_1e_2}\mathbf{T}_{e_2} \quad \mathbf{p}_{e_1} = \mathbf{Y}_{e_1}\mathbf{T}_{e_1} \quad \mathbf{Z}_{e_2} = \mathbf{T}_{e_2}\mathbf{1} \quad e_1,e_2=0,1 \tag{3.6.2}$$

Therefore, in the regular case, after estimating the parameters of the matrix process, it is necessary to find matrices $\mathbf{T}_e$ satisfying the conditions

$$\mathbf{T}_{e_1}^{-1}\mathbf{M}_{e_1e_2}\mathbf{T}_{e_2} \geq 0 \quad \mathbf{Y}_e\mathbf{T}_e \geq 0 \quad \mathbf{T}_e^{-1}\mathbf{Z}_e = \mathbf{1} \tag{3.6.3}$$

If, on the other hand, the matrix process is not equivalent to a regular Markov function, i.e., matrices $\mathbf{T}_e$ that satisfy conditions (3.6.3) do not exist, then, possibly, the process is equivalent to a function of a Markov chain with a number of states greater than r_0+r_1. In this case we might find a rectangular matrix $\mathbf{T}$ which satisfies the conditions

$$\mathbf{TP}(e) = \mathbf{M}(e)\mathbf{T} \quad \mathbf{p} = \mathbf{YT} \quad \mathbf{T1} = \mathbf{Z} \tag{3.6.4}$$

where

$$\mathbf{p} \geq 0 \quad \mathbf{P}(e) \geq 0$$

The transformations described above are quite complex, so some assumptions on the model structure are often made. It is convenient to use a multilayered model structure: Independent errors occur within bursts of errors, error bursts occur independently within chains of bursts, which in turn occur independently within larger clusters, and so on. These concepts are closely related to the method of obtaining and processing experimental data: As new experimental data are obtained, the model is refined by gradually making it more sophisticated and introducing new parameters.

Let us consider the problem of estimating parameters of an SSM model, assuming that the states of the underlying chain are semi-Markov lumpable. We partition the experimental sequence into segments of different categories, depending on the intensity of the error flow. The classification can be based on one parameter, for instance, by identifying an error burst as a sequence in which a distance between any errors does not exceed a selected value ρ while the distance between errors outside the burst is greater than ρ. We can also have a classification based on several parameters, for instance by choosing a sequence of gaps $\rho_1 < \rho_2 < \cdots < \rho_s$ and identifying bursts of type i as a sequence of bits in which the distance λ between adjacent errors lies within the limits $\rho_{i-1} < \lambda \leq \rho_i$. So-defined bursts are sometimes called *experimental bursts* in contrast with *theoretical bursts* which are defined as sequences of bits in which errors occur independently with constant conditional probability ε_c.

If so defined error bursts can be described by a semi-Markov process whose interval transition probabilities have matrix-geometric form and errors within bursts are i. i. d., then we can create the SSM error source model as is shown in Sec. 1.4. To test the validity of these assumptions, we must verify the following hypotheses:

I. The errors within bursts of the same i-th category are independent and occur with constant conditional probability ε_i.

II. Bursts of different categories constitute a semi-Markov process with the matrix-geometric interval transition distributions of form (1.4.15).

If experimental data do not reject any of these hypotheses, then by estimating parameters ε_i and parameters of distribution (1.4.15), we can, according to Theorem 1.4.3, estimate SSM model parameters. If, for a selected method of burst error classification, any of the hypotheses discussed above is rejected, then we conduct a more detailed classification based on a larger number of parameters.

Consider now the problem of verification of the hypotheses I and II in more detail.

I. *Distribution of errors inside bursts.* We would like to test the hypothesis

H_0: errors within bursts are i. i. d. against $H_1: H_0$ is not true.

The verification of this hypothesis is performed with the procedures of random number testing.[86]

Preliminarily, during error burst identification it is better to test the hypothesis using some nonparametric criterion. Let $e_1, e_2, \ldots, e_n$ be a sequence of errors belonging to bursts of the same category. If hypothesis H_0 is true, then $\xi_i = e_{2i-1} - e_{2i}$ is also i. i. d. and $Pr(\xi_i = 1) = Pr(\xi_i = -1) = \varepsilon_c(1-\varepsilon_c)$. By deleting from this sequence all the elements that are equal to zero, we create a sequence $\{\eta_i\}$ of i. i. d. random numbers whose probabilities $Pr(\eta_i = 1)$ and $Pr(\eta_i = -1)$ are equal to 0.5 and do not depend on the error source parameters.

Thus, we transformed the test of the hypothesis with unknown error probability ε_c to the similar test with probability equal to 0.5, which was considered in Example 3.1.4. The two-sided acceptance region is given by equation (3.1.22) with $p_0 = q_0 = 0.5$. If the sample size n is large, we may use the acceptance region (3.1.24), which takes the form

$$A(0.5) = (0.5-2Z_\alpha\sqrt{2/n},\ 0.5+2Z_\alpha\sqrt{2/n}) \tag{3.6.5}$$

We reject the hypothesis if the frequency of ones in the sequence $\{\eta_i\}$ lies outside this interval. This criterion is not sensitive to data dependency. The so-called series criteria are better suited to this test. [86]

We call a sequence $\eta_{i+1}=1, \eta_{i+2}=1, ..., \eta_{i+j}=1$ a length j series of pluses if $\eta_i = -1$ and $\eta_{i+j+1} = -1$. A series of minuses is defined similarly. If the hypothesis H_0 is true, the number of series should be large and there should not be very long series of pluses or minuses. Therefore, one may use a number of series and a maximal series length criteria for the hypothesis testing.

It is easy to show that the number of series r in the sequence of n samples $\{\eta_i\}$ has a binomial distribution

$$\binom{n-1}{r-1}0.5^{n-1}$$

so that the two-sided acceptance region for the number of series may be expressed similarly to (3.1.20) and (3.1.22). For large n the acceptance region is similar to (3.1.24):

$$(0.5n-2Z_\alpha\sqrt{2n},\ 0.5n+2Z_\alpha\sqrt{2n})$$

It may be shown[87] that the probability that there is at least one series whose length is greater than y is given by

$$Pr(l_{\max} > y) \sim 1 - \exp(-n\,2^{-y-1})$$

so that the significance level α acceptance interval may be expressed as

$$(0, \log(-n / \ln(1-\alpha))-1)$$

The considered nonparametric criteria are simple, but to use them we need to transform the original sequence $\{e_i\}$ into the sequence $\{\eta_i\}$, which leads to a significant loss of information about error dependencies. After we have roughly outlined burst boundaries, using these simple criteria, we need to perform more detailed analysis of the bursts, using more powerful parametric criteria.

If H_0 is true, then the error number inside bursts of the category c has a binomial distribution

$$P_{c,n}(t) = \binom{n}{t}\varepsilon_c^t(1-\varepsilon_c)^{n-t}$$

and the hypothesis can be tested using the ML estimate of the parameter ε_c and the χ^2 criterion (3.1.11) for the binomial distribution. The ML parameter estimate can be expressed as the ratio $\hat{\varepsilon}_c = t_c / N_c$ of the total number t_c of errors inside the bursts of c-th categories to the total number of bits in the bursts. The χ^2 criterion is given by

$$\chi^2 = \sum_i (\nu_i - Kp_i)^2 / Kp_i$$

where ν_i is the total number of blocks of length n that contain more than t_i, but less or equal to t_{i+1} errors, $K = \lfloor N_c / n \rfloor$ is the total number of blocks in the sequence of N_c bits, and

$$p_i = \sum_{k=t_i+1}^{t_{i+1}} \binom{n}{t}\hat{\varepsilon}_c^k(1-\hat{\varepsilon}_c)^{n-t}$$

Another distribution that we may use for hypothesis H_0 testing is the distribution of intervals between errors inside bursts. Again, replacing ε_c with its estimate we can use the χ^2 criterion to verify the agreement of the distribution

$$p_{0,c}(\lambda) = \hat{\varepsilon}_c(1 - \hat{\varepsilon}_c)^{\lambda-1}$$

with the experimental data.

We have examined just the simplest criteria for testing the hypothesis I. If this hypothesis is not true, then we must perform a more thorough sorting of bursts of the category under consideration and partition them into several types of bursts. For this we can use approximation (by polygeometric distributions) of the distribution of lengths of the intervals between adjacent errors within the bursts of the category under consideration.

Sorting can also be achieved using the following method. Let us establish level α acceptance region $(\varepsilon_{cL}, \varepsilon_{cH})$ for the hypothesis that error probability within the bursts of c-th category is equal to ε_c. If the relative frequency of errors in an actual burst falls within that acceptance region, then we assume that the burst belongs to this category. By changing the α, we can decrease or expand the acceptance interval changing the method of sorting of errors by categories.

II. *Semi-Markov process hypothesis testing*. We are ready now to test a hypothesis H_0 that bursts of different categories constitute a semi-Markov process whose interval transition distributions have form (1.4.15). The criteria developed in Sec. 3.3 enable us to perform the testing. In our case the interval transition distributions have independent parameters. The parameters of each individual distribution can be estimated by one of the methods developed in Sec. 3.1.

However, the fact that the theoretical bursts of errors of a given category are not directly observed in an experiment complicates their identification. We can observe the experimental bursts of errors, i.e., sequences beginning and ending with errors, in which the distance between adjacent errors lies within the given limits $\rho_{i-1} < \lambda \leq \rho_i$. If the conditional probability of error ε_c is large enough, then by identifying the theoretical burst with the experimental, we make a relatively small error. If, on the other hand, the probability ε_c is small, then the error in determining the boundaries of the theoretical burst with respect to the experimental burst can be large.

The solution lies in approximating distributions of lengths of experimental bursts using matrix-geometric functions and subsequent calculation of the parameters of the distribution of lengths of theoretical bursts. As noted earlier, most difficulty in estimating the parameters of a model is due to the fact that it is impossible to determine exactly a burst's boundaries (especially for small values of conditional probability ε_c of the error within the burst). We can resolve this ambiguity by considering a model in which conditional probabilities of errors ε_i are equal 1 or 0 (see Sec. 1.2.3). Then the model can be described by matrices of type (1.2.21), and the necessity to verify hypothesis I no longer exists.

The essence of the method of estimating the parameters of the model represented by matrices of type (1.2.21) consists of the following. Using certain rules, we relate some nonbinary sequence Y_t with the binary sequence e_t: $e_t = f(Y_t)$. If the constructed sequence Y_t is a Markov chain, then by estimating its parameters we can build an SSM model. In practice, it is possible to combine the above methods using Markovian nesting of the sequence of models.

3.6.2 Nested Markov Chains. Let us consider a Markov chain with s states and matrix of transition probabilities $\mathbf{P}$; we will call them macrostates. In addition, let us assume that the i-th macrostate ($i=1,2,...,s$) consists of s_i microstates n_i ($n_i=1,2,...,s_i$). If the macrostate does not change, then, within the limits of the time of stay in that macrostate, the microstates form a (conditional) Markov chain with a matrix

$$\boldsymbol{\varepsilon}_{ii} = \left[\varepsilon_{n_i m_i}\right]_{s_i, s_i} \qquad i=1,2,...,s \tag{3.6.6}$$

If, on the other hand, macrostate i changes into macrostate j, then with the probability $\varepsilon_{n_i m_j}$ and independently from past history the chain passes from microstate n_i to microstate n_j. We denote the matrix of these conditional probabilities as

$$\boldsymbol{\varepsilon}_{ij} = \left[\varepsilon_{n_i m_j}\right]_{s_i, s_j} \qquad i,j=1,2,...,s \tag{3.6.7}$$

It is obvious that the sequence of microstates forms a Markov chain with transition probabilities of the form

$$q_{n_i m_j} = p_{ij}\varepsilon_{n_i m_j} \tag{3.6.8}$$

In the matrix form this relationship can be written as follows:

$$\mathbf{Q} = \begin{bmatrix} p_{11}\boldsymbol{\varepsilon}_{11} & p_{12}\boldsymbol{\varepsilon}_{12} & \cdots & p_{1s}\boldsymbol{\varepsilon}_{1s} \\ p_{21}\boldsymbol{\varepsilon}_{21} & p_{22}\boldsymbol{\varepsilon}_{22} & \cdots & p_{2s}\boldsymbol{\varepsilon}_{2s} \\ \cdots & \cdots & \cdots & \cdots \\ p_{s1}\boldsymbol{\varepsilon}_{s1} & p_{s2}\boldsymbol{\varepsilon}_{s2} & \cdots & p_{ss}\boldsymbol{\varepsilon}_{ss} \end{bmatrix} \tag{3.6.9}$$

The matrices $\boldsymbol{\varepsilon}_{ij}$, defined by formulas (3.6.6) and (3.6.7), obviously satisfy the normalization conditions

$$\boldsymbol{\varepsilon}_{ij}\mathbf{1} = \mathbf{1} \tag{3.6.10}$$

In a more general case, when the same set of microstates corresponds to different macrostates the matrix of transition probabilities is given by

$$\mathbf{Q} = \begin{bmatrix} \mathbf{P}_{11}\boldsymbol{\varepsilon}_{11} & \mathbf{P}_{12}\boldsymbol{\varepsilon}_{12} & \cdots & \mathbf{P}_{1s}\boldsymbol{\varepsilon}_{1s} \\ \mathbf{P}_{21}\boldsymbol{\varepsilon}_{21} & \mathbf{P}_{22}\boldsymbol{\varepsilon}_{22} & \cdots & \mathbf{P}_{2s}\boldsymbol{\varepsilon}_{2s} \\ \cdots & \cdots & \cdots & \cdots \\ \mathbf{P}_{s1}\boldsymbol{\varepsilon}_{s1} & \mathbf{P}_{s2}\boldsymbol{\varepsilon}_{s2} & \cdots & \mathbf{P}_{ss}\boldsymbol{\varepsilon}_{ss} \end{bmatrix} \tag{3.6.11}$$

It is obvious that any stochastic matrix can be expressed in the form of (3.6.9) or (3.6.11) in innumerable ways. If the elements of matrices $\mathbf{P}_{ii}\mathbf{1}$ are large, then transitions from macrostates occur after relatively long time intervals. The above construction of a Markov chain using the concept of macrostates is called Markov chain nesting. If this operation is repeated m times, we call it m-th order Markov chain nesting.

Let us assume that we have been able to build an SSM model for relatively short segments of the experimental data. Then, if we approximate the segment transition process by a Markov chain and the transition probabilities (elements of matrices $\boldsymbol{\varepsilon}_{ij}$ for $i \neq j$) satisfy the conditions of independence from past history, then (3.6.11) gives the matrix of transition probabilities of the combined process. The elements of matrices $\mathbf{P}_{ij}$ are estimated through the method described above only with application to a conditional Markov function. To estimate the model parameters by multiple Markov nesting, let us partition the sequence of errors into segments of different categories and interpret them as times of stay in respective macrostates. Generally speaking, the chosen segments are different from the bursts we considered earlier (the mean lengths of these segments are usually large, and therefore they are sometimes called clusters of bursts). Next, let us partition the selected segments of each category into (second-order) segments of smaller length. We try to find a partition, such that the distribution of lengths of segments (of second order) of each type can be approximated by a geometric distribution (for polygeometric approximation one must try to extract each geometric component). We estimate matrices $\boldsymbol{\varepsilon}_{ij}$ and $\mathbf{P}_{ij}$ using distributions of lengths of segments of the first and second order and, through the process of their change, we estimate the matrix (3.6.11). If by so doing it appears that errors within each second-order segment are independent and occur with constant conditional probability, then we obtain estimates of the parameters of the SSM model. If the hypothesis of the independence of errors is rejected, then we must conduct a more detailed sorting, or partition the segments of the second order into segments of the third order, and so on.

If we assume that during the change of the macrostates the microstates of the category that follows do not depend on the microstates of the previous category, then the matrices $\boldsymbol{\varepsilon}_{ij}$ have the special structure outlined in Sec. 1.4. The sufficient condition of this is the factorization:

$$\boldsymbol{\varepsilon}_{ij} = \mathbf{M}_i\mathbf{N}_j \qquad \mathbf{M}_i = col\ \{m_{ij}\} \quad \mathbf{N}_j = row\ \{n_{ij}\}$$

EXAMPLE 3.6.1: Suppose that we have partitioned the error sequence into two categories of error clusters that constitute an alternating renewal process. Let

$$p_{ch}(\lambda) = p_1 q_1^{\lambda-1}$$

be the distribution of intervals between the clusters and

$$p_{ch}(l) = p_2 q_2^{l-1}$$

be the distribution of lengths of clusters. Then, according to (1.4.26), the macrostate transition probability matrix is equal to

$$\mathbf{P} = \begin{bmatrix} q_1 & p_1 \\ p_2 & q_2 \end{bmatrix}$$

Suppose that inside the clusters error bursts are independent and their lengths have a polygeometric distribution

$$p_b(l) = a_{21}p_{21}q_{21}^{l-1} + a_{22}p_{22}q_{22}^{l-1}$$

and intervals between bursts inside the clusters have geometric distribution

$$p_b(\lambda) = p_{11} q_{11}^{\lambda-1}$$

Then, according to (1.4.26), we obtain

$$\boldsymbol{\varepsilon}_{11} = 1 \quad \boldsymbol{\varepsilon}_{12} = \begin{bmatrix} a_{21}p_{11} & a_{22}p_{11} & q_{11} \end{bmatrix}$$

$$\boldsymbol{\varepsilon}_{21} = \begin{bmatrix} 1 \\ 1 \\ 1 \end{bmatrix} \quad \boldsymbol{\varepsilon}_{22} = \begin{bmatrix} q_{21} & 0 & p_{21} \\ 0 & q_{22} & p_{22} \\ a_{21}p_{11} & a_{22}p_{11} & q_{11} \end{bmatrix}$$

Finally, by (3.6.9), the model transition probability matrix is given by

$$\mathbf{P} = \begin{bmatrix} q_1\boldsymbol{\varepsilon}_{11} & p_1\boldsymbol{\varepsilon}_{12} \\ p_2\boldsymbol{\varepsilon}_{21} & q_2\boldsymbol{\varepsilon}_{22} \end{bmatrix}$$ ■

3.6.3 Single-Error-State Models. A single-error-state model is a particular case of a renewal process. Therefore, the intervals between consecutive errors are independent, and all we need is to estimate parameters of the matrix-geometric distributions of the intervals. If this distribution is polygeometric

$$f_\lambda(\lambda) = Pr(0^\lambda 1 \mid 1) = \sum_{i=1}^{n-1} b_i(1-q_i)q_i^\lambda \quad \text{for } \lambda > 0$$

with positive parameters, then the process can be described by Fritchman's partitioned model (see Sec. 1.5.1) with the matrix

$$\mathbf{P} = \begin{bmatrix} q_1 & 0 & \dots & 0 & 1-q_1 \\ 0 & q_2 & \dots & 0 & 1-q_2 \\ \dots & \dots & \dots & \dots & \dots \\ 0 & 0 & \dots & q_{n-1} & 1-q_{n-1} \\ b_1 & b_2 & \dots & b_{n-1} & p_{nn} \end{bmatrix}$$

where $p_{nn} = f_\lambda(0)$ and $b_j = (1-p_{nn})a_j$. This method is broadly used in practice.[5,11,12,19]

However, more stable estimates can be obtained by the experimental data sorting discussed in the previous section. Depending on error density we define different types of error bursts, thus constructing a nonbinary sequence out of binary error statistics. Then we can apply the methods considered in Sec. 3.2 for estimating model parameters. Suppose that an experimental burst of i-th category contains m_i errors and c_i correct symbols. For this burst the number of transitions from the error state n into the state i which corresponds to a correct symbol inside the burst and from the state i back into the state n is equal to $m_i - 1$; the total number of transitions for all bursts is equal to $\sum(m_i - 1)$. The total number of transitions from the state i is equal to $\sum c_i$. Using equation (3.2.10), we obtain estimates of the transition probabilities

$$\hat{p}_{in} = 1 - \hat{q}_i = \sum(m_i-1) / \sum c_i$$

If we denote m_n the length of a series of errors, then the number of transitions from the error state to itself for this series is equal to $m_n - 1$. Equation (3.2.10) gives us

$$\hat{p}_{nn} = \sum(m_n-1) / \sum_i(m_i-1)$$

$$\hat{p}_{ni} = b_i = \sum(m_i-1) / \sum_i(m_i-1)$$

To complete the statistical analysis we must test the hypothesis that the experimental data do not contradict the single-error-state model. One method of testing this hypothesis is outlined in Sec. 3.2.2. The other method consists in testing the hypothesis that the intervals between consecutive errors are statistically independent and their length distribution is polygeometric. Using this simple approach we were able to build models for the source of errors in the majority of the T1 digital repeatered lines.[122]

3.7 MONTE CARLO METHOD OF MODEL BUILDING

The idea of the Monte Carlo method is simple: find model parameters through computer simulations. If the results of simulation are close to the experimental data, the computer model parameters can be used as the estimates. To test the hypothesis that the computer model describes the real process, we need some criteria of closeness between the experimental data and the result of simulation.

Normally we can select several distributions of some random variables and compare the difference between their measured and simulated estimates. Let $p(x)$ be a variable x probability distribution and

$$\hat{P}_{i,exp} = m_{i,exp} / N_{exp} \qquad \hat{P}_{i,sim} = m_{i,sim} / N_{sim}$$

be the probability that the random variable belongs to $[x_{i-1}, x_i)$ estimated from the experimental data and simulation, respectively.

If the null hypothesis is true and the estimates $\hat{P}_{i,exp}$ and $\hat{P}_{i,sim}$ are asymptotically normal, then[53]

$$\chi^2 = N_{sim} N_{exp} \sum_{i=1}^{r} (\hat{P}_{i,exp} - \hat{P}_{i,sim})^2 / (\hat{P}_{i,exp} + \hat{P}_{i,sim})$$

and

$$2I = 2N_{exp} N_{sim} \sum_{i=1}^{r} (\hat{P}_{i,exp} - \hat{P}_{i,sim}) \ln (\hat{P}_{i,exp} / \hat{P}_{i,sim}) / (N_{exp} + N_{sim})$$

are asymptotically χ^2 distributed with $r-1$ degrees of freedom. The null hypothesis is rejected if the calculated values χ^2 or $2I$ exceed the selected critical value. We can use similar criteria for the two-dimensional distributions described in Sec 3.5.3.

We would like to point out that the Monte Carlo method is less efficient than the majority of the above-considered methods. However, this method is comparatively easy to implement. The method should also be used as an additional test of a model agreement with the experimental data, if the model parameters have been estimated using some other method. It is especially important to do this test if we intend to use the model for computer simulation.

3.8 ERROR SOURCE MODEL IN SEVERAL CHANNELS.

Let us now address the question of error source model building in several communication channels. As we mentioned before, the methodology of model parameter estimation in this case does not differ significantly from the case of a single channel. We just outline the steps in model building and emphasize only the features that are specific to multiple channels.

Consider a matrix process model of error source in h binary symmetrical channels. There are $\sigma = 2^h$ possible values of the vector error $\mathbf{e} = (e_1, e_2, ..., e_h)$ in the set of h channels. The matrix process can be described by the following block matrices:

$$\mathbf{M}_{e_j} = \begin{bmatrix} 0 & \cdots & 0 & \mathbf{M}_{\mathbf{e}_1\mathbf{e}_j} & 0 & \cdots & 0 \\ 0 & \cdots & 0 & \mathbf{M}_{\mathbf{e}_2\mathbf{e}_j} & 0 & \cdots & 0 \\ \cdots & \cdots & \cdots & \cdots & \cdots & \cdots & \cdots \\ 0 & \cdots & 0 & \mathbf{M}_{\mathbf{e}_\sigma\mathbf{e}_j} & 0 & \cdots & 0 \end{bmatrix}$$

We need to estimate the elements of these matrices. As in the case of a single channel, the diagonal blocks $\mathbf{M}_{\mathbf{e}_i\mathbf{e}_i}$ of the matrix

$$\mathbf{M} = \sum_{i=1}^{\sigma} \mathbf{M}_{\mathbf{e}_i}$$

can be estimated using the estimated parameters of the error series distributions which have the form of (3.4.7):

$$p_{\mathbf{e}_i \mathbf{e}_j}(l) = \mathbf{S}_{\mathbf{e}_i} \mathbf{M}_{\mathbf{e}_i \mathbf{e}_i}^{l-1} \mathbf{F}_{\mathbf{e}_j}$$

As in the case of a single channel, the parameters of these distributions can be estimated by the difference equations (3.4.16). The rest of the matrices can be estimated using the error series two-dimensional distributions, as described in Sec. 3.5.3.

The methodology of model parameter estimation that has been proposed in Sec. 3.5 can be modified for the case of multiple channels. The model based on the Markov chain with the matrix $\mathbf{P}$ and matrix conditional probabilities of errors $\mathbf{E}_i$, described in Sec. 1.1.2, has a special interest for applications. For this interpretation it is necessary to test the hypothesis that errors inside a burst are independent and occur with constant conditional probabilities so that the conditional probability $P_{c,n}(t_1,t_2,...,t_h)$ that in a block of n symbols inside the burst of the c-th category there are t_1 errors in the first channel, t_2 errors in the second channel, and so on, is given by

$$P_{c,n}(t_1,t_2,...,t_h) = \prod_{i=1}^{h} \binom{h}{t_i} \varepsilon_{ci}^{t_i} (1-\varepsilon_{ci})^{n-t_i}$$

where ε_{ci} is the probability of an error in the i-th channel inside the burst of the c-th category. If this hypothesis does not contradict the experimental data, then we test the hypothesis that bursts of different categories constitute a semi-Markov process with the state-holding distributions of form (3.4.7).

In general, the complexity of the model of error source in multiple channels grows exponentially with the number of channels. However, model parameter estimation can be simplified if we make some assumptions about model's structure. In many practical cases we can neglect the adjacent channel interference and therefore assume that error sources in these channels are independent. If this hypothesis is true, it is possible to build the error source models independently for each channel.

To test the hypothesis of independence consider two communication channels. Let $\mathbf{P}_1(e)$ and $\mathbf{P}_2(e)$ be the matrix probabilities of errors in the first and the second channels, respectively. If the hypothesis is true, the error source in the two channels can be described by the model with the matrix probabilities

$$\mathbf{P}(e_1,e_2) = \mathbf{P}_1(e_1) \otimes \mathbf{P}_2(e_2)$$

This equation may be treated as a matrix generalization of the condition of independence (3.1.25) of random variables. In the particular case where the model is described by multiple Markov chains this condition was used [57] to test the hypothesis that two subsets of states of the chain are independent. However, since the matrix probabilities are not defined uniquely, the above condition is not necessarily true when the hypothesis of independence is true. So this condition cannot be used for hypothesis testing in the general case.

In order to test the hypothesis in the general case, we use distributions of different sequences of errors. If the hypothesis is true, then the intervals between errors $\mathbf{e}_i$ and $\mathbf{e}_j$ have the following distribution:

$$p_{\mathbf{e}_i\mathbf{e}_j}(l) = \mathbf{F}_{\mathbf{e}_i}\mathbf{P}^{l-1}(0,0)\mathbf{S}_{\mathbf{e}_j}$$

If the chain is stationary, then, using the properties of the Kronecker product, we can rewrite the previous equation as

$$p_{\mathbf{e}_i\mathbf{e}_j}(l) = p_{e_{1i}e_{1j}}(l)p_{e_{2i}e_{2j}}(l)$$

where

$$p_{e_{ki}e_{kj}}(l) = \mathbf{F}_{\mathbf{e}_{ki}}\mathbf{P}_k^{l-1}(0)\mathbf{S}_{\mathbf{e}_{kj}}$$

Therefore, if the null hypothesis is true, then intervals between errors in different channels are independent. We can prove similarly that for the stationary models all sequences of intervals in different channels are independent.

3.9 CONCLUSION

We have considered the questions of statistical inference concerning the SSM models. Since the model has numerous representations, we proposed a variety of methods of model parameter estimation. The states of the underlying Markov chain are not uniquely identified in the general case, which creates additional difficulties in model identification and statistical hypothesis testing. We can only test hypotheses about visible sequences of errors rather than invisible states of the SSM. Therefore, any hypothesis connected with states has to be translated into a hypothesis in the error sequence space.

We have shown that the two-dimensional distributions of the series of errors and intervals between errors can be used to identify uniquely the canonical matrix process that is equivalent to a model. This canonical matrix process can be viewed as the kernel of any SSM. If two models are equivalent, their kernels must coincide. Therefore, we can translate statistical problems related to the SSM into the corresponding problems of two-dimensional distributions.

The problem of estimating interval distribution parameters is a problem of the nonlinear function minimization. It can be reduced to a problem of a linear equation parameter estimation, which has been studied by many authors. However, this approach is very formal and destroys the clear relationship between the model parameters and statistical data. The alternative approach, which uses the notion of error bursts and clusters of bursts, allows us not only to have a "physical" model but often is more efficient. In many cases a model based on bursts of errors and clusters of bursts better suits applications. For example, if we calculate the error number distribution in a short block we may make an assumption that this block either lies inside a cluster of bursts or outside the clusters, which simplifies the distribution calculation.

We have considered questions of estimating SSM model parameters on the basis of experimental data of fixed size. The problems of estimating confidence intervals and experiment planning are not treated in detail. This can be explained by the nature of statistical data that is usually available. Normally, the experimental data are collected for some fixed period of time and then processed, so that it is impossible to set the sample size depending on experiment results.

CHAPTER 4

PERFORMANCE OF FORWARD ERROR CORRECTION SYSTEMS

4.1 BASIC CHARACTERISTICS OF ONE-WAY SYSTEMS

In one-way systems the information flow is strictly unidirectional: from transmitter to receiver. Such systems can be described in terms of their input and output process characterization. However, in practical applications only certain characteristics of this process are usually considered. Some basic performance characteristics that are used for comparing these systems are:

p_s	the symbol-error probability on a decoder output;
p_*	the symbol-erasure probability (the probability of receiving a symbol with detected errors);
p_c	the probability of receiving a message without errors;
p_u	the probability of receiving a message with undetected errors;
p_d	the probability of receiving a message with detected errors (obviously $p_c+p_u+p_d = 1$);
t_d	the average path delay;
R	the average information rate (the average ratio of the number of information symbols to the total number of transmitted symbols);
$P(EFS)$	the average percent of error-free seconds (the average percentage of the one-second intervals that do not have errors);
$P_e(L)$	the probability that a series of errored seconds (one-second intervals containing at least one error) has length L.

These parameters, along with the system's complexity and implementation cost, are taken into consideration when the system is designed. The method of comparison depends upon the system's designation.

Selecting a method of forward error correction (FEC) or error detection is the important part of system design. The optimal FEC scheme takes into account real-channel error statistics and therefore corrects the most probable sequences of errors. Normally, it is assumed that a channel is the random-error channel or memoryless channel when errors occur independently with constant bit-error probability. This assumption often leads to selecting an inappropriate code for the channels with memory. For example, a very popular extended Hamming cyclic code with the generator polynomial[88] $x^{16} + x^{15} + x^2 + 1$, broadly known as CRC-16, † is able to detect up to three errors or correct a single error in a block.[89] If channel errors are bursty, then the probability of having more than one error in a block might be higher than the probability of a single error. In this case, the redundancy of 16 bits is not used for the blocks without errors, but this redundancy is not sufficient to correct the most probable multiple errors.

To be able to select an optimal code one must have an adequate error source description. As we saw in Chapter 1, the SSM model is general enough to describe error statistics in real channels. In this chapter we show that the model is also convenient to use for calculating system performance parameters. We consider first a general approach to computing FEC efficiency, then analyze particular coding algorithms and derive the numerical results.

4.2 ELEMENTS OF ERROR-CORRECTING CODING

Our goal is to demonstrate the application of methods developed in the previous chapters to evaluating error correction and error detection performance. We assume that the reader is familiar with the basics of coding theory; however, we present here the background necessary for understanding performance evaluation of communication systems. We present the basic results without proofs, since excellent treatments of the subject can be found in texts by W. W. Peterson and E. J. Weldon,[88] E. R. Berlekamp,[75] and R. G. Gallager.[90]

4.2.1 Field Galois Arithmetic. Digital communication devices perform operations on bits using finite-length registers. The logical operations between these registers may be interpreted as some mathematical operations between the integers that those registers represent. If we impose certain rules on these operations, we may call them addition, subtraction, multiplication, and division. The "usual" mathematical operations performed by digital computers suffer from a deficiency that is not tolerable in data communications: The result of an operation may not fit into the finite-length register, and thus register overflow and loss of accuracy occur. We want the operations to be defined in such a way that for any two numbers of the set the result of any operation (except for division by zero) is a number of the same set that fits into a finite-length register. Such number sets are called *finite fields* if the operations satisfy some standard conditions (axioms) that

† The acronym CRC-16 means Cyclic Redundancy Check with 16 redundant bits.

are familiar from elementary mathematics. We present here a method of finite field construction.

Let p be a prime number. A *ground field*, or *Galois field* of p elements GF(p) may be defined as a set of non-negative integers that are less than p with the modulo p operations between them. The modulo p operation on an integer gives the remainder of division of the number by p. The modulo p addition of two numbers $(a+b) \bmod p$ is equal to the remainder of division of $(a+b)$ by p. The product $ab \bmod p$ is also the remainder of division of ab by p. The subtraction and division modulo p are defined as the inverse operations to the addition and multiplication respectively.

EXAMPLE 4.2.1: Let us illustrate the operations in GF(5), which is the set of numbers 0, 1, 2, 3, 4 with the modulo 5 operations:

addition $(2 + 3) \bmod 5 = 5 \bmod 5 = 0$, $(4 + 3) \bmod 5 = 7 \bmod 5 = 2$, $(1 + 2) \bmod 5 = 3$.

subtraction is inverse to addition. To find $(2-4) \bmod 5 = x$ we must solve the equation $(x+4) \bmod 5 = 2$. By testing all five elements of the field we see that $x=3$ is the only solution of this equation. Therefore, $(2 - 4) \bmod 5 = 3$. This result may be obtained faster if we make use of the identity $2 = 7 \bmod 5$, so that $(2 - 4) \bmod 5 = (7 - 4) \bmod 5 = 3$.

multiplication $2 \cdot 3 \bmod 5 = 6 \bmod 5 = 1$, $4^2 \bmod 5 = 16 \bmod 5 = 1$.

division is inverse to multiplication. To find $(2 / 3) \bmod 5 = x$ we must solve the equation $3x \bmod 5 = 2$. By testing all five elements of the field we determine that $x = 4$ is the only solution of this equation. Therefore, $(2 / 3) \bmod 5 = 4$. It is convenient to find the inverse of each field element (except for 0): $1^{-1} = 1 / 1 = 1$, $2^{-1} = 1 / 2 \bmod 5 = 3$, $3^{-1} = 1 / 3 \bmod 5 = 2$, $4^{-1} = 1 / 4 \bmod 5 = 4$. Using these results as a table and the identity $a / b \bmod 5 = a \cdot b^{-1}$ we find that $2 / 3 \bmod 5 = 2 \cdot 3^{-1} = 2 \cdot 2 = 4$, etc.

■

Using the ground field we can define a more general GF(p^n). Let $q(x)$ be n-th degree polynomial with the coefficient from the ground field GF(p). Then the field GF(p^n) can be constructed as the set of all polynomials whose degree is less than n

$$a(x) = \sum_{i=0}^{n-1} a_i x^i$$

with the coefficients a_i from the ground field GF(p), and modulo $q(x)$ operations between the polynomials. The polynomial $q(x)$ is assumed to be *irreducible* (that is, not factorable) over GF(p) and is called the *field generator polynomial*. According to this definition, the sum (or difference) of two polynomials is the polynomial whose coefficients are modulo p sums (or differences) of their corresponding coefficients.

The definition of the product of two polynomials is more complex. The formal product of two polynomials can be presented as

$$\sum_{i=0}^{n-1} a_i x^i \cdot \sum_{i=0}^{n-1} b_i x^i = \sum_{i=0}^{2n-2} c_i x^i$$

where

$$c_i = \sum_{k=0}^{i} a_k b_{i-k} \bmod p$$

The degree of the product $c(x)$ may be greater than $n-1$, so that it may not belong to the set of polynomials with degrees less than n. However, the remainder after division of $a(x)b(x)$ by $q(x)$ has a degree less than n, which we define as the product $a(x)b(x) \bmod q(x)$ in GF(p^n).

Since a polynomial

$$a(x) = \sum_{i=0}^{n-1} a_i x^i$$

is uniquely represented by its coefficients, we sometimes use vector notation to represent field elements:

$$\mathbf{a} = (a_0, a_1, \ldots, a_{n-1}) = row\{a_i\} \in \mathrm{GF}(p^n).$$

The correspondence between these two equivalent representations is denoted as $a(x) \sim \mathbf{a}$.

The most important fields used in practice are GF(2^k), because binary operations are the basic operations of the vast majority of digital devices. In this case, the ground field FG(2) operations have special notations. The modulo 2 addition or subtraction is called exclusive-OR and is denoted as $\oplus$: $0 \oplus 0=0$, $1 \oplus 0=1$, $0 \oplus 1=1$, $1 \oplus 1=0$. The modulo 2 multiplication is normally called the AND operation and is denoted as $\&$: $0 \& 0 = 0 \& 1 = 1 \& 0 = 0$, $1 \& 1=1$. The field GF(2^k) addition or subtraction is usually called bitwise exclusive-OR. However, we will not use these special notations in the future if it is clear that the operations are carried out in GF(2^k).

EXAMPLE 4.2.2: Consider the field GF(2^3) with the generator polynomial $q(x) = 1 + x + x^3$. This polynomial is irreducible, which can be proved by trying to divide it by all possible linear and quadratic polynomials having binary coefficients. The field consists of all polynomials whose degree is less than 3: $a(x) = a_0 + a_1 x + a_2 x^2$ or, equivalently, all possible three-bit-long sequences $\mathbf{a} = (a_0, a_1, a_2)$.

Let us illustrate the field GF(2^3) operations between two numbers $\mathbf{a} = (0,1,0)$ and $\mathbf{b}=(0,1,1)$. Employing the bitwise exclusive-OR we obtain

$(0,1,0)+(0,1,1) = (0,0,1)$. To find the product of these numbers, we represent them as polynomials: $a(x) = x$, $b(x) = x+x^2$. Then

$$a(x)b(x) = x\cdot(x + x^2) \bmod q(x) = (x^2 + x^3) \bmod q(x)$$

The polynomial division †

$$\begin{array}{r l}
 & \quad 1 \quad \text{(quotient)} \\
x^3 + x + 1 &)\, x^3 + x^2 \\
 & \underline{x^3 \qquad + x + 1} \\
 & \qquad x^2 + x + 1 \quad \text{(remainder)}
\end{array}$$

gives the remainder $1 + x + x^2$, and we can write $(0,1,0)\cdot(0,1,1) = (1,1,1)$.

In every GF(p^n) there is a so-called *primitive element* α such that any nonzero element of the field can be expressed as a power of α. We can illustrate this statement by using the Galois field of Example 4.2.2. Indeed, let $\boldsymbol{\alpha} = (0,1,0)$ which corresponds to the polynomial $\alpha = x$.

$$\begin{array}{ll}
\alpha^1 = x \sim & (0,1,0) \\
\alpha^2 = x^2 \sim & (0,0,1) \\
\alpha^3 = x^3 \bmod (1 + x + x^3) = 1 + x \sim & (1,1,0) \\
\alpha^4 = \alpha^3 x = x + x^2 \sim & (0,1,1) \\
\alpha^5 = \alpha^4 x \bmod (1 + x + x^3) = 1 + x + x^2 \sim & (1,1,1) \\
\alpha^6 = 1 + x^2 \sim & (1,0,1) \\
\alpha^7 = 1 \sim & (1,0,0)
\end{array}$$

■

Galois fields are important in practice, because they enable us to apply mathematical methods to optimize the performance of digital communication systems. Many familiar mathematical methods are directly applicable to these fields. For example, we can solve a system of linear equations by using the standard Gaussian algorithm; a linear system has a unique solution if its determinant is not equal to zero. In the next section we illustrate applications of matrix theory and linear algebra to channel-coding algorithms based on the Galois field notion.

4.2.2 Linear Codes. Suppose that elements of a finite field represent symbols $s_1, s_2, \ldots, s_k$ that are transmitted over a noisy channel. Let us also assume that any element of the field can be transmitted so that there is no redundancy in the

† Note that addition and subtraction of polynomials with binary coefficients is the same operation.

alphabet. Then it is impossible to detect an error, since a corrupted symbol is still a legitimate element of the alphabet. In order to be able to detect and/or correct errors we need to transmit some redundant symbols $s_{k+1}, s_{k+2}, ..., s_n$ that depend on the previously transmitted symbols. The initial k symbols in this case are called the *information* symbols and the added $r = n - k$ symbols are called the *parity check* symbols. The set of all possible combinations $s_1, s_2, ..., s_n$ of information and parity symbols is called a *systematic* (n,k) *code*.

We consider only the case where information and parity symbols satisfy a system of *linear* equations

$$\sum_{i=1}^{n} s_i h_{ji} = 0 \pmod{p} \quad j=1,2,..,r$$

This system can be rewritten in the matrix form:

$$\mathbf{sH}' = 0 \quad \mathbf{s} = row\{s_i\} \quad \mathbf{H} = \left[h_{ij}\right]_{r,n} \tag{4.2.1}$$

The matrix **H** is called a *parity-check matrix.* The set of all possible solutions of this system is called the *linear* (n,k) *code* that corresponds to the parity-check matrix **H**; any element **s** of the code is called a *code word.*

A well-known result from linear algebra says that all possible solutions of a linear homogeneous system constitute a linear vector space.[90] Thus, this code represents a linear vector subspace of the n-dimensional vector space. If we denote $\mathbf{g}_1, \mathbf{g}_2, ..., \mathbf{g}_k$ the code basis, then any code word can be represented as a linear combination of these vectors:

$$\mathbf{s} = \sum_{i=1}^{k} c_i \mathbf{g}_i = \mathbf{cG}$$

where

$$\mathbf{G}' = \begin{bmatrix} \mathbf{g}_1 & \mathbf{g}_2 & \cdots & \mathbf{g}_k \end{bmatrix}$$

is the matrix whose rows are the k basis vectors of the code, which is called the *code generator matrix*. From the definitions of **G** and **H** it follows that

$$\mathbf{GH}' = 0 \tag{4.2.2}$$

Suppose that, because of the channel errors, the transmitted code word **s** is received as $\tilde{\mathbf{s}}$. If $\tilde{\mathbf{s}}$ is not a code word, that is

$$\tilde{\mathbf{s}}\mathbf{H}' = \mathbf{S} \neq 0 \tag{4.2.3}$$

then the received information contains errors. We may try to correct the errors by selecting a code word that is closest to the received word $\tilde{\mathbf{s}}$, or we may just decide not to try to recover the information, but instead ask to retransmit the whole message.

The vector $\mathbf{S}$ in the right-hand side of the previous equation is called the received message $\tilde{\mathbf{s}}$ *syndrome*, because if $\mathbf{S} \neq 0$, it indicates the presence of errors. However, when $\mathbf{S} = 0$ the received information is not necessarily correct: the error pattern might be such that $\tilde{\mathbf{s}}$ is also a valid code word that differs from the transmitted $\mathbf{s}$. In this case the code is not able to detect the error pattern.

The difference between the received and transmitted words is called an *error word* $\mathbf{e} = \tilde{\mathbf{s}} - \mathbf{s}$. Since $\mathbf{sH} = 0$, the syndrome can be expressed through the error word

$$\mathbf{eH}' = \mathbf{S} \tag{4.2.4}$$

If errors are additive, the syndrome probability distribution depends only on the error word distribution. In this case, for each syndrome we can define the most probable error pattern $\mathbf{e}(\mathbf{S})$. Then the most probable transmitted word can be obtained by subtracting the most probable error pattern from the received word

$$\hat{\mathbf{s}} = \tilde{\mathbf{s}} - \mathbf{e}(\mathbf{S})$$

This technique is called *maximum likelihood syndrome decoding*.

EXAMPLE 4.2.3: Consider the binary (7,4) Hamming code whose parity-check matrix is given by

$$\mathbf{H} = \begin{bmatrix} 1 & 1 & 1 & 0 & 1 & 0 & 0 \\ 0 & 1 & 1 & 1 & 0 & 1 & 0 \\ 0 & 0 & 1 & 1 & 1 & 0 & 1 \end{bmatrix}$$

This code consists of all possible solutions of equation (4.2.1) and is the four-dimensional subspace of the seven-dimensional vector space. The basis of the code consists of four linearly independent vectors. The vectors

$$\begin{aligned} \mathbf{s}_1 &= [1\ 0\ 1\ 1\ 0\ 0\ 0] \\ \mathbf{s}_2 &= [0\ 1\ 0\ 1\ 1\ 0\ 0] \\ \mathbf{s}_3 &= [0\ 0\ 1\ 0\ 1\ 1\ 0] \\ \mathbf{s}_4 &= [0\ 0\ 0\ 1\ 0\ 1\ 1] \end{aligned}$$

constitute the code basis, since the matrix has full rank (see Appendix 5):

$$rank \begin{bmatrix} 1 & 0 & 1 & 1 & 0 & 0 & 0 \\ 0 & 1 & 0 & 1 & 1 & 0 & 0 \\ 0 & 0 & 1 & 0 & 1 & 1 & 0 \\ 0 & 0 & 0 & 1 & 0 & 1 & 1 \end{bmatrix} = 4$$

This is true because the determinant of the leftmost square sub-block equals 1:

$$\det \begin{bmatrix} 1 & 0 & 1 & 1 \\ 0 & 1 & 0 & 1 \\ 0 & 0 & 1 & 0 \\ 0 & 0 & 0 & 1 \end{bmatrix} = 1$$

Thus, the matrix whose rows are $\mathbf{s}_1$, $\mathbf{s}_2$, $\mathbf{s}_3$, and $\mathbf{s}_4$ can be selected as the code-generator matrix

$$\mathbf{G} = \begin{bmatrix} 1 & 0 & 1 & 1 & 0 & 0 & 0 \\ 0 & 1 & 0 & 1 & 1 & 0 & 0 \\ 0 & 0 & 1 & 0 & 1 & 1 & 0 \\ 0 & 0 & 0 & 1 & 0 & 1 & 1 \end{bmatrix}$$

Every code word can be represented as a linear combination (with binary coefficients) of the basis vectors

$$\mathbf{s} = c_1\mathbf{s}_1 + c_2\mathbf{s}_2 + c_3\mathbf{s}_3 + c_4\mathbf{s}_4 = \mathbf{cG}$$

Let us consider now the problem of error correction. Suppose that single errors in the block of seven bits have the highest probabilities. The syndrome of the first bit error $\mathbf{e}_1 = (1\ 0\ 0\ 0\ 0\ 0\ 0)$ is equal to $\mathbf{e}_1\mathbf{H}' = (1\ 0\ 0)$, which is the transposed first column of $\mathbf{H}$. Similarly, one can easily see that the i-th bit error syndrome is the transposed i-th column of the matrix $\mathbf{H}$. Therefore, each syndrome uniquely identifies the position of a single error. If we know the position of the error, we can correct it by inverting the corresponding received bit. More specifically, if the syndrome of the received word is equal to the i-th column of the matrix $\mathbf{H}$, then the i-th bit of the received word should be inverted.

By permuting columns of $\mathbf{H}$ it is possible to obtain the parity-check matrix, whose i-th column is the binary representation of i:

$$\mathbf{H}_p = \begin{bmatrix} 0 & 0 & 0 & 1 & 1 & 1 & 1 \\ 0 & 1 & 1 & 0 & 0 & 1 & 1 \\ 1 & 0 & 1 & 0 & 1 & 0 & 1 \end{bmatrix}$$

then the decoding rule can be simplified: Invert i-th bit of the received word if the syndrome's binary representation is i. For instance, if $\mathbf{S}=(1\ \ 0\ \ 1)$, then $i=5$ and $\mathbf{e}(\mathbf{S})=(0\ \ 0\ \ 0\ \ 0\ \ 1\ \ 0\ \ 0)$. We can use a so-called 3-to-8 decoder to implement this algorithm in hardware.

■

An important requirement of code design consists of finding a parity-check matrix $\mathbf{H}$ such that the most probable error patterns have different syndromes. Another important property of a code is the complexity of its implementation. From this point of view, the class of cyclic codes is the most attractive class of codes.

4.2.3 Cyclic Codes. A linear code is called *cyclic* if a cyclic shift of any code word results in another code word. It is convenient to use polynomial notation when dealing with cyclic codes. The cyclic shift, for example, can be represented analytically as modulo x^n-1 multiplication of a polynomial by x:

$$xa(x) \bmod (x^n-1) = a_0x + a_1x^2 + \cdots + a_{n-2}x^{n-1} + a_{n-1}$$

since $x^n = 1 \bmod (x^n-1)$. Thus, if $a(x)$ is a code word, then modulo x^n-1 products $xa(x), x^2a(x), \ldots, x^{n-1}a(x)$ are code words. Since any code is a linear vector space, the product of a code word by a number from the ground field is a code word, and the sum of any two code words is a code word. In terms of polynomials this means that, if $a(x)$ is a code word and $b(x)$ is an arbitrary polynomial, then $b(x)a(x) \bmod (x^n-1)$ is also a code word.

Every cyclic code has a code word $g(x) = g_0 + g_1x + \ldots + g_{r-1}x^{r-1} + x^r$ of the minimal degree, which is called the *code-generator polynomial*, such that the whole code can be generated by multiplying this polynomial by arbitrary polynomials modulo x^n-1:

$$c(x) = b(x)g(x) \bmod (x^n-1) \tag{4.2.5}$$

Since any code word can be expressed as a linear combination of $g(x)$, $xg(x), \ldots, x^{n-r-1}g(x)$, these polynomials represent a basis of the code, and the code-generator matrix can be written as

$$\mathbf{G} = \begin{bmatrix} x^{n-r-1}g(x) \\ x^{n-r-2}g(x) \\ \cdots \\ xg(x) \\ g(x) \end{bmatrix} = \begin{bmatrix} 1 & g_{r-1} & g_{r-2} & \cdots & g_0 & 0 & \cdots & 0 \\ 0 & 1 & g_{r-1} & \cdots & g_1 & g_0 & \cdots & 0 \\ \multicolumn{8}{c}{\cdots\cdots\cdots\cdots\cdots\cdots\cdots\cdots} \\ 0 & 0 & \cdots & 1 & g_{r-1} & \cdots & g_0 & 0 \\ 0 & 0 & \cdots & 0 & 1 & \cdots & g_1 & g_0 \end{bmatrix} \tag{4.2.6}$$

The rank of this matrix is equal to $n-r$, because the determinant of its leftmost square sub-block is equal to 1.

The corresponding parity-check matrix can be found by solving equation (4.2.2). One of the solutions may be expressed in the polynomial form

$$\mathbf{H} = \begin{bmatrix} x^{r-1}h(x) \\ x^{r-2}h(x) \\ \cdots \\ xh(x) \\ h(x) \end{bmatrix}$$

where

$$h(x) = (x^n-1)/g(x) \tag{4.2.7}$$

is the so-called *parity-check polynomial.* Indeed, the product of the i-th row of the matrix $\mathbf{G}$ by the j-th column of the matrix $\mathbf{H}'$ is equal to zero:

$$x^{n-r-i}g(x)x^{r-j}h(x) = x^{n-i-j}(x^n-1) = 0 \mod (x^n-1)$$

Cyclic codes are popular because the above-described polynomial operations can easily be implemented in hardware by using shift registers.

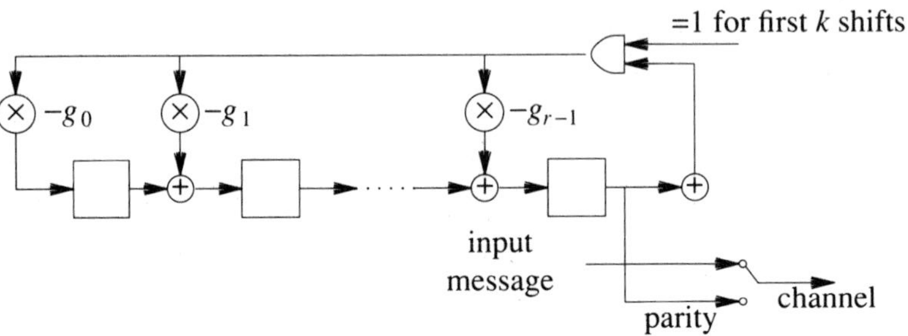

Figure 4.1. Encoder for cyclic (n,k) code.

For example, we can use the shift register shown in Fig. 4.1 to encode the information with the cyclic code whose generator polynomial is

$$g(x) = g_0 + g_1 x + \cdots + g_{k-1} x^{k-1} + x^k$$

This encoder works as follows: Initially the shift registers are empty. The AND gate is open, and k information symbols are shifted simultaneously into the circuit and the communication channel. After that the AND gate is closed. The shift register contents are the r parity symbols, which are shifted into the channel.

It is not difficult to prove[75,88] that the shift register contents after k shifts can be expressed as $-r(x)$, where $r(x) = b_0 + b_1 x + ... + b_{r-1} x_{r-1}$ is the remainder of division of the information polynomial

$$i(x) = s_1 x^{n-1} + s_2 x^{n-2} + \cdots + s_k x^{n-k}$$

by the generator polynomial $g(x)$. If we denote $q(x)$ the quotient of the division, then the remainder is given by $r(x) = i(x) - q(x)g(x)$. The sequence transmitted over the channel consists of the information symbols that were shifted when the AND gate was open and the remainder multiplied by (-1) when the gate was closed. Using polynomial notation, the transmitted sequence can be expressed as

$$i(x) - r(x) = q(x)g(x)$$

which, according to (4.2.5), is a code word.

In the case of the binary cyclic code, the encoder implementation is a fairly simple task, because the ground field elements are either 0 or 1. The circuit shown in Fig. 4.1 consists only of binary adders and binary shift registers.

EXAMPLE 4.2.4: Consider the (7,4) Hamming code with the generator polynomial $g(x) = 1 + x + x^3$. According to (4.2.6), its generator matrix is given by

$$\mathbf{G} = \begin{bmatrix} 1 & 0 & 1 & 1 & 0 & 0 & 0 \\ 0 & 1 & 0 & 1 & 1 & 0 & 0 \\ 0 & 0 & 1 & 0 & 1 & 1 & 0 \\ 0 & 0 & 0 & 1 & 0 & 1 & 1 \end{bmatrix}$$

The parity-check polynomial equals $h(x) = (x^7 - 1)/(x^3 + x + 1) = 1 + x + x^2 + x^4$. Parity-check matrix (4.2.7) has the following form

$$\mathbf{H} = \begin{bmatrix} 1 & 1 & 1 & 0 & 1 & 0 & 0 \\ 0 & 1 & 1 & 1 & 0 & 1 & 0 \\ 0 & 0 & 1 & 1 & 1 & 0 & 1 \end{bmatrix}$$

The encoder can be implemented using the circuit shown in Fig. 4.2. Suppose that the information sequence is 1010. After this sequence is shifted

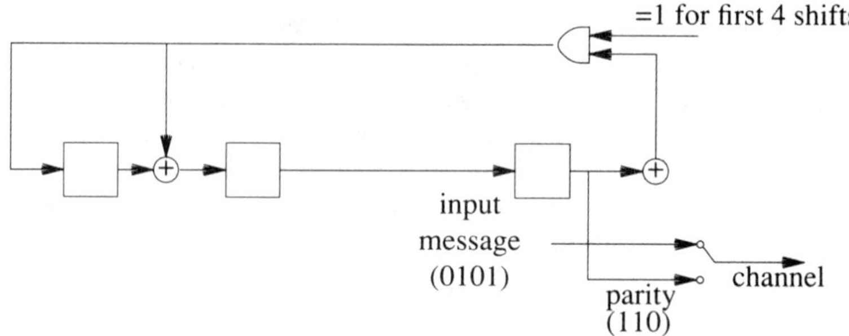

Figure 4.2. Encoder for the (7,4) Hamming code.

simultaneously into the channel and the shift register, the AND gate is closed and the shift register's contents, 110, are shifted into the channel. Thus, the transmitted code word is 1010011. The first four bits of this word represent the information, and the remaining three bits represent the parity.

■

Let us address now the problem of detecting and correcting errors by cyclic codes. If the transmitted code word is

$$s(x) = s_1 x^{n-1} + s_2 x^{n-2} + \cdots + s_n$$

then errors are detected if the received word $\tilde{s}(x)$ does not belong to the code, or, in other words, it is not divisible by the code generator $g(x)$. The same circuit that we used for encoding can be used for detecting errors, since it produces the division remainder. The information portion of the received word is shifted into the shift register, where its contents are compared with the parity-check portion of the received word. If they are different, the received word contains errors.

We can also shift the whole received word $\tilde{s}(x)$ into the shift register. Since the register's contents represent the remainder, the received polynomial is not divisible by the code-generator polynomial if the register's contents are not equal to zero. The polynomial that corresponds to the shift register's contents, after the whole received block is shifted into it is called the *syndrome polynomial*. The syndrome polynomial represents the syndrome vector defined by (4.2.3).

If channel errors $e(x)$ are additive

$$\tilde{s}(x) = s(x) + e(x)$$

then the syndrome polynomial

$$S(x) = \text{remainder}\{\tilde{s}(x) / g(x)\} = \text{remainder}\{e(x) / g(x)\}$$

depends only on errors, since the code word $s(x)$ is divisible by $g(x)$.

Thus, cyclic codes that are used only for detecting errors can be easily implemented. The same shift registers can perform encoding and decoding. The decoder implementation is usually much more complex when the codes are used for correcting errors. One method of error correction consists of selecting (decoding) a code word whose conditional probability, given the received word $\tilde{s}(x)$, is highest:

$$Pr\,[\,\hat{s}(x) | \tilde{s}(x)\,] = \max$$

In general, this method can be implemented only for short codes, because the number of code words grows exponentially with the code length. A very attractive class of codes that are both powerful and implementable in a simple manner are the codes considered in the next section.

4.2.4 Bose-Chaudhuri-Hocquenghem Codes. Let channel symbols be elements of GF(q) and α be an element of GF(q^m). Then the Bose-Chaudhuri-Hocquenghem (BCH) codes are defined by the lowest degree generator polynomial $g(x)$ over GF(q) whose roots are $\alpha^{m_0}, \alpha^{m_0+1}, ..., \alpha^{m_0+d-2}$, where m_0 and d are fixed numbers.

Any code word also has these roots, since it is divisible by $g(x)$:

$$s(\alpha^j) = \sum_{i=1}^{n} s_i \alpha^{(n-i)j} = 0 \quad j = m_0, m_0+1, ..., m_0+d-2 \tag{4.2.8}$$

These equations suggest that the code parity-check matrix is

$$\mathbf{H} = \begin{bmatrix} \alpha^{(n-1)m_0} & \alpha^{(n-2)m_0} & \cdots & \alpha^{m_0} & 1 \\ \alpha^{(n-1)(m_0+1)} & \alpha^{(n-2)(m_0+1)} & \cdots & \alpha^{m_0+1} & 1 \\ \cdots & \cdots & \cdots & \cdots & \cdots \\ \alpha^{(n-1)(m_0+d-2)} & \alpha^{(n-2)(m_0+d-2)} & \cdots & \alpha^{m_0+d-2} & 1 \end{bmatrix} \tag{4.2.9}$$

Suppose that channel errors $e(x)$ are additive: $\tilde{s}(x) = s(x) + e(x)$. Then the received message syndrome may be expressed in the polynomial form:

$$\tilde{s}(\alpha^j) = e(\alpha^j) = S_j \quad j = m_0, m_0+1, ..., m_0+d-2$$

These relations can be treated as equations for the unknown errors that are the coefficients of the polynomial $e(x)$:

$$e(\alpha^j) = \sum_{i=1}^{n} e_i \alpha^{(n-i)j} = S_j \qquad j=m_0,m_0+1,...,m_0+d-2 \tag{4.2.10}$$

If the number of errors t is not greater than $(d-1)/2$, equation (4.2.10) has a unique solution that determines the error values. In this case only t coefficients of the polynomial $e(x)$ differ from zero. If we denote them $Y_k = e_{i_k}$, $k=1,2,...,t$, the previous system can be rewritten as

$$\sum_{k=1}^{t} Y_k X_k^j = S_j \qquad j=m_0,m_0+1,...,m_0+d-2 \tag{4.2.11}$$

where $X_k = \alpha^{(n-i_k)}$. This system contains $2t$ unknowns $X_1,X_2,...,X_t$, $Y_1,Y_2,...,Y_t$ and is equivalent to the previous system. The values X_k uniquely identify error locations i_k and are called the *error locators*, while Y_k give the corresponding *error values*.

The problem of solving system (4.2.11) is similar to the problem of approximating a probability distribution by the polygeometric distribution (3.4.8). Applying Prony's method (described in Sec 3.4.5) to the system solution,[75,81] we obtain a linear difference equation

$$S_l + \sum_{j=1}^{t} \sigma_j S_{l-j} = 0 \tag{4.2.12}$$

whose characteristic equation

$$x^t + \sum_{j=1}^{t} \sigma_j x^{t-j} = 0 \tag{4.2.13}$$

roots are the error locators X_k. The corresponding error values are found by solving system (4.2.11).

These results can be summarized as the following algorithm [75] of BCH code decoding:

- Find the coefficients σ_i of difference equation (4.2.12) that generates the syndrome sequence.
- Find the error locators X_k as the roots of characteristic equation (4.2.13).
- Find the error values Y_k by solving equation (4.2.11). This step is not needed in decoding the binary BCH codes, since then $Y_k = 1$.
- Correct errors by subtracting the error values Y_k from the received symbols $\tilde{s}_{n-i_k}$.

System (4.2.12) is a Toeplitz system, therefore the fast recursive algorithms discussed in the previous chapter can be used to solve it. However, not all the recursive methods are applicable in this case, because the majority of them assume that the Toeplitz matrix principal determinants are not equal to zero. This is not a significant restriction with complex numbers, but it is intolerable with finite fields. Berlekamp's algorithm[75] or Euclid's algorithm[91] are free of these restrictions.

We can also improve the algorithm's performance if we notice that equations (4.2.10) can be treated as a part of the discrete Fourier transform of the received sequence. For the complete Fourier transform we need n equations rather than the $d-1$ equations that we have. If the number of errors is not greater than $(d-1)/2$, we can obtain as many syndromes as we need by making use of difference equation (4.2.12). Therefore, instead of solving the characteristic equation to find the error locators and subsequently solving system (4.2.11) to find the error values, we may proceed as follows:

- Find the linear difference equation (4.2.12) coefficients σ_i by using Berlekamp's or Euclid's algorithms.
- Compute the error sequence Fourier transform by using recurrence equation (4.2.12) for $l=m_0+d-1,...,m_0+n-1$.
- Find the error polynomial estimate $\hat{e}(x)$ by applying the inverse Fourier transform.
- Subtract the estimated error sequence from the received sequence to correct errors.

The complexity of this method is proportional to $n\log^2 n$, whereas the complexity of the previous method is proportional to n^2.[92,93]

EXAMPLE 4.2.5: The Hamming code considered in Example 4.2.4 can be treated as a BCH code. Its generator polynomial $g(x)=1+x+x^3$ is the lowest degree polynomial whose root α is the primitive element of GF(2^3) of Example 4.2.2. Indeed,

$$g(\alpha)=1+\alpha+\alpha^3=(1,0,0)+(0,1,0)+(1,1,0)=0$$

and no linear or quadratic polynomial with binary coefficients has a root α. The code is capable of correcting a single bit error because the condition $\alpha=\alpha^{m_0+d-2}$ is satisfied by $m_0=1$ and $d=2$. Suppose that a single bit error occurred in the first information bit. Equation (4.2.11) then becomes $\alpha^6=S_1$, so that the syndrome uniquely identifies the error location: bit $i=1$ must be complemented, since $\alpha^{7-i}=S_1$.

If we decide to use the Fourier transform method, we first find difference equation (4.2.12), which in our case takes the form $S_l+\sigma_1 S_{l-1}=0$ with $\sigma_1=\alpha^6$. Using this equation we calculate $S_2=\alpha^{12}=\alpha^5$, $S_3=\alpha^{18}=\alpha^4$, $S_4=\alpha^{24}=\alpha^3$, $S_5=\alpha^{30}=\alpha^2$, $S_6=\alpha^{36}=\alpha^1$, $S_7=\alpha^{42}=1$. We find the error bit pattern by using the inverse Fourier transform

$$e_k = \sum_{j=1}^{7} S_j \alpha^{-(7-k)j}$$

which is equal to 1 when $k=1$ and to 0 otherwise. By adding (1 0 0 0 0 0 0) to the received word we correct the error.

■

4.2.5 Reed-Solomon Codes. The Reed-Solomon (RS) code is the particular case of the BCH codes when $\alpha \in \mathrm{GF}(q)$. These codes are becoming increasingly popular, because of their efficiency in correcting bursts of errors.

It is obvious that the RS code generator polynomial is given by

$$g(x) = (x-\alpha^{m_0})(x-\alpha^{m_0+1}) \cdots (x-\alpha^{m_0+d-2})$$

since this polynomial is the minimum degree polynomial that has the roots $\alpha^{m_0}, \alpha^{m_0+1}, \ldots, \alpha^{m_0+d-2}$. Being a particular case of the BCH codes, this code is able to correct up to $\lfloor (d-1)/2 \rfloor$. Since the generator polynomial degree $r=d-1$ is equal to the number of the parity symbols, this code requires only $d-1$ parity check symbols to be able to correct up to $\lfloor (d-1)/2 \rfloor$ symbol errors, whereas a general BCH code requires more than $d-1$ parity-check symbols to be able to correct up to $\lfloor (d-1)/2 \rfloor$ symbol errors.

4.2.6 Convolutional Codes. In contrast with the block (n,k) codes, convolutional codes continuously encode information so that parity symbols of any block depend not only on the information symbols of this block but also on some symbols of previous blocks.

A *convolutional* (n,k,m) *code* can be defined by a difference equation

$$\mathbf{y}_j = \sum_{i=0}^{m} \mathbf{x}_{j-i} \mathbf{G}_i \tag{4.2.14}$$

where , $\mathbf{G}_i$ are the $k \times n$ matrices, vectors $\mathbf{x}_i = [x_{i1} \; x_{i2} \; \ldots \; x_{ik}]$ represent the information symbols, and $\mathbf{y}_j = [y_{j1} \; y_{j2} \; \ldots \; y_{jn}]$ represent the encoded symbols. These matrices are used to perform the encoding and are called the *code generator matrices*. We assume also that the elements of the matrices belong to a Galois field GF(q).

It is convenient to describe convolutions using polynomials. Denote the encoder input sequence

$$\mathbf{x}(D) = \sum_{i=0}^{\infty} \mathbf{x}_i D^i \tag{4.2.15}$$

the encoder output sequence

$$\mathbf{y}(D) = \sum_{i=0}^{\infty} \mathbf{y}_i D^i \tag{4.2.16}$$

and the code generator polynomial

$$\mathbf{G}(D) = \sum_{i=0}^{m} \mathbf{G}_i D^i \tag{4.2.17}$$

Then the convolution (4.2.14) can be expressed as

$$\mathbf{y}(D) = \mathbf{x}(D)\mathbf{G}(D) \tag{4.2.18}$$

This equation is merely a compressed version of the convolution (4.2.14). If we *formally* multiply the polynomials and equate the coefficients of the same powers of D, we obtain equation (4.2.14).

Equation (4.2.14) defines nonsystematic codes in the general case, since the encoded information $\mathbf{y}_j$ might not contain the original information $\mathbf{x}_j$. In the particular case, when the code generator polynomial has the special form

$$\mathbf{G}(D) = \begin{bmatrix} \mathbf{I} & \mathbf{G}_P(D) \end{bmatrix} \tag{4.2.19}$$

the code is systematic, because the encoder output vector can be decomposed into the information part

$$\mathbf{y}_I(D) = \mathbf{x}(D) \tag{4.2.20}$$

and the parity part

$$\mathbf{y}_P(D) = \mathbf{x}(D)\mathbf{G}_P(D) \tag{4.2.21}$$

According to (4.2.14), convolutional encoders can be implemented using Galois field multipliers and adders. In the case of binary convolutional codes we need only binary adders.

EXAMPLE 4.2.6: Let $\mathbf{G}(D) = [1+D+D^2 \quad 1+D^2]$ be a convolutional code generator polynomial. Then the encoder input–output relationship can be expressed by the convolution (4.2.14), which takes the form

$$\mathbf{y}_j = x_j[1 \quad 1] + x_{j-1}[1 \quad 0] + x_{j-2}[1 \quad 1] \tag{4.2.22}$$

and can be rewritten as

$$\begin{aligned} y_{j1} &= x_j + x_{j-1} + x_{j-2} \\ y_{j2} &= x_j + x_{j-2} \end{aligned} \tag{4.2.23}$$

This encoder is shown in Fig. 4.3. One bit at a time is shifted into the shift register. For each input bit x_j, the encoder outputs two bits: y_{j1} and y_{j2} when the switch is in positions 1 and 2, respectively.

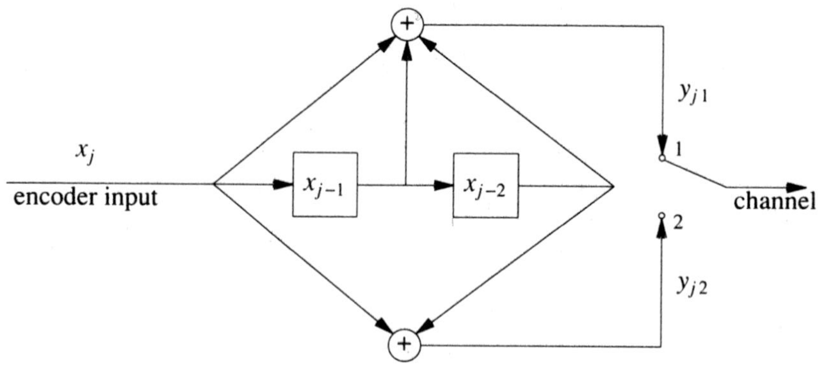

Figure 4.3. Encoder for the (2,1,2) convolutional code.

■

A convolutional encoder can be described as a linear sequential circuit.[94] Indeed, it is easy to verify that the convolution (4.2.14) can be expressed as

$$\begin{aligned} \mathbf{s}_{j+1} &= \mathbf{s}_j\mathbf{A} + \mathbf{x}_j\mathbf{B} \\ \mathbf{y}_j &= \mathbf{s}_j\mathbf{C} + \mathbf{x}_j\mathbf{G}_0 \end{aligned} \tag{4.2.24}$$

where $\mathbf{s}_j = [\mathbf{x}_{j-1}\ \mathbf{x}_{j-2}\ \dots\ \mathbf{x}_{j-m}]$ is the *encoder state* vector. (Sometimes the encoder state is defined as the number of shift registers used in its implementation: only the components of $\mathbf{s}_j$ that are actually used in the system (4.2.24) constitute the state.) Matrices **A**, **B**, and **C** are given by

$$\mathbf{A} = \begin{bmatrix} 0\ \mathbf{I}\ 0 \dots 0 \\ 0\ 0\ \mathbf{I} \dots 0 \\ \dots\dots\dots \\ 0\ 0\ 0 \dots \mathbf{I} \\ 0\ 0\ 0 \dots 0 \end{bmatrix} \quad \mathbf{B} = \begin{bmatrix} \mathbf{I}\ 0 \dots 0 \end{bmatrix} \quad \mathbf{C}' = \begin{bmatrix} \mathbf{G}_1\ \mathbf{G}_2 \dots \mathbf{G}_m \end{bmatrix} \tag{4.2.25}$$

In this interpretation the encoder output $\mathbf{y}_j$ and the next state $\mathbf{s}_{j+1}$ depend on its input $\mathbf{x}_j$ and state $\mathbf{s}_j$. The encoder state diagram is a flow graph whose nodes represent the encoder states, and the symbols I/O along the edges represent the encoder input and output symbols. The state diagram of the encoder of Fig. 4.3 is shown in Fig. 4.4. Sometimes we want to show a convolutional code state transition in time. In this case we treat states $\mathbf{s}_j$ and $\mathbf{s}_{j-1}$ as elements of different sets of

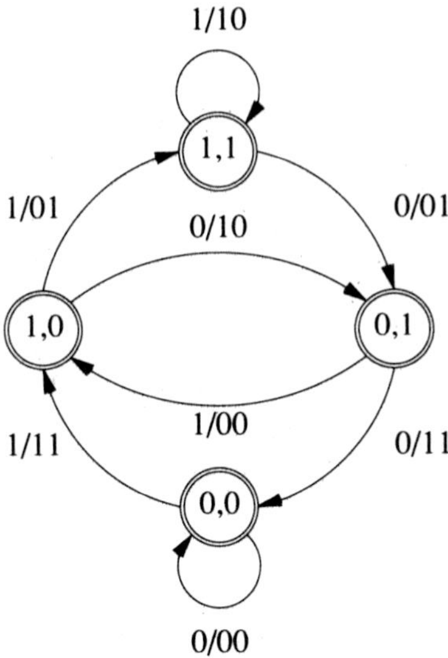

Figure 4.4. Encoder state diagram.

states. The state diagram of this process is called a *trellis diagram*. The trellis diagram that corresponds to the state diagram of Fig. 4.4 is shown in Fig. 4.5.

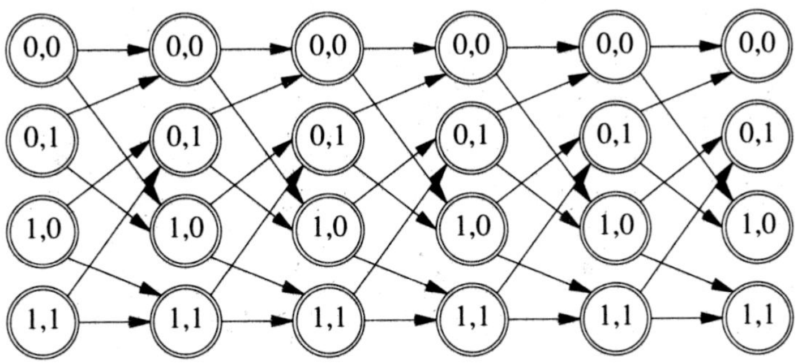

Figure 4.5. Encoder trellis diagram.

State and trellis diagrams are useful in describing convolutional code operation and evaluating its performance.

Let us address now the problem of decoding cyclic codes. We consider here only the algorithms whose performance will be evaluated later in this chapter.

4.2.6.1 Maximum Likelihood Decoding. A maximum likelihood decoder (MLD) outputs the information sequence for which the received sequence is most probable. Suppose that an information sequence $\mathbf{x}_0, \mathbf{x}_1, ..., \mathbf{x}_l$ is encoded into a sequence $\mathbf{y}_0, \mathbf{y}_1, ..., \mathbf{y}_l$ and that a sequence $\mathbf{z}_0, \mathbf{z}_1, ..., \mathbf{z}_l$ is received over a noisy channel. The MLD outputs the sequence $\hat{\mathbf{x}}_0, \hat{\mathbf{x}}_1, ..., \hat{\mathbf{x}}_l$ that maximizes the probability

$$Pr(\mathbf{z} \mid \mathbf{y}) = Pr(\mathbf{z}_0, \mathbf{z}_1, ..., \mathbf{z}_l \mid \mathbf{y}_0, \mathbf{y}_1, ..., \mathbf{y}_l) \tag{4.2.26}$$

If information symbols appear independently and the channel is memoryless, then the MLD output also maximizes the log-likelihood function

$$\log Pr(\mathbf{z} \mid \mathbf{y}) = \sum_{i=0}^{l} Pr(\mathbf{z}_i \mid \mathbf{y}_i) \tag{4.2.27}$$

4.2.6.2 Viterbi Algorithm. The Viterbi algorithm finds the MLD output recursively. If a maximum-likelihood estimate $\hat{\mathbf{s}}_i$ of the encoder state sequence has been found, then the information sequence $\hat{\mathbf{x}}_i$ can be uniquely identified:

$$\hat{\mathbf{x}}_j = \hat{\mathbf{s}}_{j+1} [\mathbf{I} \ 0 \ 0 ... 0]' \tag{4.2.28}$$

Each sequence of states can be depicted as a path on a trellis diagram. The maximum-likelihood state sequence estimate can be interpreted as a selection of the path that has the highest probability. Finding the true maximum seems to be impossible, since the number of paths grows exponentially with the sequence length. However, if the log-likelihood function is expressed by equation (4.2.27), the solution can be found recursively. Indeed, the log-likelihood maximum corresponds to the minimum of the function

$$M(\mathbf{z}, \mathbf{s}) = \sum_{i=0}^{l} m(\mathbf{z}_i, \mathbf{s}_i)$$

where $m(\mathbf{z}_i, \mathbf{s}_i) = -\log Pr(\mathbf{z}_i \mid \mathbf{s}_i)$ may be interpreted as a trellis diagram branch metric. The problem of maximum-likelihood decoding can be interpreted as a problem of finding the shortest path on the trellis diagram, which can be solved by the Viterbi recursive algorithm:[95]

> Let $L(\mathbf{s}_t)$ be the length of the shortest path that ends at the node $\mathbf{s}_t$ at the moment t. Then, because of metric additivity, the shortest path that ends at a node $\mathbf{s}_{t+1}$ at the moment $t+1$ can be obtained from one of the shortest paths at the moment t, and its length is given by

$$L(\mathbf{s}_{t+1}) = \min_{\mathbf{s}_t}\{L(\mathbf{s}_t) + m(\mathbf{z}_t, \mathbf{s}_t)\}$$

We retain only the shortest paths at every trellis node. These paths are called the *survivors.*

EXAMPLE 4.2.7: Let us illustrate the Viterbi algorithm for the code of Example 4.2.6, assuming that channel errors are independent. The likelihood function can be expressed as

$$Pr(\mathbf{z}|\mathbf{y}) = p^e q^{2N-e}$$

where p is the bit error probability, N is the total number of transmitted symbols $\mathbf{y}_i$ (each consisting of two bits), and e is the total number of bit errors. The log-likelihood function has the form

$$\log Pr(\mathbf{z}|\mathbf{y}) = 2N \log q + e \log (p/q)$$

From this equation it follows that the number e of different bits between received $\mathbf{z}$ and hypothetical sequences $\mathbf{y}$ (the Hamming distance) can be used as a path metric.

Suppose that the sequence 0 0 0 0 0 0 entered the encoder when it was in the state (0,0). The encoder sends 00 00 00 00 00 00 (see Fig. 4.4), and the decoder receives the sequence 00 01 00 00 00 00, which contains a single error.

If we assume that initially the decoder was in the state (0,0), then the branch (0,0)→(0,0) length is equal to zero, since the received 00 coincides with this transition output (see Fig. 4.4). The branch (0,0)→(1,0) has length 2, since the received 00 is different from this transition output (11) in both positions. Repeating these calculations we see that the path (0,0)→(0,0)→(0,0)→(0,0) has length 1, whereas the path (0,0)→(1,0)→(0,1)→(0,0) that terminates at the same state has length 4. Therefore, the path (0,0)→(0,0)→(0,0)→(0,0) is a survivor. Similarly, we select survivors for each state in the trellis diagram (see Fig. 4.6). Finally, we discover that the path

(0,0)→(0,0)→(0,0)→(0,0)→(0,0)→(0,0)

is the shortest (its length is equal to 1), and the corresponding all-zero sequence is decoded, thus correcting the single error.

■

We described the algorithm by assuming that the information symbols are independent and the channel is memoryless. However, this algorithm is applicable to more general cases. The metric does not have to be the log-likelihood function, and the states are not necessarily defined by the contents of the encoder shift registers. In general, the algorithm finds the shortest path with respect to some metrics

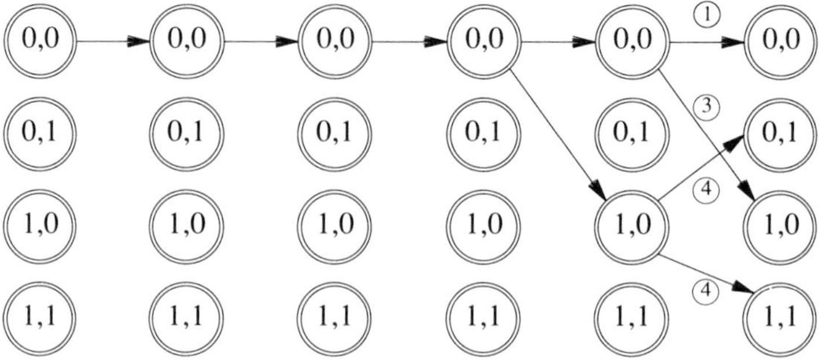

Figure 4.6. The survivors.

if the metrics are additive:[96]

$$M(\mathbf{z},\mathbf{s}) = \sum_{i=0}^{l} m(\mathbf{z}_i, \mathbf{s}_{i+1}, \mathbf{s}_i)$$

Numerous applications of the algorithm use this approach.

If a convolutional code is used in a channel with an SSM error source, we define the composite state as a Kronecker product of the encoder state and the SSM state: if the encoder state is $\mathbf{u}$ and the SSM state is $\boldsymbol{\sigma}$, then the composite state is $\mathbf{s} = (\mathbf{u},\boldsymbol{\sigma})$. Since the composite states constitute a Markov chain, the likelihood function can be expressed as

$$Pr(\mathbf{z}|\,\mathbf{s}) = p_{\mathbf{s}_0} \prod_{i=1}^{l} p_{\mathbf{s}_i,\mathbf{s}_{i+1}}(\mathbf{e}_i)$$

where $p_{\mathbf{s}_0}$ is the state $\mathbf{s}_0$ initial probability and $p_{\mathbf{s}_i,\mathbf{s}_{i+1}}(\mathbf{e}_i)$ is the composite state transition probability of error $\mathbf{e}_i = \mathbf{z}_i - \mathbf{y}_i$ that transforms an encoder output symbol $\mathbf{y}_i$ into received symbol $\mathbf{z}_i$. The MLD outputs the sequence of states $\hat{\mathbf{s}}_0, \hat{\mathbf{s}}_1, ..., \hat{\mathbf{s}}_l$ for which the received sequence $\mathbf{z}_0, \mathbf{z}_1, ..., \mathbf{z}_l$ has the highest probability.

The log-likelihood function is an additive function

$$\log Pr(\mathbf{z}|\,\mathbf{s}) = \log p_{\mathbf{s}_0} + \sum_{i=1}^{l} \log p_{\mathbf{s}_i,\mathbf{s}_{i+1}}(\mathbf{e}_i) \tag{4.2.29}$$

We can neglect the initial distribution by assuming that the encoder initial state is always known to the decoder, or by assuming that the process started long time ago ($t_0 = -\infty$). Thus, we can use the Viterbi algorithm with the branch metrics

$$m(\mathbf{z}_i, \mathbf{s}_{i+1}, \mathbf{s}_i) = -\log p_{\mathbf{s}_i, \mathbf{s}_{i+1}}(\mathbf{e}_i) \tag{4.2.30}$$

to decode a convolutional code when the channel error source is described by the SSM. The number of composite states which is equal to the product of the number of the encoder states and the number of the SSM states may be large. This would require us to perform some approximations, such as replacing a group of states with an aggregate state.

Ideally, at the end of communication the survivor $\hat{\mathbf{s}}_0, \hat{\mathbf{s}}_1, ..., \hat{\mathbf{s}}_T$ that has the shortest length is selected, and the decoder outputs the corresponding information symbols $\hat{\mathbf{x}}_0, \hat{\mathbf{x}}_1, ..., \hat{\mathbf{x}}_T$. This method would require a very large memory to store all the survivors. In practice it is necessary to perform information decoding based on truncated survivors. If the truncation depth Δ is large enough, then it is highly probable that the initial part (outside the truncation interval) of all the survivors is the same so that we can decode $\hat{\mathbf{x}}_{t-\Delta}$ at the moment t that corresponds to the merged part of all the survivors (the common part of all the survivors in Fig. 4.6). If the survivors disagree outside the truncated interval at the moment $t-\Delta$, then we select $\hat{\mathbf{x}}_{t-\Delta}$, which corresponds to the shortest path.

4.2.6.3 Syndrome Decoding. Let us consider now syndrome decoding of convolutional codes. Suppose that we have a systematic convolutional code and that the channel errors are additive. Then the transmitted information is expressed by equations (4.2.20) and (4.2.21) and the received information is given by

$$\mathbf{r}(D) = \mathbf{x}(D)\mathbf{G}(D) + \mathbf{e}(D) \tag{4.2.31}$$

where $\mathbf{e}(D)$ is the channel error polynomial. Since the code is systematic, the information part of the received data is given by

$$\mathbf{r}_I(D) = \mathbf{x}_I(D) + \mathbf{e}_I(D) \tag{4.2.32}$$

and the parity-check information by

$$\mathbf{r}_P(D) = \mathbf{x}_P(D)\mathbf{G}_P(D) + \mathbf{e}_P(D) \tag{4.2.33}$$

The difference

$$\mathbf{s}(D) = \mathbf{r}_I(D)\mathbf{G}_P(D) - \mathbf{r}_P(D) \tag{4.2.34}$$

between the parity check of the received information and the received parity check indicates the presence of errors and is called the *error syndrome*. This syndrome does not depend on the transmitted information but only on errors:

$$\mathbf{s}(D) = \mathbf{e}_I(D)\mathbf{G}_P(D) - \mathbf{e}_P(D) \tag{4.2.35}$$

since

$$\mathbf{y}_I(D)\mathbf{G}_P(D) - \mathbf{y}_P(D) = 0 \tag{4.2.36}$$

If $\hat{\mathbf{e}}(D)$ is a solution of equation (4.2.35), then the received information is corrected by subtracting the estimated information error sequence from the received information:

$$\hat{\mathbf{x}}(D) = \mathbf{y}_I(D) - \hat{\mathbf{e}}_I(D) \tag{4.2.37}$$

Denote $\mathbf{s}_j$, $\mathbf{e}_{Ij}$, and $\mathbf{e}_{Pj}$ the coefficients of $\mathbf{s}(D)$, $\mathbf{e}_I(D)$, and $\mathbf{e}_p(D)$, respectively. Then equation (4.2.35) can be rewritten as a system of linear equations for the unknown channel errors,

$$\mathbf{s}_j = \sum_{i=0}^{m} \mathbf{e}_{I(j-i)} \mathbf{G}_{Pi} - \mathbf{e}_{Pj} \tag{4.2.38}$$

This system may be rewritten as

$$\mathbf{s}_j = \sum_{i=0}^{m} \mathbf{e}_{j-i} \mathbf{H}_i$$

where $\mathbf{e}_i = (\mathbf{e}_{Ii}, \mathbf{e}_{Pi})$ and $\mathbf{H}_i$ are the rows of the matrix

$$\mathbf{H}' = [\mathbf{G}_P \quad -\mathbf{I}]$$

which is called a parity-check matrix. Any method of solving (4.2.38) represents an algorithm of error correction.

Usually this system is broken down into subsystems. The subsystem that checks $\mathbf{e}_{Ij}$ has the form

$$\mathbf{s}_{j+k} = \sum_{i=0}^{m} \mathbf{e}_{I(j+k-i)} \mathbf{G}_{Pi} - \mathbf{e}_{P(j+k)} \qquad k=1,2,\ldots,m \tag{4.2.39}$$

The decoding algorithm that estimates

$$\hat{\mathbf{e}}_{Ij} = \mathbf{f}(\mathbf{s}_j, \mathbf{s}_{j+1}, \ldots, \mathbf{s}_{j+m}) \tag{4.2.40}$$

from this subsystem is called *definite decoding*. It can be implemented using shift registers and a circuit that computes the above function (see Fig. 4.7).

The received information sequence $\mathbf{r}_I(D)$ is re-encoded by using the encoder replica shown in upper part of Fig. 4.7 to produce $\mathbf{r}_I(D)\mathbf{G}_P(D)$. Then the received parity sequence $\mathbf{r}_P(D)$ is subtracted from $\mathbf{r}_I(D)\mathbf{G}_P(D)$, which, according to (4.2.34), yields the syndrome sequence $\mathbf{s}(D)$. The syndrome registers remember m most recent syndromes, and the decision circuit estimates error sequence according to (4.2.40).

The other technique, which is called *feedback decoding*, uses the estimates from the previous subsystems. If error estimates from the previous subsystems are subtracted from the actual errors, the system (4.2.39) becomes

$$\boldsymbol{\sigma}_j^{(k)} = \sum_{i=0}^{m} \boldsymbol{\varepsilon}_{I(j-i)}^{(k)} \mathbf{G}_{Pi} - \mathbf{e}_{P(j+k)} \qquad k=1,2,\ldots,m$$

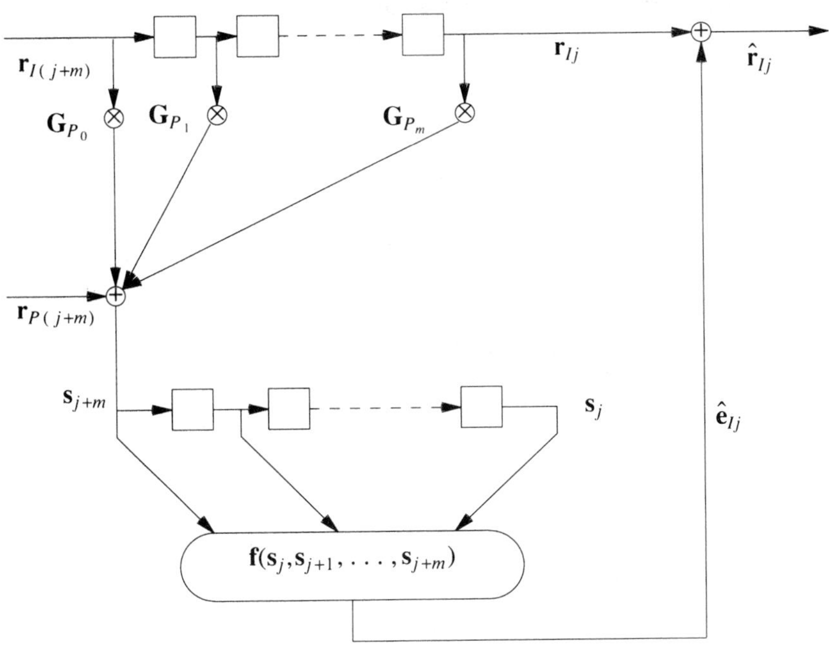

Figure 4.7. Decoder for the convolutional (n,k,m) code.

where

$$\boldsymbol{\varepsilon}_{Ij}^{(k)} = \begin{cases} \mathbf{e}_{Ij} - \hat{\mathbf{e}}_{Ij} & \text{for } j<k, \\ \mathbf{e}_{Ij} & \text{for } j \geq k \end{cases}$$

If the estimated error values from the previous subsystems were correct, then $\boldsymbol{\varepsilon}_{Ij}^{(k)} = 0$ for $j<k$ and the system becomes significantly simpler.

The feedback decoding rule

$$\hat{\mathbf{e}}_{Ij} = \mathbf{f}(\boldsymbol{\sigma}_j^{(j)}, \boldsymbol{\sigma}_{j+1}^{(j)}, \ldots, \boldsymbol{\sigma}_{j+m}^{(j)})$$

is based on the values $\boldsymbol{\sigma}_j^{(k)}$ that are called the *modified syndromes*. Note that $\boldsymbol{\sigma}_j^{(k)} = \mathbf{s}_j$ for $j \geq k+m$.

This technique is called feedback decoding because it can be implemented by feeding the estimated errors back into the syndrome registers. Indeed, one can easily verify by direct substitutions that the modified syndromes used for the $\hat{\mathbf{e}}_{I(j+1)}$ estimation satisfy the following recurrent equations:

$$\begin{aligned}
\boldsymbol{\sigma}_{j+1}^{(j+1)} &= \boldsymbol{\sigma}_{j+1}^{(j)} - \hat{\mathbf{e}}_{Ij}\mathbf{G}_{P_1}, \\
\boldsymbol{\sigma}_{j+2}^{(j+1)} &= \boldsymbol{\sigma}_{j+2}^{(j)} - \hat{\mathbf{e}}_{Ij}\mathbf{G}_{P_2}, \\
&\cdots\cdots\cdots\cdots\cdots\cdots \\
\boldsymbol{\sigma}_{j+m}^{(j+1)} &= \mathbf{s}_{j+m} - \hat{\mathbf{e}}_{Ij}\mathbf{G}_{P_m}.
\end{aligned} \tag{4.2.41}$$

These equations can be implemented by adding a feedback to the syndrome registers of the definite decoder, as shown in Fig. 4.8.

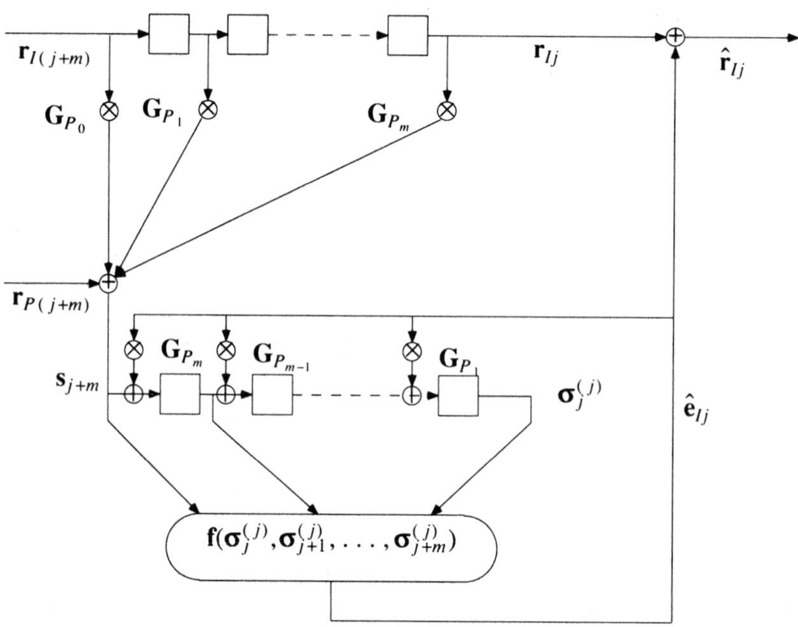

Figure 4.8. Feedback decoder for the (n,k,m) code.

Like equations (4.2.24), equations (4.2.41) may be rewritten in the more compact form

$$\mathbf{u}_{j+1} = \mathbf{u}_j\mathbf{A} + \mathbf{s}_{j+m}\mathbf{B} - \hat{\mathbf{e}}_j\mathbf{G}_P \tag{4.2.42}$$

where

$$\mathbf{u}_j = (\boldsymbol{\sigma}_{j+m-1}^{(j)}, \boldsymbol{\sigma}_{j+m-2}^{(j)}, \ldots, \boldsymbol{\sigma}_j^{(j)})$$

is the syndrome register state, $\mathbf{G}_P = (\mathbf{G}_{P_m}, \mathbf{G}_{P_{m-1}}, \ldots, \mathbf{G}_{P_1})$, and matrices $\mathbf{A}$ and $\mathbf{B}$ are defined by equations (4.2.25).

If the code has a special structure, the decoder decision element can be implemented by using simple majority-logic circuits. The most popular codes of this type are the so-called *convolutional self-orthogonal codes* (CSOCs). [97—100]

In the case of a binary convolutional code, elements of system (4.2.38) are zeroes and ones. We say that e_{Ij} is checked by J equations of system (4.2.38) if this system has j equations of the form

$$s_{j_k} = e_{Ij} + \delta_{j_k} \qquad k=1,2,\ldots,J \tag{4.2.43}$$

where δ_{j_k} denotes the sum of other than e_{Ij} terms. Let us assume now that δ_{j_k} and δ_{j_l} for $k \neq l$ do not have any common terms e_{Ik}. The convolutional codes that possess this property are called self-orthogonal. They can be decoded by a simple majority-logic rule.

If the number of errors that are checked by (4.2.43) is not greater than $\lfloor J/2 \rfloor$ and e_{Ij}=1, then $\delta_{j_k} = 0$ in more than $\lfloor J/2 \rfloor$ of the equations so that the majority of them have the form

$$s_{j_k} = 1 \tag{4.2.44}$$

If the number of errors that are checked by (4.2.43) is not greater than $\lfloor J/2 \rfloor$ and e_{Ij}=0, then $\delta_{j_k} = 1$ in no more than $\lfloor J/2 \rfloor$, so that the majority of equations (4.2.43) cannot have the form of (4.2.44). Therefore, if the number of errors that are checked by the equations (4.2.43) is not greater than $\lfloor J/2 \rfloor$, the following algorithm corrects the error: Decode $e_{Ij} = 1$ when the majority of the syndromes in (4.2.43) are equal to 1, and decode 0 otherwise.

EXAMPLE 4.2.8: Consider decoding of the systematic code with

$$\mathbf{G} = \begin{bmatrix} 1 & 1+D+D^4+D^6 \end{bmatrix}$$

The consecutive equations of system (4.2.38) containing e_{Ij} are

$$s_j = e_{I(j-6)} + e_{I(j-4)} + e_{I(j-1)} + e_{Ij} + e_{Pj}$$

$$s_{j+1} = e_{I(j-5)} + e_{I(j-3)} + e_{Ij} + e_{I(j+1)} + e_{P(j+1)}$$
$$s_{j+2} = e_{I(j-4)} + e_{I(j-2)} + e_{I(j+1)} + e_{I(j+2)} + e_{P(j+2)}$$
$$s_{j+3} = e_{I(j-3)} + e_{I(j-1)} + e_{I(j+2)} + e_{I(j+3)} + e_{P(j+3)}$$
$$s_{j+4} = e_{I(j-2)} + e_{Ij} + e_{I(j+3)} + e_{I(j+4)} + e_{P(j+4)}$$
$$s_{j+5} = e_{I(j-1)} + e_{I(j+1)} + e_{I(j+4)} + e_{I(j+5)} + e_{P(j+5)}$$
$$s_{j+6} = e_{Ij} + e_{I(j+2)} + e_{I(j+5)} + e_{I(j+6)} + e_{P(j+6)}$$

The part of this system that checks e_{Ij} is given by

$$s_j = e_{I(j-6)} + e_{I(j-4)} + e_{I(j-1)} + e_{Ij} + e_{Pj}$$
$$s_{j+1} = e_{I(j-5)} + e_{I(j-3)} + e_{Ij} + e_{I(j+1)} + e_{P(j+1)}$$
$$s_{j+4} = e_{I(j-2)} + e_{Ij} + e_{I(j+3)} + e_{I(j+4)} + e_{P(j+4)}$$
$$s_{j+6} = e_{Ij} + e_{I(j+2)} + e_{I(j+5)} + e_{I(j+6)} + e_{P(j+6)}$$

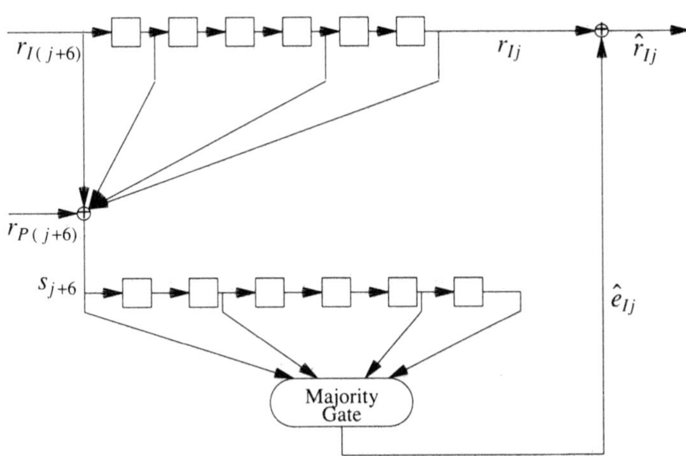

Figure 4.9. Decoder for the convolutional (2,1,6) code.

Each of these equations contains e_{Ij}, but they don't have any other common elements. Therefore, the code is self-orthogonal. If the number of errors in the total of seventeen symbols that are checked by these equations is not greater than two, then $e_{Ij} = 1$ only when the majority of the syndromes s_j, s_{j+1}, s_{j+4} and s_{j+6} are equal to 1. This method does not take into account the solution of the rest of the system and therefore represents definite decoding.

The decoder diagram is shown in Fig. 4.9. The received information symbols r_{Ik} are re-encoded, and the received parity symbols r_{Pk} are subtracted from the calculated parity of the received information symbols, thus producing syndromes s_k according to (4.2.38). The syndromes are shifted into the syndrome registers. Then the majority-logic element selects $\hat{e}_{Ij} = 1$ if at least three out of four syndromes that enter it are equal to 1, otherwise it selects $\hat{e}_{Ij}$=0. The estimated error is added to the received information bit, thus correcting the error. To illustrate feedback decoding, let us consider the first four equations of the modified system:

$$
\begin{aligned}
s_0 &= e_{I0} + e_{P0} \\
s_1 &= e_{I0} + e_{I1} + e_{P1} \\
s_4 &= e_{I0} + e_{I3} + e_{I4} + e_{P4} \\
s_6 &= e_{I0} + e_{I2} + e_{I5} + e_{I6} + e_{P6}
\end{aligned}
$$

If the number of errors in the total of eleven symbols that are checked by these equations is not greater than two, the error e_{I0} is corrected. If the estimated error $\hat{e}_{I0}$ is equal to the actual error ($e_{I0} = \hat{e}_{I0}$), subtracting of the error from the previous equations eliminates it from the rest of the system.

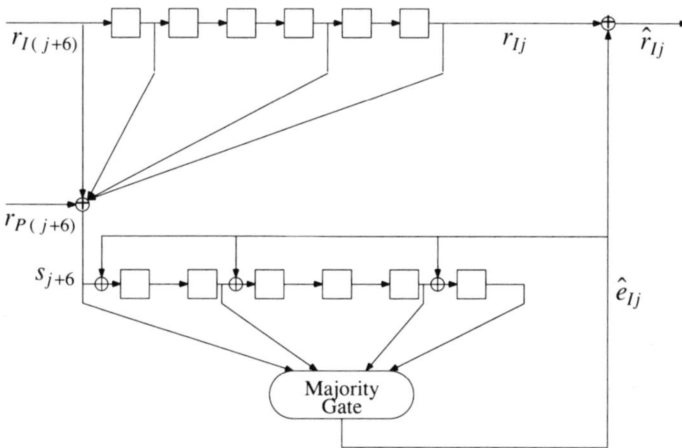

Figure 4.10. Feedback decoder for the (2,1,6) code.

The subtraction can be performed by feeding the estimated error $\hat{e}_{I0}$ back into syndrome registers (see Fig. 4.10). The contents of the syndrome registers after subtraction are the modified syndromes. The system that checks e_{I1} then becomes

$$\sigma_1 = e_{I1} + e_{P1}$$
$$s_2 = e_{I1} + e_{I2} + e_{P2}$$
$$s_5 = e_{I1} + e_{I4} + e_{I5} + e_{P5}$$
$$s_7 = e_{I1} + e_{I3} + e_{I6} + e_{I7} + e_{P7}$$

where $\sigma_1 = s_1 + e_{I0} + \hat{e}_{I0}$ is the modified syndrome. Again, if no more than two out of eleven symbols have errors and our previous estimation $\hat{e}_{I0}$ was correct, then e_{I1} is corrected. It is obvious that the process can be repeated. If the probability of two errors in eleven is smaller than the probability of having two errors in seventeen symbols, feedback decoding might be better than definite decoding.

The problem arises when an error is estimated incorrectly. Then the incorrect estimate is shifted into the syndrome registers, which could lead to additional decoding errors (error propagation of feedback decoding), even in the absence of channel errors. We investigate the performance merits of feedback decoding later in this chapter.

■

4.3 BLOCK CODE PERFORMANCE CHARACTERIZATION

In any one-way communication system, three mutually exclusive events can happen after a message $\mathbf{s} = [s_1, s_2, ..., s_n]$ decoding:

- C_s The message has been decoded without errors (that is, the message either did not have errors or the errors have been corrected).
- U_s The message has been decoded with errors (that is, it contains undetected errors or an attempt to correct errors was unsuccessful).
- D_s The message contains errors that the decoder can detect but is not able to correct.

The probabilities of these events characterize the message transmission correctness in one-way systems.

Suppose that messages are encoded by a linear (n,k) block code and that channel errors are additive. Then the above-described events do not depend on the transmitted message, and their probabilities can be expressed as

$$p_u = \sum_{\mathbf{e} \in U} Pr(\mathbf{e}) \tag{4.3.1}$$

$$p_d = \sum_{\mathbf{e} \in D} Pr(\mathbf{e}) \tag{4.3.2}$$

$$p_c = \sum_{\mathbf{e} \in C} Pr(\mathbf{e}) \tag{4.3.3}$$

where **e** is the difference between received and transmitted messages; sets C, U, and D correspond to above-defined events. Obviously, we need to compute only two of these probabilities, since their sum is equal to 1. The use of formulas (4.3.1), (4.3.2), and (4.3.3) is often accompanied by many difficulties due to the necessity of having to perform summation over all the elements of sets C, D, and U. This process can be systematized by using matrix generating functions.

4.3.1 The Probability of Undetected Error. Suppose that a linear (n,k) code with the code-generator matrix $\mathbf{G}$ is used for error detection only. An error $\mathbf{e} = (e_1, e_2, ..., e_n) \neq 0$ where $e_i \in \mathrm{GF}(p^m)$ is not detected if it is a code word:

$$\mathbf{e} = \mathbf{cG}$$

Thus, according to (4.3.1),

$$p_u = \sum_{\mathbf{c} \neq 0} Pr(\mathbf{cG}) = \sum_{\mathbf{c}} Pr(\mathbf{cG}) - p_n(0) \tag{4.3.4}$$

where $p_n(0)$ is the probability that the block does not contain errors.

The probability of undetected errors can also be found by equation (4.3.1) if we describe the set U of all undetected errors as the set of all nonzero vectors whose syndromes (4.2.4) are equal to zero:

$$U = \{\mathbf{e}\text{: } \mathbf{eH}' = 0 \quad \mathbf{e} \neq 0\}$$

It is convenient to use generating functions to perform summation in the right-hand side of equation (4.3.1). As we have seen in Sec 2.2.1, elements of $\mathrm{GF}(p^m)$ may be interpreted as m-tupples with components from the ground field $\mathrm{GF}(p)$. Let $(a)_i$ denote an i-th component of an element. Then error syndrome components can be expressed in the form

$$(S_j)_i = \sum_{k=1}^{n} (e_k h_{jk})_i \pmod{p} \quad j=1,2,...,r;\ i=1,2,...,m \tag{4.3.5}$$

This equation is the expanded version at the level of the ground field $\mathrm{GF}(p)$ of equation (4.2.4), which uses matrices whose elements belong to $\mathrm{GF}(q)$ with $q=p^m$.

The probability of undetected errors is equal to the sum of probabilities of nonzero errors, for which the right-hand side of equation (4.3.5) is equal to zero. To find this sum, we introduce the generating function

$$f(\mathbf{z}) = \sum_{\mathbf{e}} Pr(\mathbf{e}) \mathbf{z}^{\mathbf{eH}'} \tag{4.3.6}$$

where $\mathbf{z}^{\mathbf{eH}'} = \prod_{i,j=1}^{m,r} z_{ij}^{w_{ij}}$ and $w_{ij} = \sum_{k=1}^{n} (e_k h_{jk})_i$ for $j=1,2,...,r$; $i=1,2,...,m$.

The probability of undetected errors p_u is equal to the sum of probabilities $Pr(\mathbf{e})$, for which $w_{ij} = 0 \pmod p$ minus the probability $p_n(0)$ of the absence of errors in the block. This probability can be determined from generating function (4.3.6) by using formula (2.2.29):

$$p_u = q^{-r} \sum_{\mathbf{v}} f(\boldsymbol{\zeta}^{\mathbf{v}}) - p_n(0) \tag{4.3.7}$$

where $\mathbf{v}$ is an rm-dimensional vector whose coordinates v_{ij} vary from 0 to $p-1$, $\boldsymbol{\zeta}^{\mathbf{v}}$ denotes a vector with the coordinates $\zeta^{v_{ij}}$, and $\zeta = e^{-2\pi j/p}$. This equation expresses the probability of undetected errors for any additive error source model. It can be simplified if the error source is described by an SSM model, because we can make use of the properties of matrix probabilities. The matrix probability of absence of errors in the block is equal to $\mathbf{P}^n(0)$. The generating function $\mathbf{F}(\mathbf{z})$ that corresponds to (4.3.6) is given by

$$\mathbf{F}(\mathbf{z}) = \sum_{\mathbf{e}} \prod_{k=1}^{n} \mathbf{P}(\mathbf{e}) \mathbf{z}^{\mathbf{eH}'} \tag{4.3.8}$$

Since

$$\mathbf{z}^{\mathbf{eH}'} = \prod_{k=1}^{n} \prod_{ij} z_{ij}^{(e_k h_{jk})_i}$$

and the matrix probability

$$\mathbf{P}(\mathbf{e}) = \prod_{k=1}^{n} \mathbf{P}(e_k)$$

generating function (4.3.8) can be expressed as

$$\mathbf{F}(\mathbf{z}) = \prod_{k=1}^{n} \sum_{e_k} \mathbf{P}(e_k) \mathbf{z}^{e_k \mathbf{h}_k} \tag{4.3.9}$$

where

$$\mathbf{z}^{e_k \mathbf{h}_k} = \prod_{ij} z_{ij}^{(e_k h_{jk})_i}$$

Therefore, as in (4.3.7), we can write the matrix probability of undetected errors

$$\mathbf{P}_u = q^{-r}\sum_{\mathbf{v}}\mathbf{F}(\zeta^{\mathbf{v}}) - \mathbf{P}^n(0) \tag{4.3.10}$$

In the most important case of binary codes ($q=2$, $\zeta=-1$), generating function (4.3.9) becomes

$$\mathbf{F}(\mathbf{z}) = \prod_{k=1}^{n}[\mathbf{P}(0) + \mathbf{P}(1)\mathbf{z}^{\mathbf{h}_k}] \tag{4.3.11}$$

where $\mathbf{z}^{\mathbf{h}_k} = z_1^{h_{1k}} z_2^{h_{2k}} \cdots z_r^{h_{rk}}$. Thus, the matrix probability of undetected errors (4.3.10) is given by

$$\mathbf{P}_u = 2^{-r}\sum_{\mathbf{v}}\prod_{k=1}^{n}[\mathbf{P}(0) + \mathbf{P}(1)(-1)^{\mathbf{vh}_k}] - \mathbf{P}^n(0) \tag{4.3.12}$$

where $\mathbf{vh}_k = v_1 h_{1k} + v_2 h_{2k} + \ldots + v_r h_{rk}$ is a scalar product of binary vectors. The scalar probability of undetected errors can be found in the usual way:

$$p_u = \mathbf{p}\mathbf{P}_u\mathbf{1} \tag{4.3.13}$$

Since the code is used only for detecting errors, the matrix probability $\mathbf{P}_c$ of correct reception of a combination and the matrix probability $\mathbf{P}_d$ of error detection are equal to

$$\mathbf{P}_c = \mathbf{P}^n(0) \tag{4.3.14}$$

and

$$\mathbf{P}_d = \mathbf{P}^n - \mathbf{P}^n(0) - \mathbf{P}_u \tag{4.3.15}$$

respectively.

EXAMPLE 4.3.1:
Let us consider the simple $(n,n-1)$ parity-check code. The parity-check matrix of this code has the form $\mathbf{H}=[1\ 1\ 1\ \ldots\ 1]$. Equation (4.3.11) now takes the form $\mathbf{F}(z) = [\mathbf{P}(0) + \mathbf{P}(1)z]^n$ and, according to (4.3.12), the matrix probability of undetected errors is given by

$$\mathbf{P}_u = 2^{-1}\mathbf{P}^n + 2^{-1}[\mathbf{P}(0) - \mathbf{P}(1)]^n - \mathbf{P}^n(0)$$

The matrix probability of detected errors, according to (4.3.15), is equal to

$$\mathbf{P}_d = 2^{-1}\mathbf{P}^n - 2^{-1}[\mathbf{P}(0) - \mathbf{P}(1)]^n \qquad \blacksquare$$

Consider now a more complex case.

EXAMPLE 4.3.2: For the binary (7,4) Hamming code with the following parity-check matrix:

$$\mathbf{H} = \begin{bmatrix} 1 & 1 & 1 & 0 & 1 & 0 & 0 \\ 0 & 1 & 1 & 1 & 0 & 1 & 0 \\ 0 & 0 & 1 & 1 & 1 & 0 & 1 \end{bmatrix}$$

find the probability of undetected errors.

In this case equation (4.3.11) becomes

$$\mathbf{F}(z_1,z_2,z_3) = [\mathbf{P}(0)+\mathbf{P}(1)z_1][\mathbf{P}(0)+\mathbf{P}(1)z_1z_2][\mathbf{P}(0)+\mathbf{P}(1)z_1z_2z_3]\times$$
$$[\mathbf{P}(0)+\mathbf{P}(1)z_2z_3][\mathbf{P}(0)+\mathbf{P}(1)z_1z_3][\mathbf{P}(0)+\mathbf{P}(1)z_2][\mathbf{P}(0)+\mathbf{P}(1)z_3]$$

Then by formula (4.3.12) we can find

$$\mathbf{P}_u = 2^{-3}\sum_{i,j,k=0}^{1} \mathbf{F}((-1)^i,(-1)^j,(-1)^k) - \mathbf{P}^7(0)$$

In the case of the BSC with independent errors and $\mathbf{P}(0) = p_0$ and $\mathbf{P}(1) = p_1$, these equations are simplified:

$$p_u = 2^{-3}[1+7(p_0-p_1)^4] - p_0^7 \qquad \blacksquare$$

Note that the MacWilliams formula[101] follows from the results of this section. Indeed, if we substitute $f(\boldsymbol{\zeta}^{\mathbf{v}})$ from equation (4.3.6) into (4.3.7), we obtain

$$p_u = q^{-r}\sum_{\mathbf{v}}\sum_{\mathbf{e}} Pr(\mathbf{e})\zeta^{(\mathbf{eH}')\mathbf{v}} - p_n(0) \tag{4.3.16}$$

This equation expresses the probability of undetected errors directly through the code parity-check matrix.

Comparing equations (4.3.16) and (4.3.4), we obtain the identity

$$\sum_{\mathbf{c}} Pr(\mathbf{cG}) = q^{-r}\sum_{\mathbf{v}}\sum_{\mathbf{e}} Pr(\mathbf{e})\zeta^{(\mathbf{eH}')\mathbf{v}} \tag{4.3.17}$$

which is the generalization of the MacWilliams identity. The latter is obtained if we assume that errors are independent

$$Pr(\mathbf{e}) = \prod_{k=1}^{n} Pr(e_k)$$

and have the same probability $Pr(e) = P/(q-1)$ for $e \neq 0$.[102]

4.3.2 Performance of Forward Error-Correcting Codes. When a block code is used for correcting and detecting errors, the basic probabilities that characterize the system's performance depend not only on the code-generator (or parity-check) matrix but also on the method of correcting errors.

Suppose that maximum-likelihood syndrome decoding is being used. Then for each syndrome $\mathbf{S} = \mathbf{eH}'$ we select the most probable error word $\mathbf{e}(\mathbf{S})$ and subtract it from the received word. In this case, the full power of the code is used to correct errors so that the probability of detecting errors is equal to zero. Sometimes not the full code power is used to correct errors. If a syndrome belongs to a subset C, errors are corrected by the maximum-likelihood rule; otherwise errors are just detected. The detected error patterns are either suppressed or flagged for subsequent application of some other error correction scheme.

The probability of correct decoding is the probability that channel error words are equal to $\mathbf{e}(\mathbf{S})$ when syndromes belong to C:

$$p_c = \sum_{\mathbf{S} \in C} Pr(\mathbf{e}(\mathbf{S})) \tag{4.3.18}$$

The probability of detected errors is equal to

$$p_d = \sum_{\mathbf{S} \in \bar{C}} Pr(\mathbf{S}) \tag{4.3.19}$$

where $\bar{C}$ is the complement of C.

The probability of detecting errors can be found using generating function (4.3.6). Indeed, using equation (2.2.29) we obtain the probability that a syndrome is equal to $\mathbf{S}$:

$$Pr(\mathbf{S}) = q^{-r} \sum_{\mathbf{v}} \zeta^{-\mathbf{S}\mathbf{v}} f(\zeta^{\mathbf{v}}) \tag{4.3.20}$$

The probability of detected errors is found by substituting these values into (4.3.19):

$$p_d = q^{-r} \sum_{\mathbf{S} \in \bar{C}} \sum_{\mathbf{v}} \zeta^{-\mathbf{S}\mathbf{v}} f(\zeta^{\mathbf{v}})$$

In calculating characteristics of FEC performance, we can always make use of the identity $p_c + p_u + p_d = 1$ to find one of the probabilities when the other two are known. In many practical cases it is simpler to find p_c and p_u than p_d because it is difficult to describe the set $\overline{C}$. The probability p_u is equal to the probability that an error syndrome $\mathbf{S} = \mathbf{eH}'$ belongs to C, but the error is not equal to $\mathbf{e}(\mathbf{S})$:

$$p_u = Pr(\mathbf{S} \in C) - p_c \tag{4.3.21}$$

Using equation (4.3.20), the previous formula can be rewritten as

$$p_u = q^{-r} \sum_{\mathbf{S} \in C} \sum_{\mathbf{v}} \zeta^{-\mathbf{S}\mathbf{v}} f(\zeta^{\mathbf{v}}) - p_c \tag{4.3.22}$$

Note that this equation is more general than (4.3.7), which is obtained when C consists of only one element $\mathbf{S} = 0$.

In the majority of cases, the error patterns that correspond to syndromes are selected according to the *minimum Hamming weight rule*. To describe it we define a block $\mathbf{a}$ of n symbols Hamming weight $|\mathbf{e}|$ as a total number of its nonzero symbols and define the Hamming distance between two vectors $\mathbf{a}$ and $\mathbf{b}$ as the Hamming weight of their difference: $d(\mathbf{a},\mathbf{b}) = |\mathbf{a}-\mathbf{b}|$. Now the minimum Hamming weight rule of decoding defines $\mathbf{e}(\mathbf{S})$ as the word with the smallest number of nonzero components whose syndrome is $\mathbf{S}$:

$$|\mathbf{e}(\mathbf{S})| = \min_{\mathbf{eH}' = \mathbf{S}} |\mathbf{e}|$$

If a code is able to correct up to t errors and minimum Hamming weight decoding is used, the set C consists of syndromes of all error sequences that contain up to t nonzero components.

The alternative decoding procedure consists of selecting the code word $\mathbf{c}$ whose distance from the received word $\mathbf{e}$ does not exceed t. This procedure is equivalent to the above-described syndrome decoding. Indeed, suppose that $\mathbf{c}$ is the code word whose distance from $\mathbf{e}$ does not exceed t: $|\mathbf{e}-\mathbf{c}| \le t$. Denote $\mathbf{e}_1 = \mathbf{e}-\mathbf{c}$. Since $|\mathbf{e}_1| \le t$ and $\mathbf{e}_1\mathbf{H}' \in C$, the syndrome decoder outputs the same vector $\mathbf{e}+\mathbf{e}_1 = \mathbf{c}$.

The converse statement is also true. If $\mathbf{eH}' \in C$, the syndrome decoder finds a vector $\mathbf{e}_1$ that has the same syndrome as $\mathbf{e}$ (i. e., $\mathbf{e}_1\mathbf{H}' = \mathbf{eH}'$) and weight $|\mathbf{e}_1| \le t$. Then the decoder outputs a code word $\mathbf{c} = \mathbf{e}-\mathbf{e}_1$ whose distance from $\mathbf{e}$ does not exceed t: $|\mathbf{e}-\mathbf{c}| = |\mathbf{e}_1| \le t$.

The alternative decoding rule can be interpreted geometrically. A set of points $\mathbf{x}$ that satisfy the equation $|\mathbf{x}-\mathbf{c}| \le r$ is called a sphere of radius r with the center in $\mathbf{c}$. Thus, we may say that a received message $\mathbf{e}$ is decoded as $\mathbf{c}$ if it falls inside the sphere whose center is defined by the code word and has radius t. The code is able to correct up to t errors if and only if these spheres do not have common points, meaning that the Hamming distance between their centers is at least $2t+1$. The minimum distance between code words of a code is called the *code minimum distance*. A code is able to correct up to t errors if and only if its minimum distance $d_{\min} \geq 2t+1$.

If a code corrects up to t errors and does not correct errors of higher multiplicity, the probability p_c of correct decoding is equal to the probability that the number of errors in a word is not greater than t. For the SSM error source model we have

$$p_c = \mathbf{p}\mathbf{P}_c\mathbf{1} \qquad \mathbf{P}_c = \mathbf{P}_n(\leq t) \tag{4.3.23}$$

where $\mathbf{P}_n(\leq t)$ is the matrix probability that the number of errors does not exceed t and is defined in Sec 2.2.

The probability of incorrect decoding may be found using equation (4.3.22) and generating function (4.3.9):

$$p_u = \mathbf{p}\mathbf{P}_u\mathbf{1} \tag{4.3.24}$$

where

$$\mathbf{P}_u = q^{-r} \sum_{\mathbf{S}\in C} \sum_{\mathbf{v}} \zeta^{-\mathbf{S}\mathbf{v}} \mathbf{F}(\zeta^{\mathbf{v}}) - \mathbf{P}_n(\leq t) \tag{4.3.25}$$

Since syndromes may be expressed as $\mathbf{S} = \mathbf{e}\mathbf{H}'$, we can replace the index of summation $\mathbf{S}$ by $\mathbf{e}$:

$$\mathbf{P}_u = q^{-r} \sum_{|\mathbf{e}|\leq t} \sum_{\mathbf{v}} \zeta^{-(\mathbf{e}\mathbf{H}')\mathbf{v}} \mathbf{F}(\zeta^{\mathbf{v}}) - \mathbf{P}_n(\leq t) \tag{4.3.26}$$

In the case of binary codes these equations become

$$\mathbf{P}_u = 2^{-r} \sum_{\mathbf{S}\in C} \sum_{\mathbf{v}} (-1)^{-\mathbf{S}\mathbf{v}} \prod_{k=1}^{n} [\mathbf{P}(0) + \mathbf{P}(1)(-1)^{\mathbf{v}\mathbf{h}_k}] - \mathbf{P}_n(\leq t) \tag{4.3.27}$$

and

$$\mathbf{P}_u = 2^{-r} \sum_{|\mathbf{e}|\leq t} \sum_{\mathbf{v}} (-1)^{-(\mathbf{e}\mathbf{H}')\mathbf{v}} \prod_{k=1}^{n} [\mathbf{P}(0) + \mathbf{P}(1)(-1)^{\mathbf{v}\mathbf{h}_k}] - \mathbf{P}_n(\leq t) \tag{4.3.28}$$

EXAMPLE 4.3.3: Suppose that the binary (7,4) Hamming code with the parity-check matrix

$$\mathbf{H} = \begin{bmatrix} 1 & 1 & 1 & 0 & 1 & 0 & 0 \\ 0 & 1 & 1 & 1 & 0 & 1 & 0 \\ 0 & 0 & 1 & 1 & 1 & 0 & 1 \end{bmatrix}$$

is used for correcting errors. For minimum Hamming weight decoding, let us find the probability of undetected errors.

First we define the decoding rule $\mathbf{e}(\mathbf{S})$ for error words with the smallest Hamming weight. The syndrome for the zero-weight error $\mathbf{e}_0 = (0\ 0\ 0\ 0\ 0\ 0\ 0\ 0\ 0)$ is the zero syndrome $\mathbf{e}_0\mathbf{H}' = \mathbf{S}_0 = (0\ 0\ 0)$. Thus, we decide that no errors occurred, if the received message syndrome equals $\mathbf{S}_0$. Let us now compute syndromes for the single bit errors. The syndrome for $\mathbf{e}_1 = (1\ 0\ 0\ 0\ 0\ 0\ 0\ 0\ 0)$ is $\mathbf{e}_1\mathbf{H}' = \mathbf{S}_1 = (1\ 0\ 0)$, which coincides with the transposed first column of $\mathbf{H}$. Quite analogously we find that the syndrome for $\mathbf{e}_2 = (0\ 1\ 0\ 0\ 0\ 0\ 0\ 0\ 0)$ is $\mathbf{S}_2 = (1\ 1\ 0)$, which is the transposed second column of $\mathbf{H}$, and so on. Since all the columns of matrix $\mathbf{H}$ are different, we obtain all possible syndromes $\mathbf{S}_0, \mathbf{S}_1, ..., \mathbf{S}_7$, because there are only eight different combinations of three bits. This means that the set C consists of all possible sequences of three bits. Thus, the minimum Hamming weight decoding rule can be expressed as $\mathbf{e}(\mathbf{S}_i) = \mathbf{e}_i$. In other words, if a received message syndrome is equal to the i-th column of the matrix $\mathbf{H}$, then the i-th bit of the received block must be corrected (see also Example 4.2.3).

The probability of correct decoding, according to (4.3.18), is the probability that the number of errors in the block is not greater than 1. For the SSM model this probability is equal to $p_c = \mathbf{pP}^7(0)\mathbf{1}+\mathbf{pP}_7(1)\mathbf{1}$ where matrix probabilities $\mathbf{P}_n(t)$ of t errors in a block of n symbols are defined in Sec 2.3. The probability of incorrect decoding is equal to $p_u = 1-p_c$, since $p_d = 0$ in this case.

■

4.3.3 Symmetrically Dependent Errors. The exact computations by formulas developed in the previous section are, in general, quite complex, so that we need to find their approximations. For short codes a good approximation can be found by assuming the symmetric dependence of errors,[103] meaning that the probabilities of different permutations of the same combination of errors are equal:

$$Pr(e_1, e_2, ..., e_n) = Pr(e_1^{(1)} e_2^{(1)}, ..., e_n^{(1)}) \tag{4.3.29}$$

where $e_i, e_i^{(1)} \in \mathrm{GF}(q)$ and $(e_1, e_2, ..., e_n)$ is a permutation of $(e_1^{(1)}, e_2^{(1)}, ..., e_n^{(1)})$. This assumption agrees reasonably well with experimental data when n and t are small and in effect is very close to the assumption that error source is memoryless.[103] According to this assumption we need not be concerned with the order of matrices in matrix products, and a sum of products of probabilities (4.3.29) can be replaced by the number of addends times the probability. The matrix probability $\mathbf{P}_c$ in (4.3.23) is equivalent to

$$\mathbf{P}_c = \sum_{i=0}^{t} \binom{n}{i} \mathbf{P}_0^{n-i} \mathbf{Q}_0^i \tag{4.3.30}$$

where $\mathbf{Q}_0 = \mathbf{P} - \mathbf{P}_0$.

The matrix probability of incorrect decoding is given by the equation

$$\mathbf{P}_u = \mathbf{P}(\mathbf{S} \in C) - \mathbf{P}_c \tag{4.3.31}$$

where $\mathbf{P}(\mathbf{S} \in C)$ is the matrix probability that error syndrome belongs to C. This probability is equal to the sum of probabilities that an error vector $\mathbf{e}$ falls into a radius-t sphere with the center at a code vector $\mathbf{c}$:

$$\mathbf{P}(\mathbf{S} \in C) = \sum_{\mathbf{c}} \sum_{|\mathbf{e}-\mathbf{c}| \leq t} \mathbf{P}(\mathbf{e}) \tag{4.3.32}$$

This sum can be rearranged into sums according to Hamming weights:

$$\mathbf{P}(\mathbf{S} \in C) = \sum_{w=0}^{n} \sum_{\mathbf{c}_w} \sum_{i=0}^{t} \sum_{j=0}^{n} \sum_{|\mathbf{e}_j - \mathbf{c}_w| = i} \mathbf{P}(\mathbf{e}_j) \tag{4.3.33}$$

where $\mathbf{c}_w$ denotes a weight w code word and $\mathbf{e}_j$ denotes a weight j error vector. After changing the order of summation this equation can be rewritten in the following form:

$$\mathbf{P}(\mathbf{S} \in C) = \sum_{i=0}^{t} \sum_{j=0}^{n} A_{ij} \mathbf{P}_1^j \mathbf{P}_0^{n-j} \tag{4.3.34}$$

where $\mathbf{P}_1 = (q-1)^{-1} \mathbf{Q}_0$ is the average symbol-error probability, and A_{ij} is the number of vectors of Hamming weight j whose distance from some code word is equal to i. This number is a coefficient of $x^i y^j$ in the generating function expansion

$$\sum_{ij} A_{ij} x^i y^j = A(x+y+(1-\gamma)xy,\ 1+\gamma xy) \tag{4.3.35}$$

where $\gamma = q-1$

$$A(x,y) = \sum_{i=0}^{n} A_i x^i y^{n-i} \tag{4.3.36}$$

and A_i is the number of code words whose Hamming weight is equal to i; $A_i = 0$ for $i = 1, 2, \ldots, 2t$.[75,101] The function

$$A(z,1) = A(z) = \sum_{i=0}^{n} A_i z^i \tag{4.3.37}$$

is called the *code-weight generator* polynomial. Obviously, $A(x,y) = y^n A(x/y)$.

It is easy to express probability (4.3.34) in terms of these generating functions. Indeed, from (4.3.35) it follows that

$$\sum_{ij} A_{ij} \mathbf{P}_1^j \mathbf{P}_0^{n-j} x^i y^j = A(x\mathbf{P}_0 + y\mathbf{P}_1 + (1-\gamma)xy\mathbf{P}_1,\ \mathbf{P}_0 + \gamma xy\mathbf{P}_1) \tag{4.3.38}$$

Comparing this equation with (4.3.34), we see that $\mathbf{P}(\mathbf{S} \in C)$ is equal to the sum of the coefficients of x^i for $i=0,1,\ldots,t$ in the expansion of function (4.3.38) when $y=1$:

$$A(x\mathbf{P}_0 + \mathbf{P}_1 + (1-\gamma)x\mathbf{P}_1,\ \mathbf{P}_0 + \gamma x\mathbf{P}_1) \tag{4.3.39}$$

But the probability in (4.3.30) can also be found by summing the coefficients of x^i for $i=0,1,\ldots,t$ in the expansion of the binomial

$$(\mathbf{P}_0 + \gamma x\mathbf{P}_1)^n$$

Thus, the probability of incorrect decoding equals the sum of the coefficients of x^i for $i=0,1,\ldots,t$ in the expansion of the function

$$\mathbf{V}(x) = A(x\mathbf{P}_0 + \mathbf{P}_1 + (1-\gamma)x\mathbf{P}_1,\ \mathbf{P}_0 + \gamma x\mathbf{P}_1) - (\mathbf{P}_0 + \gamma x\mathbf{P}_1)^n \tag{4.3.40}$$

This equation can be simplified if we use the generator polynomial

$$W(x,y) = y^n [A(x/y) - 1] \tag{4.3.41}$$

of all nonzero code weights. It is easy to verify that

$$\mathbf{V}(x) = W(x\mathbf{P}_0 + \mathbf{P}_1 + (1-\gamma)x\mathbf{P}_1,\ \mathbf{P}_0 + \gamma x\mathbf{P}_1) \tag{4.3.42}$$

Using Taylor's formula for the coefficients of a power series, we can express the probability of incorrect decoding as

$$\mathbf{P}_u = \sum_{i=0}^{t} \frac{1}{i!} \frac{d^i}{dx^i} \mathbf{V}(x) \Big|_{x=0} \tag{4.3.43}$$

This probability can also be found as a coefficient of x^t in the expansion of the generating functions of sums:

$$\mathbf{P}_u = \frac{1}{t!}\frac{d^t}{dx^t}\mathbf{R}(x)\mid_{x=0} \tag{4.3.44}$$

$$\mathbf{R}(x) = W(x\mathbf{P}_0+\mathbf{P}_1+(1-\gamma)x\mathbf{P}_1,\ \mathbf{P}_0+\gamma x\mathbf{P}_1)(1-x)^{-1} \tag{4.3.45}$$

After performing the expansion we obtain

$$\mathbf{P}_u = \sum_{i=1}^{n} A_i \sum_{k=0}^{t} \sum_{h=h_1}^{h_2} \binom{i}{k-h}\binom{n-i}{h}\gamma^h \mathbf{P}_0^{n-i-h}\mathbf{P}_1^{2h+i-k}(\mathbf{P}_0+(\gamma-1)\mathbf{P}_1)^{k-h} \tag{4.3.46}$$

where $h_1 = \max\{0,k-i\}$ and $h_2 = \min\{n-i,k\}$. In the case of binary codes $\gamma = 1$, the previous expression becomes

$$\mathbf{P}_u = \sum_{i=1}^{n} A_i \sum_{k=0}^{t} \sum_{h=h_1}^{h_2} \binom{i}{k-h}\binom{n-i}{h}\mathbf{P}_0^{n-i+k-2h}\mathbf{P}_1^{2h+i-k} \tag{4.3.47}$$

We would like to emphasize once more that the above approximation is based on the assumption that errors are symmetrically dependent which allowed us to change the order in matrix products. In the particular case of the channel with independent errors, the matrices are replaced with the probabilities, and the above results coincide with well-known equations. [75,101]

EXAMPLE 4.3.4: Let us illustrate the calculations for the CRC-16 code with the generator polynomial

$$x^{16} + x^{15} + x^2 + 1$$

assuming that channel errors are independent and occur with the probability p_1.

Since $x^{16} + x^{15} + x^2 + 1 = (x+1)(x^{15}+x+1)$ and $x^{15}+x+1$ is irreducible, this code is an extended $(n,n-16)$ Hamming code (see Ref. [88] p. 220) with the block length $n=2^{15}-1$. Its code-weight generator polynomial (see problem 5. of Ref. [88], p. 142) is equal to

$$A(x) = \frac{1}{2(n+1)}[(1+x)^n+(1-x)^n+2n(1-x^2)^{(n-1)/2}]$$

Therefore,

$$W(x,y) = \frac{1}{2(n+1)}[(y+x)^n+(y-x)^n+2n(y^2-x^2)^{(n-1)/2}] - y^n$$

Generating function (4.3.45) then takes the form

$$R(x) = W(xp_0+p_1, p_0+xp_1)(1-x)^{-1}$$

where $p_0 = 1-p_1$.

If the code is used only for detecting errors, then the probability of correct decoding is equal to p_0^n. The probability of undetected errors is equal to the coefficient of x^0 in the generating function expansion

$$p_u = R(0) = \frac{1}{2(n+1)}[1+g^n+2ng^{(n-1)/2}] - p_0^n$$

where $g = 1-2p_1$.

Suppose now that the code is being used to correct single errors. In this case the probability of correct decoding is the probability of receiving a block with no more than one error:

$$p_c = p_0^n + np_1p_0^{n-1}$$

The probability of incorrect decoding is equal to the coefficient of x in the expansion of the generating function $R(x)$:

$$p_u = \frac{d}{dx}R(x)\,|_{\,x=0} = R(0)+(1-g^n)n/2(n+1) - np_1p_0^{n-1}$$ ■

EXAMPLE 4.3.5: Consider now an (n,k) Reed-Solomon code whose symbols belong to GF(q) where $q=n+1$. Suppose that symbol-errors are independent and the probability of correct symbol reception is equal to P_0.

Let us evaluate the code's performance characteristics when it is used for error detection only. In this case the probability of correct decoding is equal to $p_c = P_0^n$. The code Hamming weight generator polynomial (see Appendix 4.2) is equal to

$$A(x) = q^{-r}(1+nx)^n+\sum_{i=0}^{r-1}\binom{n}{i}x^i(1-x)^{n-i}(1-q^{i-r})$$

so that

$$W(x,y) = q^{-r}(y+nx)^n + \sum_{i=0}^{r-1} \binom{n}{i} x^i (y-x)^{n-i}(1-q^{i-r}) - y^n$$

where $r=n-k$. Generating function (4.3.42) is given by

$$W(P_1+xQ_1, P_0+nxP_1) = q^{-r}(1+nx)^n - (P_0 + nxP_1)^n +$$
$$\sum_{i=0}^{r-1} \binom{n}{i} (P_1+xQ_1)^i (1-x)^{n-i} (P_0-P_1)^{n-i}(1-q^{i-r})$$

where $Q_1 = 1-P_1$ and generating function (4.3.45) takes the form

$$R(x) = q^{-r}(1+\gamma x)^n (1-x)^{-1} - (P_0+nxP_1)^n (1-x)^{-1} +$$
$$\sum_{i=0}^{r-1} \binom{n}{i} (P_1+xQ_1)^i (1-x)^{n-i-1} (P_0-P_1)^{n-i}(1-q^{i-r})$$

The probability of undetected errors is equal to

$$p_u = R(0) = q^{-r} - P_0^n + \sum_{i=0}^{r-1} \binom{n}{i} P_1^i (P_0-P_1)^{n-i}(1-q^{i-r})$$

If the code redundancy r is an even number, then this code is able to correct up to $r/2$ symbol errors (see Sec 4.2.5). Suppose that the code is used to correct up to $t \le r/2$ errors. Then the probability of correct decoding is given by (4.3.30):

$$p_c = \sum_{i=0}^{t} \binom{n}{i} P_0^{n-i} P_1^i n^i$$

The probability of incorrect decoding is the coefficient of x^t in the expansion of $R(x)$:

$$p_u = q^{-r} \sum_{i=0}^{t} \binom{n}{i} \gamma^i - \sum_{i=0}^{t} P_0^{n-i} P_1^i n^i +$$
$$\sum_{i=0}^{r-1} \binom{n}{i} (P_0-P_1)^{n-i}(1-q^{i-r}) \sum_{j=j_1}^{j_2} \binom{i}{j} \binom{n-i-1}{t-j} (-1)^{t-j} P_1^{i-j} Q_1^j$$

where $j_1 = \max\{0, t+i-n\}$ and $j_2 = \min\{i,t\}$. ■

4.3.4 Upper and Lower Bounds. In many applications of probability theory it is often easier to find bounds on some probability than to find the probability itself. This observation is especially true for the probability of incorrect decoding, even in the case of independent errors, since the code-weight generator polynomial is known for only a small number of codes.

Upper bounds are often used improperly to compare the performance of different systems. If a performance parameter upper bound of a system A is less than that of a system B, it is not necessarily true for the real values of the parameters. To prove a system's superiority one must compare the upper bound of the parameter of one system with the lower bound of the parameter of the other system. However, if the upper bounds are "tight" they may be used for comparison, but in this case they are used not as upper bounds but rather as good approximations of real parameters (since tightness means that they are close to the bounds).

4.3.4.1 Superset Bound. A simple technique of obtaining the upper bound is based on the inequality

$$Pr(A) \leq Pr(B) \quad \text{if } A \subset B$$

For example, if a code can correct up to t errors and is used for error correction and detection, then the probability of incorrect decoding is upper bounded by the probability that the number of errors is greater than t.

$$p_u \leq \mathbf{p}\mathbf{P}_n(>t)\mathbf{1} \tag{4.3.48}$$

This upper bound may be exact in some cases.

4.3.4.2 Union Bound. Another commonly used method of finding bounds is based on the fact that the probability of the union of events does not exceed the sum of the probabilities of the events

$$Pr(\bigcup_i A_i) \leq \sum_i Pr(A_i)$$

The right-hand side of this inequality is called the *union bound* of its left-hand side.

By using this method we can obtain a more accurate upper bound than in (4.3.48). Let us consider the case when the maximum-likelihood decoding is used for correcting errors: for each received word the decoder outputs the closest code word and, in the case of a tie, each code word is selected with the same probability by some random-number generator. If we assume that the all-zero code word has been transmitted, then the received block $\mathbf{e}$ is decoded incorrectly if its Hamming distance from the transmitted word is larger than from some other code word $|\mathbf{c}-\mathbf{e}| < |\mathbf{e}|$ or $|\mathbf{c}-\mathbf{e}| = |\mathbf{e}|$, but the random-number generator selected $\mathbf{c}$

rather than **e**. The sum of the probabilities of these inequalities gives us the (union) upper bound of the probability of incorrect decoding:

$$p_u \leq \mathbf{pU1}$$

$$\mathbf{U} = \sum_{\mathbf{c}\neq 0}\sum_{\mathbf{e}}\mathbf{P}(\,|\,\mathbf{c}-\mathbf{e}\,|<|\,\mathbf{e}\,|\,) + 0.5\,\mathbf{P}(\,|\,\mathbf{c}-\mathbf{e}\,| = |\,\mathbf{e}\,|\,)$$

For the SSM model we can calculate this upper bound by using generating functions. Let $\mathbf{c} = (c_1, c_2, ..., c_n)$ be a code word and $\mathbf{e} = (e_1, e_2, ..., e_n)$ denote an error sequence. Then the Hamming distance between the code word and the received word can be expressed as

$$|\,\mathbf{c}-\mathbf{e}\,| = \sum_{i=1}^{n}\Delta(c_i, e_i)$$

and the distance between the received word and the transmitted word

$$|\,\mathbf{e}\,| = \sum_{i=1}^{n}\Delta(0, e_i)$$

where $\Delta(x,y) = 0$ if $x=y$ and otherwise $\Delta(x,y) = 1$. The generating function of the distance difference distribution is given by

$$\mathbf{\Phi}(z) = \sum_{\mathbf{c}\neq 0}\sum_{\mathbf{e}}\mathbf{P}(\mathbf{e})z^{|\,\mathbf{e}\,|-|\,\mathbf{c}-\mathbf{e}\,|}$$

If we denote $\delta(c_i, e_i) = \Delta(0, e_i) - \Delta(c_i, e_i)$, then this generating function may be expressed as

$$\mathbf{\Phi}(z) = \sum_{\mathbf{c}}\prod_{i=1}^{n}\mathbf{\Psi}(c_i, z) \tag{4.3.49}$$

where

$$\mathbf{\Psi}(c_i, z) = \sum_{e_i}\mathbf{P}(e_i)z^{\delta(c_i, e_i)}$$

Since $\delta(c_i, e_i) = 1$ when $c_i = e_i \neq 0$; $\delta(c_i, e_i) = -1$ when $c_i \neq 0, e_i = 0$, and $\delta(c_i, e_i) = 0$ in all other cases, we have

$$\mathbf{\Psi}(c_i;z) = \begin{cases} \mathbf{P}(c_i)z + \mathbf{P}(0)z^{-1} + \mathbf{P}-\mathbf{P}(c_i)-\mathbf{P}(0) & \text{when } c_i \neq 0 \\ \mathbf{P} & \text{otherwise} \end{cases} \tag{4.3.50}$$

The matrix probability **U** is equal to the sum of the coefficients of the positive power terms and one-half the probability of the zero power term in the expansion of the generating function $\mathbf{\Phi}(z)$. This sum can be expressed using contour integral representation of the Laurent series [46] coefficients:

$$\mathbf{P}_u \leq \frac{1}{4\pi j} \oint_{\gamma} \frac{z+1}{z(z-1)} \mathbf{\Phi}(z) dz$$

This upper bound does not look simpler than exact formula (4.3.25). However, it becomes much simpler when channel errors are symmetrically dependent, and in particular for channels without memory. Indeed, generating function (4.3.50) in this case takes the form

$$\mathbf{\Psi}(c_i,z) = \mathbf{P}_1 z + (q-2)\mathbf{P}_1 + \mathbf{P}_0 z^{-1}$$

for $c_i \neq 0$. Generating function (4.3.49) is equal to

$$\mathbf{\Phi}(z) = \sum_{i=1}^{n} A_i [\mathbf{P}_1 z + (q-2)\mathbf{P}_1 + \mathbf{P}_0 z^{-1}]^i \mathbf{P}^{n-i}$$

where A_i is the number of the code words whose Hamming weight is equal to i. Since $\mathbf{P}^{n-i}\mathbf{1} = \mathbf{1}$, this generating function is equivalent to

$$\mathbf{\Phi}(z) = \sum_{i=1}^{n} A_i [\mathbf{P}_1 z + (q-2)\mathbf{P}_1 + \mathbf{P}_0 z^{-1}]^i = W(\mathbf{P}_1 z + (q-2)\mathbf{P}_1 + \mathbf{P}_0 z^{-1}, 1)$$

In the case of a BSC without memory it takes the form

$$\Phi(z) = \sum_{i=1}^{n} A_i (P_1 z + P_0 z^{-1})^i = W(P_1 z + P_0 z^{-1}, 1) \tag{4.3.51}$$

Collecting all the positive power terms and one-half of the zero power term we obtain

$$p_u \le \sum_{i=1}^{n} A_i U_i$$

$$U_i = \begin{cases} \displaystyle\sum_{j=(i+1)/2}^{i} \binom{i}{j} P_1^j P_0^{i-j} & \text{when } i \text{ is odd} \\ \displaystyle 0.5\binom{i}{i/2}(P_0 P_1)^{i/2} + \sum_{j=i/2+1}^{i} \binom{i}{j} P_1^j P_0^{i-j} & \text{when } i \text{ is even} \end{cases}$$

In conclusion, we would like to note that the zero power term in the generating function expansion is needed only when the random-number generator is used to resolve a tie. If the code is being used to correct up to t errors, then we can include into the union bound only the terms whose power is greater than $|\mathbf{c}| - 2t$.

4.3.4.3 Lower Bound. If we know a sum $S = X + Y$ of two non-negative variables and a upper bound of one of the addends ($X \le U$), then we can obtain a lower bound of the other ($Y \ge S - U$). Because of the sum's symmetry, we can also find a upper bound on one of the addends if we know a lower bound on the other.

This primitive technique is often applied to the identity

$$Pr(A) = Pr(A \cup B) - Pr(B) \quad \text{where } A \cap B = \varnothing$$

If the probability $Pr(A \cup B)$ is known and the probability $Pr(B)$ is upper bounded by some value U, then the lower bound is given by

$$Pr(A) \ge Pr(A \cup B) - U$$

In particular, if A is a complement of B, $Pr(A \cup B) = 1$, and the previous bound takes the form

$$Pr(A) \ge 1 - U$$

To illustrate this technique, suppose that we know a distribution-generating function

$$f(z) = \sum_{i=0}^{\infty} p_i z^{-i}$$

and we want to find a lower bound on the sum

$$p_t = \sum_{i=0}^{t} p_i .$$

This sum can be expressed through the distribution tail:

$$p_t = 1 - \sum_{i=t+1}^{\infty} p_i$$

Using the Chernoff inequality (see Appendix 2.2) for the distribution tail

$$\sum_{i=t+1}^{n} p_i \le z^{t+1} f(z)$$

where $z \ge 1$, we obtain the lower bound:

$$p_t \ge 1 - z^{t+1} f(z)$$

The best bound of this form is obtained if z delivers the global minimum to the function $z^{t+1} f(z)$.

4.3.5 Postdecoding Error Probability. In many cases it is necessary to evaluate the message-processing algorithm after decoding is done. The processing may include additional decoding, descrambling, decryption, and so on. It is therefore important to be able to describe the symbol sequence on the decoder output. Normally, only the information part of a message appears on the decoder output, so that errors in the parity portion of a message can be disregarded. However, in some cases the whole message is passed over to a processor that follows the decoder. Generalized concatenated codes represent an example of such processing.[104] We will consider only this case in the future.

The simplest characteristic of the sequence on the decoder output is the symbol-error probability p_{se}:

$$p_{se} = P \lim_{T \to \infty} \frac{N_{se}}{T}$$

where N_{se} is the total number of errors and T is the total number of symbols on the decoder output. If we assume that the all-zero code word has been transmitted and the received word syndrome $\mathbf{eH}' \in C$, then the decoder outputs the closest to $\mathbf{e}$ code word $\mathbf{c}$. The symbol-error probability on the output of a decoder of a block code used only for correcting errors is

$$p_{se} = q_c + q_e \tag{4.3.52}$$

where

$$q_c = \sum_{\mathbf{eH}' \in C} \frac{|\,\mathbf{c}(\mathbf{e})\,|}{n} Pr(\mathbf{e}) \quad q_e = \sum_{\mathbf{eH}' \notin C} \frac{|\,\mathbf{e}\,|}{n} Pr(\mathbf{e})$$

In this equation $\mathbf{c}(\mathbf{e})$ denotes the code word whose distance from the received word $\mathbf{e}$ does not exceed t.

If errors are symmetrically dependent, the above sums can be rearranged according to vectors weights as in equation (4.3.34):

$$q_c = \mathbf{p}\mathbf{Q}_c\mathbf{1}$$

$$\mathbf{Q}_c = \sum_{w=0}^{n} \sum_{\mathbf{c}_w} \sum_{i=0}^{t} \sum_{j=0}^{n} \sum_{|\,\mathbf{e}_j - \mathbf{c}_w\,| = i} \frac{w}{n} \mathbf{P}(\mathbf{e}_j)$$

After changing the order of summation we obtain

$$\mathbf{Q}_c = \sum_{i=0}^{t} \sum_{j=0}^{n} B_{ij} \mathbf{P}_1^j \mathbf{P}_0^{n-j}$$

where B_{ij} has the generating function

$$\sum_{ij} B_{ij} x^i y^j = B(x+y+(1-\gamma)xy,\ 1+\gamma xy)$$

and

$$B(x,y) = \sum_{i=0}^{n} \frac{i}{n} A_i x^i y^{n-i} = \frac{x}{n} \frac{\partial}{\partial x} A(x,y) \tag{4.3.53}$$

$\mathbf{Q}_c$ is equal to the sum of coefficients of x^i for $i \le t$ in the power series expansion of the function

$$B(x\mathbf{P}_0 + \mathbf{P}_1 + (1-\gamma)x\mathbf{P}_1,\ \mathbf{P}_0 + \gamma x \mathbf{P}_1) \tag{4.3.54}$$

It is equal to the coefficient of x^t in the expansion of the function

$$B(x\mathbf{P}_0 + \mathbf{P}_1 + (1-\gamma)x\mathbf{P}_1,\ \mathbf{P}_0 + \gamma x \mathbf{P}_1)(1-x)^{-1}$$

After performing the expansion, we obtain

$$\mathbf{Q}_c = \sum_{i=1}^{n} iA_i \sum_{k=0}^{t} \sum_{h=h_1}^{h_2} \binom{i}{k-h}\binom{n-i}{h} \gamma^h \mathbf{P}_0^{n-i-h} \mathbf{P}_1^{2h+i-k} [\,\mathbf{P}_0+(\gamma-1)\mathbf{P}_1\,]^{k-h}$$

where $h_1 = \max\{0, k-i\}$ and $h_2 = \min\{n-i, k\}$.

The second term of equation (4.3.52) can be found similarly if we make use of the identity

$$\sum_{\mathbf{eH}' \notin C} |\,\mathbf{e}\,|\, Pr(\mathbf{e}) = \sum_{\mathbf{e}} |\,\mathbf{e}\,|\, Pr(\mathbf{e}) - \sum_{\mathbf{eH}' \in C} |\,\mathbf{e}\,|\, Pr(\mathbf{e})$$

This can be rewritten as

$$q_e = \gamma \mathbf{p} \mathbf{P}_1 \mathbf{1} - \mathbf{p} \mathbf{Q}_e \mathbf{1}$$

$$\mathbf{Q}_e = \sum_{\mathbf{eH}' \in C} \frac{|\,\mathbf{e}\,|}{n} \mathbf{P}(\mathbf{e}) = \sum_{w=0}^{n} \sum_{\mathbf{c}_w} \sum_{i=0}^{t} \sum_{j=0}^{n} \sum_{|\,\mathbf{e}_j - \mathbf{c}_w\,| = i} \frac{j}{n} \mathbf{P}(\mathbf{e}_j)$$

The sum on the right-hand side of this equation can be found by using the generating function

$$C[\,x\mathbf{P}_0+\mathbf{P}_1+(1-\gamma)x\mathbf{P}_1,\ \mathbf{P}_0+\gamma x\mathbf{P}_1\,](1-x)^{-1}$$

where

$$C(x,y) = \frac{y}{n} \frac{\partial}{\partial y} A(x,y)$$

However, in the majority of cases $\mathbf{Q}_e$ is small compared to $\mathbf{Q}_c$, and in the particular case when all the syndromes are used for correcting errors $\mathbf{Q}_e = 0$.

EXAMPLE 4.3.6: Let us calculate the postdecoder error probability for the (7,4) Hamming code (see Example 4.3.3) in the case of BSC with independent errors.

$$A(x,y) = y^7+7x^3y^4+7x^4y^3+x^7$$

This generating function can be determined directly by listing all 16 code words, or using the general equation (see Ref. [88] p. 142). Equation (4.3.53) gives

$$B(x,y) = 3x^3y^4+4x^4y^3+x^7$$

After substituting $\mathbf{P}_0 = p_0$, $\mathbf{P}_1 = p_1$, and $\gamma = 1$ into (4.3.54), we obtain

$$B(p_1+p_0x, p_0+p_1x) = 3(p_1+p_0x)^3(p_0+p_1x)^4+4(p_1+p_0x)^4(p_0+p_1x)^3+(p_1+p_0x)^7$$

Finally, the probability of postdecoding error is found by summing the coefficients of x^0 and x^1:

$$p_{se} = 9p_1^2p_0^5 + 19p_1^3p_0^4 + 16p_1^4p_0^3 + 12p_1^5p_0^2 + 7p_1^6p_0 + p_1^7$$

It is easy to prove that, for the general $(n,n-r)$ Hamming code,

$$p_{se} = \frac{1}{n+1}[1+\frac{n-1}{2}g^{(n+1)/2}-\frac{n+1}{2}g^{(n-1)/2}] + p_1\frac{(n-1)}{n+1}[1-g^{(n+1)/2}]$$

where $g = 1 - 2p_1$. ■

4.3.6 Soft Decision Decoding. So far we have considered the performance of systems in which the transmitted and received symbols belong to the same alphabet GF(q). This happens if the so-called *hard decision demodulator* described in Sec 1.1.6 is used. In the case of a binary channel, the decision circuit in the demodulator outputs bit 1 if at the sampling instant the received signal voltage is positive and outputs 0 otherwise. However, a more sophisticated decision circuit can output more than two alternative values that depend on the received signal voltage. The received analog signal is quantized and a decoder receives non-binary symbols. This type of demodulation is called a *soft decision demodulation*. Soft decision demodulator provides more information about the received signal than hard decision demodulator. The decoder can use this information to improve system performance.

Let $\mathbf{s}=(s_1,s_2,...,s_n)$ be a transmitted block and $\boldsymbol{\theta}(\mathbf{s})=(\theta_1,\theta_2,...,\theta_n)$ be the received block of n symbols, assuming that there were no errors. Suppose that the transmitted symbols belong to GF(q) and are encoded by an (n,k) code. The received symbols, in general, do not belong to the field of transmitted symbols. We assume also that all the messages are equally probable and that channel errors $\mathbf{e}$ are additive, so that

$$Pr(\boldsymbol{\theta}(\mathbf{s})+\mathbf{e}\,|\,\mathbf{s}) = Pr(\mathbf{e})$$

The maximum-likelihood decoder outputs the code word $\mathbf{s}^*$ for the received message $\mathbf{r}^*$ if the received message has the largest probability:

$$Pr(\mathbf{r}^* \mid \mathbf{s}^*) = \max_{\mathbf{s}} Pr(\mathbf{r}^* \mid \mathbf{s})$$

If this equation has several solutions, a unique solution is selected from the set of solutions with equal probabilities by using a random number generator. For each code word we define the decision region as the set $\Theta(\mathbf{s})$ of all $\mathbf{r}^*$ that satisfy the previous equation. The decoder outputs $\mathbf{s}$ when the received block $\boldsymbol{\theta} \in \Theta(\mathbf{s})$.

Let us evaluate the performance of a soft decision decoder. Since errors do not depend on the transmitted symbols, we can assume that the all-zero code word $\mathbf{0}$ has been transmitted. The receiver decodes it correctly if the soft decision demodulator outputs $\mathbf{e} \in \Theta(\mathbf{0})$, so that

$$p_c = Pr(\mathbf{e} \in \Theta(\mathbf{0}))$$

The probability of incorrect decoding is equal to

$$p_u = \sum_{\mathbf{s} \neq \mathbf{0}} Pr(\mathbf{e} \in \Theta(\mathbf{s}))$$

The probability of detected errors is equal to zero in this case, since every received block $\boldsymbol{\theta}$ is decoded into a code word $\mathbf{s}$. If we want to correct only the most probable errors, then the decision region can be narrowed by eliminating points $\mathbf{r}^*$ for which the probability $Pr(\mathbf{r}^* \mid \mathbf{s}^*)$ is small. In this case the probability of detected errors $p_d = 1 - p_c - p_u$.

In many cases when it is difficult to calculate $Pr(\mathbf{e} \in \Theta(\mathbf{s}))$, the following union bound may be handy:

$$Pr(\mathbf{e} \in \Theta(\mathbf{s})) \leq \sum_{U(\mathbf{s})} Pr(\mathbf{e}) + 0.5 \sum_{T(\mathbf{s})} Pr(\mathbf{e})$$

where $U(\mathbf{s}) = \{\mathbf{e}\text{: } Pr(\mathbf{e} - \boldsymbol{\theta}(\mathbf{s})) > Pr(\mathbf{e})\}$, $T(\mathbf{s}) = \{\mathbf{e}\text{: } Pr(\mathbf{e} - \boldsymbol{\theta}(\mathbf{s})) = Pr(\mathbf{e})\}$. In order to calculate these probabilities we need to have an error source model.

EXAMPLE 4.3.7: Consider the satellite channel described in Sec 1.1.6. For the sake of simplicity we assume that the channel is linear, ideally equalized, and that there is no differential encoder. In this case the signal on the input to the data detector is given by equation (1.1.22). Let us assume now that the data detector makes a "soft" decision about the received signal: it does not change the signal. This assumption is valid when the number of quantization levels is large.

Binary information is encoded by the (n,k) code, demultiplexed into the inphase and quadrature sequences, and transmitted over the channel. The receiver multiplexes the samples $\mathrm{Re}\, u_3(iT)$ and $\mathrm{Im}\, u_3(iT)$, which are given by equation (1.1.22) and outputs them to the soft decision decoder. Our goal is to evaluate the decoder's performance.

Suppose that a code word $\mathbf{s} = (s_1, s_2, ..., s_n)$ has been transmitted and the block $\boldsymbol{\theta}(\mathbf{s}) = (\theta_1, \theta_2, ..., \theta_n)$ has been received in the absence of noise. Then, according to the channel description, $\boldsymbol{\theta}(\mathbf{s}) = 2\mathbf{s}-\mathbf{1}$, where $\mathbf{1} = (1,1,...,1)$. In the presence of noise, however, we receive $\mathbf{r} = \boldsymbol{\theta}+\mathbf{e}$, where $\mathbf{e}$ is a zero-mean Gaussian variable whose covariance matrix is equal to $diag\{\sigma^2\}$ where σ^2 is the noise power on the output of the receiver filter.

Let us define now the maximum-likelihood decoder decision regions. Since $\mathbf{r}$ is a Gaussian variable whose mean is equal to $\boldsymbol{\theta}$, the probability density of receiving $\mathbf{r}$ if $\mathbf{s}$ was transmitted is given by

$$f(\mathbf{r}|\mathbf{s}) = A e^{-||\mathbf{r}-\theta(\mathbf{s})||^2/2\sigma^2}$$

where $A = (2\pi)^{-n/2}\sigma^{-n}$ and

$$||\mathbf{r}-\theta(\mathbf{s})||^2 = \sum_{i=1}^{n}(\theta_i - r_i)^2$$

is the squared Euclidean distance between vectors $\mathbf{r}$ and $\boldsymbol{\theta}$. From this equation it follows that the maximum of $f(\mathbf{r}|\mathbf{s})$ corresponds to the minimum of the Euclidean distance. Therefore, the maximum-likelihood decoding rule can be formulated as follows: Decode the code word $\mathbf{s}$ if $\boldsymbol{\theta}(\mathbf{s})$ is the closest to the received vector $\mathbf{r}$. This rule is similar to the minimum Hamming distance decoding considered in the previous section.

In order to evaluate the decoder's performance suppose that the all-zero code word $\mathbf{0} = (0,0,...,0)$ has been transmitted. In the absence of noise the receiver outputs the sequence $\boldsymbol{\theta}(\mathbf{0}) = -\mathbf{1}$. A received block $\mathbf{r}$ is decoded correctly if it is closer to $\boldsymbol{\theta}(\mathbf{0})$ than to some other $\boldsymbol{\theta}(\mathbf{s})$, so that

$$p_c = \int_{\Theta(\mathbf{0})} f(\mathbf{r}|\mathbf{0})d\mathbf{r}$$

where the integration is performed over all $\mathbf{r}$ that are closer to $\boldsymbol{\theta}(\mathbf{0})$ than to $\boldsymbol{\theta}(\mathbf{s})$ with $\mathbf{s}\neq\mathbf{0}$.

The probability of incorrect decoding is

$$p_u = \sum_{\mathbf{s}\neq\mathbf{0}} \int_{\Theta(\mathbf{s})} f(\mathbf{r}|\mathbf{0})d\mathbf{r}$$

It is difficult to evaluate the above integrals, because the integration regions are complex. The union bound can be expressed using simpler integrals:

$$p_u \leq \sum_{\mathbf{s}\neq\mathbf{0}} \int_{U(\mathbf{s})} f(\mathbf{r}|\mathbf{0})d\mathbf{r} \tag{4.3.55}$$

where $U(\mathbf{s}) = \{\mathbf{r}: \| \mathbf{r}-\boldsymbol{\theta}(\mathbf{0}) \| > \| \mathbf{r}-\boldsymbol{\theta}(\mathbf{s}) \| \}$ is the set of all $\mathbf{r}$ that are closer to $\boldsymbol{\theta}(\mathbf{s})$ than to $\boldsymbol{\theta}(\mathbf{0})$. The integration region $U(\mathbf{s})$ is the semi-space bounded by the hyperplane which is equidistant from the points $\boldsymbol{\theta}(\mathbf{0})$ and $\boldsymbol{\theta}(\mathbf{s})$. The integral over $U(\mathbf{s})$ is equal to the probability that the projection of $\mathbf{r}$ onto $\boldsymbol{\theta}(\mathbf{s})-\boldsymbol{\theta}(\mathbf{0})$ is greater than $\| \boldsymbol{\theta}(\mathbf{s})-\boldsymbol{\theta}(\mathbf{0}) \| /2$, which is equal to half of the distance between the points. Since the projection of $\mathbf{r}$ is a Gaussian variable, we obtain

$$\int_{U(\mathbf{s})} f(\mathbf{r} \mid \mathbf{0}) d\mathbf{r} = 0.5 erfc\,(\beta \| \boldsymbol{\theta}(\mathbf{s})-\boldsymbol{\theta}(\mathbf{0}) \| /2)$$

It is easy to verify that the relation between the Euclidean and Hamming distances can be expressed as $\| \boldsymbol{\theta}(\mathbf{s})-\boldsymbol{\theta}(\mathbf{0}) \| /2 = \sqrt{\lceil \mathbf{s} \rceil}$. If we rearrange the summation in (4.3.55) according to Hamming weights, we obtain

$$p_u \leq \sum_{i=1} A_i erfc\,(\beta\sqrt{i}) \tag{4.3.56}$$

where A_i is the number of Hamming weight i code words.

This bound can be expressed through the code-weight generating function if we make use of the identity [105]

$$\int_a^\infty \frac{e^{-pt}dt}{t\sqrt{t-a}} = \frac{\pi}{\sqrt{a}} erfc\,(\sqrt{ap})$$

which after substitutions $\sqrt{a} = \beta$ and $\sqrt{t-a} = v$ takes the form

$$erfc\,(\beta\sqrt{p}) = \frac{2\beta}{\pi} \int_0^\infty (\beta^2+v^2)^{-1} e^{-p(\beta^2+v^2)} dv$$

Using this identity we can perform the summation in (4.3.56) and obtain the following representation of the union bound for the probability of incorrect decoding:

$$p_u \leq \frac{2\beta}{\pi} \int_0^\infty (\beta^2+v^2)^{-1} A_1(e^{-(\beta^2+v^2)}) dv$$

where $A_1(x) = W(x, 1)$ is defined by equation (4.3.41).

In particular, for the (7,4) Hamming code (see Example 4.3.6)

$$A_1(x) = 7x^3+7x^4+x^7$$

and the union bound of (4.3.56) takes the form

$$p_u \le 3.5erfc(\beta\sqrt{3}) + 3.5erfc(2\beta) + 0.5erfc(\beta\sqrt{7})$$

Let us compare the performance of the soft decision decoder with that of the hard decision decoder. For the hard decision decoder the probability of incorrect decoding is defined by equation (4.3.43), which in the case of the (7,4) Hamming code gives

$$p_u = 7(3p_1^2p_0^5 + 5p_1^3p_0^3 + 3p_1^5p_0^2 + p_1^6p_0) + p_1^7 \qquad (4.3.57)$$

where $p_1 = 0.5erfc(\beta)$ and $p_0 = 1 - p_1$. The union bound for this probability can be obtained by using generating function (4.3.51), which takes the form

$$\Phi(z) = 7(p_1z + p_0z^{-1})^3 + 7(p_1z + p_0z^{-1})^4 + (p_1z + p_0z^{-1})^7$$

Since the code corrects all single errors and does not correct or detect errors of higher multiplicity, we need to collect only the positive power terms of this function expansion:

$$p_u \le 7(3p_1^2p_0 + p_1^3 + 4p_1^3p_0 + p_1^4 + 5p_1^4p_0^3 + 3p_1^5p_0^2 + p_1^6p_0) + p_1^7$$

This bound is very tight for small values of p. It is approximately equal to $7(3p_1^2 + 5p_1^3)$, whereas the exact value of the probability of incorrect decoding (4.3.61) is approximated by $7(3p_1^2 - 10p_1^3)$ if p is small.

■

4.3.7 Correcting Errors and Erasures. Let us now consider a channel with errors and erasures. Erasures can be generated by a soft decision demodulator when it cannot make a decision about received symbol value. They can also appear on the outer decoder input of a concatenated code when the inner decoder detects symbol-errors and passes them along to the outer decoder as erasures.[106] In this case in addition to symbols belonging to GF(q) we receive a $q+1$-th symbol called an *erasure*. This symbol contains no information about the value of the transmitted symbol it only marks the location of an indefinite symbol. It is possible to use this information to improve the code's performance.

Suppose that the code minimum distance is equal to $d_{\min}$. Then the code can correct t errors and s erasures if and only if $2t + s < d_{\min}$. It follows from the fact that for correct decoding a code word with s erased symbols it is necessary and sufficient that this code word be different from any other code word in the remaining $n-s$ nonerased positions. The minimum Hamming distance decoding consists in finding a code word that is closest to the received word in the non-erased positions.

One could use a decoder that corrects errors to correct errors and erasures by trying all possible combinations of symbols in the places of erasures until the syndrome of the combination of the errors and the selected symbols belongs to the set C of the syndromes of correctable errors. Let $\mathbf{r}_0$ be a word obtained from the received word by replacing the erasures with zeroes, and let $\mathbf{e}_s$ be a word that has zeroes on the positions different from erasures. We vary $\mathbf{e}_s$ until we find the one which satisfies the condition $(\mathbf{r}_0+\mathbf{e}_s)\mathbf{H}' = \mathbf{S}\in C$, and then we decode $\mathbf{c} = \mathbf{r}_0+\mathbf{e}(\mathbf{S})$.

In the case of BCH codes, the decoding procedure can be significantly simplified. Indeed, the erasure locations are the roots of the characteristic polynomial in (4.2.13). We can factor the polynomial into the product of two polynomials $\Lambda(x)\lambda(x)$, where $\Lambda(x)$ is the known polynomial whose roots are the erasure locators, and $\lambda(x)$ is the unknown error locator polynomial. Since the syndromes satisfy difference equation (4.2.12), the syndrome-generating function is a rational function:

$$S(x) = \frac{\phi(x)}{\Lambda(x)\lambda(x)} \tag{4.3.58}$$

Multiplying both sides of this by $\Lambda(x)$, we obtain

$$S(x)\Lambda(x) = \frac{\phi(x)}{\lambda(x)}$$

The coefficients T_j of the generating function $T(x) = S(x)\Lambda(x)$ are called *modified syndromes*.[106] According to the convolution theorem they can be expressed as

$$T_j = \sum_{i=0}^{s} \Lambda_i S_{j-i} \qquad j=m_0+s, m_0+s+1, \ldots, m_0+d$$

On the other hand, since the generating function $T(x)$ is the rational function given by the right-hand side of equation (4.3.58), the modified syndromes satisfy the finite difference equation whose characteristic polynomial is equal to $\lambda(x)$. Thus, by using the modified syndromes T_j instead of the original syndromes S_j, we can use the decoding algorithm described in Sec 1.1.4: Find a difference equation for the modified syndromes; the characteristic polynomial of this equation is the error locator polynomial $\lambda(x)$. The error locators are the roots of this polynomial. Using the error locators and the known erasure locators, we find the error and erasure values by solving system (4.2.11), in which t is replaced with $t+s$.

Let us calculate the code performance characteristics by assuming that the error and erasure source is described by the SSM model with the matrix probability of error $\mathbf{P}_e$ and the matrix probability of erasure $\mathbf{P}_*$. The probability of correct decoding is equal to the sum of the probabilities of all error and erasure sequences that satisfy the condition $2t+s<d$:

$$p_c = \mathbf{p} \sum_{2t+s<d} \mathbf{P}_n(t,s)\mathbf{1}$$

where $\mathbf{P}_n(t,s)$ is the matrix probability that a block of n symbols contains t errors and s erasures. Being a matrix multinomial probability, it has generating function (2.3.3), which can be written as

$$\mathbf{\Phi}_n(1,z,w) = \sum_{t,s} \mathbf{P}_n(t,s) z^t w^s = (\mathbf{P}_0 + z\mathbf{P}_e + w\mathbf{P}_*)^n \tag{4.3.59}$$

where $\mathbf{P}_0 = \mathbf{P}-\mathbf{P}_e-\mathbf{P}_*$. The generating function $\mathbf{\Psi}_n(z)$ of the probabilities $\mathbf{\Psi}_{m,n} = \mathbf{P}_n(2t+s = m)$ that $2t+s=m$ can be obtained from the generating function $\mathbf{\Phi}_n(1,z,w)$ with the simple variable substitution

$$\mathbf{\Psi}_n(z) = \mathbf{\Phi}_n(1,z^2,z) = (\mathbf{P}_0 + z^2\mathbf{P}_e + z\mathbf{P}_*)^n \tag{4.3.60}$$

Since

$$\mathbf{\Psi}_n(z) = \mathbf{\Psi}_k(z)\, \mathbf{\Psi}_{n-k}(z)$$

the probabilities $\mathbf{\Psi}_{m,n}$ can be found from the recursive equation

$$\mathbf{\Psi}_{m,n} = \sum_{i=0}^{m} \mathbf{\Psi}_{i,k} \mathbf{\Psi}_{m-i,n-k}$$

For fast calculations it is advisable to perform the recursion by using the binary representation of n, as in evaluating the matrix binomial distribution discussed in Sec 2.6.

The matrix probability of correct decoding is equal to the sum of the coefficients of power terms z^m for $m=0,1,...,d-1$ in the function $\mathbf{\Psi}_n(z)$ expansion:

$$\mathbf{P}_c = \sum_{m=0}^{d-1} \mathbf{\Psi}_{m,n} \tag{4.3.61}$$

This probability can also be found as a coefficient of z^{d-1} in the expansion of the generating function of sums:

$$\mathbf{\Psi}_n(z)/(1-z) = \sum_{i=0}^{\infty} z^i \sum_{m=0}^{i} \mathbf{\Psi}_{m,n}$$

Since the generating function (4.3.59) is a polynomial, we may use the Fourier transform inversion formula in (2.2.25) to calculate the coefficients $\Psi_{m,n}$ and the probability of correct decoding:

$$\mathbf{\Psi}_{m,n} = p^{-1} \sum_{k=0}^{p-1} \alpha^{-mk} (\mathbf{P}_0 + \mathbf{P}_e \alpha^{2k} + \mathbf{P}_* \alpha^k)^n$$

$$\mathbf{P}_c = p^{-1} \sum_{k=0}^{p-1} \eta(\alpha^{-k}) (\mathbf{P}_0 + \mathbf{P}_e \alpha^{2k} + \mathbf{P}_* \alpha^k)^n$$

where $\alpha = e^{-2\pi j/p}$, $p > 2n$, and $\eta(x) = (1-x^d)/(1-x)$.

The lower bound on the probability of correct decoding can be obtained using Chernoff's inequality

$$p_c \geq 1 - z_0^{-d} \mathbf{p} \mathbf{\Psi}(z_0) \mathbf{1}$$

where $z_0 \geq 1$ minimizes $z^{-d} \mathbf{p} \mathbf{\Psi}(z) \mathbf{1}$.

The matrix probability of incorrect decoding can be found similarly to the case when no erasures were present. If channel errors are symmetrically dependent, the equations can be simplified. The probability of incorrect decoding can be expressed using the total probability theorem

$$p_u = \sum_{s=0}^{d-1} Pr(s) \tilde{p}_u(s) \tag{4.3.66}$$

where $\tilde{p}_u(s)$ is the average conditional probability of incorrectly decoding codes that are obtained from the original code by erasing any s symbols. The conditional probability of incorrect decoding can be determined in a manner similar to that of (4.3.47).

EXAMPLE 4.3.8: Suppose that the (7,4) Hamming code is used to correct errors and erasures in a channel with independent errors and erasures. This code minimum Hamming weight is equal to $d=3$ (see Example 4.3.6), so that it is able to correct any combination of errors and erasures that satisfy the inequality $2t+s<3$; that is, it can correct a single bit error with no erasures, or up to two erasures with no errors. The decoding procedure is simple: Correct a single error according to its syndrome in the absence of erasures, or correct no more than two erasures by finding the code word that matches the received word in unerased positions (or by varying bits in erased positions until the word becomes a code word).

Let us now evaluate the code performance characteristics. Generating function (4.3.61) takes the form

$$\Psi_7(z) = (p_0 + z^2 p_e + z p_*)^7$$

and the probability of correct decoding is equal to the sum of the coefficients of z^0, z^1, and z^2 in this function expansion:

$$p_c = p_0^7 + 7p_0^6(1-p_0) + 21p_0^5 p_*^2$$

The probability of incorrect decoding can be found using equation (4.3.66). The probability of having s=0 erasures is $Pr(s{=}0) = (1-p_*)^7$. The conditional probability $p_u(0)$ of incorrect decoding in this case is given by the ratio of the right-hand side of equation (4.3.61), in which p_1 is replaced with p_e, to $(1-p_*)^7$.

The probability of having one erasure is $Pr(s{=}1) = 7(1-p_*)^6 p_*$. To calculate the corresponding conditional probability of incorrect decoding, we have to find the average weight distribution for all the codes that can be obtained from the original (7,4) code by deleting one position. If we write down all sixteen code words and delete a bit in all of them, we obtain one all-zero word, three words whose weight is 2, eight words whose weight is 3, three words whose weight is 4, and one word whose weight is 6. Thus, the average weight-generating function is

$$A^{(1)}(x) = 1+3z^2+8z^3+3z^4+z^6$$

As we see, this code has a minimum distance d=2 and is not able to correct errors; it can only detect them. Therefore, the incorrect decoding occurs when the code does not detect errors. Using equation (4.3.43) we obtain

$$(1-p_*)^6 p_u(1) = W(p_e, p_0) = 3p_e^2 p_0^4 + 8p_e^3 p_0^3 + 3p_e^4 p_0^2 + p_e^6$$

Quite analogously we find $Pr(s{=}2){=}21p_*^2(1-p_*)^5$ and

$$(1-p_*)^5 p_u(2) = p_e p_0^4 + 6p_e^2 p_0^3 + 6p_e^3 p_0^2 + p_e^4 p_0 + p_e^5$$

Finally, the probability of incorrect decoding is given by

$$p_u = p_u(0) + 7p_* p_u(1) + 21p_*^2 p_u(2)$$ ■

4.3.8 Product Codes. Product (or iterated) codes are long codes that are formed from two or more shorter codes. Suppose that information has been encoded using an (n_1,k_1) code. A superblock consisting of k_2 code words may be viewed as a $k_2 \times n_1$ matrix, each row of which is a code word

$$\mathbf{X}_1 = \begin{bmatrix} \mathbf{X}_{11} & \mathbf{X}_{12} \end{bmatrix}$$

where matrices

$$\mathbf{X}_{11} = \begin{bmatrix} x_{11} & x_{12} & \dots & x_{1,k_1} \\ x_{21} & x_{22} & \dots & x_{2,k_1} \\ \dots & \dots & \dots & \dots \\ x_{k_2 1} & x_{k_2 2} & \dots & x_{k_2 k_1} \end{bmatrix} \quad \text{and} \quad \mathbf{X}_{12} = \begin{bmatrix} x_{1,k_1+1} & \dots & x_{1,n_1} \\ x_{2,k_1+1} & \dots & x_{2,n_1} \\ \dots & \dots & \dots \\ x_{k_2 k_1+1} & \dots & x_{k_2 n_1} \end{bmatrix}$$

represent the information and parity-check symbols, respectively. We can improve the protection of information by encoding each column of the matrix with some (n_2,k_2) block code so that the encoded information is presented by the matrix

$$\mathbf{X} = \begin{bmatrix} \mathbf{X}_{11} & \mathbf{X}_{12} \\ \mathbf{X}_{21} & \mathbf{X}_{22} \end{bmatrix}$$

where $\mathbf{X}_{21}$ represents the parity-check symbols for the columns of the matrix $\mathbf{X}_{11}$, while $\mathbf{X}_{22}$ represents parity-check symbols for columns of $\mathbf{X}_{12}$ or for rows of $\mathbf{X}_{21}$.[88] Thus we have constructed a $(n_1 n_2,\ k_1 k_2)$ code called a *product code*. We can construct products of more than two codes using multidimensional arrays.

For several reasons, product codes are widely used in practice. One of the reasons is the relative simplicity of their implementation: Very long codes can be constructed out of several short codes. The other reason is that they are capable of fighting against bursts of errors because of their built-in interleaving. If a code matrix row length n_1 is large compared to the average error burst length, then errors in the same column can be treated as independent. Also, if the (n_1,k_1) code detects noncorrectable errors in a row, it may erase the whole row, and the (n_2,k_2) code, will correct errors and erasures.

Let $\mathbf{H}_1$ and $\mathbf{H}_2$ be the parity-check matrices of the (n_1,k_1) and (n_2,k_2) code respectively. Then $\mathbf{X}$ is a code word of the product code if and only if

$$\mathbf{X}\mathbf{H}'_1 = 0 \quad \text{and} \quad \mathbf{H}_2\mathbf{X} = 0$$

Using rectangular matrices simplifies product code description. To present a code word in the conventional vector form, we may scan the matrix $\mathbf{X}$ and then string all the rows in one long vector. This operation can be expressed in the matrix form [107]

$$\mathbf{s} = (x_{11}, x_{12}, \dots, x_{1,n_1}, x_{21}, \dots, x_{n_2,n_1}) = \sum_{i=1}^{n_2} \mathbf{v}_i \mathbf{X} \mathbf{N}_i$$

where $\mathbf{v}_i$ is the row matrix whose i-th element is equal to 1 and the rest of the elements are zeroes, and $\mathbf{N}_i$ is the block-row matrix whose i-th sub-block is equal to $\mathbf{I}$ and the rest of the sub-blocks are zeroes:

$$\mathbf{v}_i = [\ 0 \ldots\ 0\ 1\ 0 \ldots\ 0\ ]$$
$$\mathbf{N}_i = [\ 0 \ldots\ 0\ \mathbf{I}\ 0 \ldots\ 0\ ]$$

The inverse relation has the form

$$\mathbf{X} = \sum_{i=1}^{n_2} \mathbf{v}'_i \mathbf{s} \mathbf{N}'_i$$

Using these equations, it is easy to verify that the product code parity-check matrix can be expressed as

$$\mathbf{H} = \begin{bmatrix} \mathbf{H}_1 \otimes \mathbf{I} & \mathbf{I} \otimes \mathbf{H}_2 \end{bmatrix}$$

To evaluate product code performance we can use its matrix $\mathbf{H}$, which contains $n_1(n_1-k_1)n_2(n_2-k_2)$ elements in all the previously derived equations. However, if we use the code matrix representation, these equations can be simplified, since the component code parity-check matrices have fewer elements $[n_1(n_1-k_1)+n_2(n_2-k_2)]$. It is convenient to use matrix notations for transmitted, received, and error sequences when dealing with product codes. Let

$$\mathbf{E} = \begin{bmatrix} \mathbf{e}_{ij} \end{bmatrix}_{n_1, n_2}$$

be the error matrix that corresponds to the transmitted code matrix $\mathbf{X}$. Matrices

$$\mathbf{S}_1 = \mathbf{E}\mathbf{H}'_1 \quad \text{and} \quad \mathbf{S}_2 = \mathbf{H}_2\mathbf{E}$$

are called *received message error syndromes.*

Using these notations, we can rewrite syndrome distribution generating function (4.3.6) as

$$f(\mathbf{z}, \mathbf{w}) = \sum_{\mathbf{E}} Pr(\mathbf{E}) \mathbf{z}^{\mathbf{E}\mathbf{H}'_1} \mathbf{w}^{\mathbf{H}'_2\mathbf{E}}$$

All the equations that use generating function (4.3.6) should be modified accordingly. Instead of doing the straightforward modifications, we illustrate their applications in the following example.

EXAMPLE 4.3.9: Consider the product of two simple parity-check codes that check the code $m \times n$ matrix rows and columns, respectively. For an SSM model, let us find the probability of correct decoding and the probability of undetected errors if we assume that the code is used for error detection only.

The generating function of the syndrome matrix distribution is similar to (4.3.11) and can be written in the form

$$\mathbf{F}(\mathbf{z},\mathbf{w}) = \prod_{i=1}^{n}\prod_{j=1}^{m}[\mathbf{P}(0) + \mathbf{P}(1)z_i w_j]$$

The matrix probability of correct decoding is equal to the zero power term of the function

$$\mathbf{P}_c = \mathbf{P}^{mn}(0)$$

The matrix probability of undetected errors is given by equation (4.3.12), which in our case has the form

$$\mathbf{P}_u = 2^{-n-m} \sum_{\mathbf{v},\boldsymbol{\mu}} \prod_{i=1}^{n}\prod_{j=1}^{m}[\mathbf{P}(0) + \mathbf{P}(1)(-1)^{\nu_i+\mu_j}] - \mathbf{P}^{mn}(0)$$

where $\nu_i, \mu_j = 0, 1$. The calculations are simplified if we replace variables z_i with $(-1)^{\nu_i}$ to obtain the generating function

$$\boldsymbol{\Psi}(\mathbf{w}) = 2^{-m}\sum_{\mathbf{v}}\prod_{i=1}^{n}\prod_{j=1}^{m}[\mathbf{P}(0) + \mathbf{P}(1)(-1)^{\nu_i} w_j] =$$

$$\left[\ \prod_{j=1}^{m}(\ \mathbf{P}(0)+\mathbf{P}(1)w_j\ ) + \prod_{j=1}^{m}(\ \mathbf{P}(0)-\mathbf{P}(1)w_j\ )\ \right]^n$$

and then determine the probability of undetected errors using the generating function

$$\mathbf{P}_u = 2^{-n}\sum_{\boldsymbol{\mu}}\boldsymbol{\Psi}((-1)^{\boldsymbol{\mu}}) - \mathbf{P}^{mn}(0)$$

where $(-1)^{\boldsymbol{\mu}} = [(-1)^{\mu_1},(-1)^{\mu_2},\ldots,(-1)^{\mu_m}]$. ■

4.3.9 Concatenated Codes. Concatenated codes are similar to product codes. The first code that is used to encode the information matrix $\mathbf{X}_{11}$ rows is called the *inner code* and is defined the same way as in product codes. The second code that is used to encode matrix $\mathbf{X}_{11}$ columns is defined differently: The rows of matrix

$\mathbf{X}_{11}$ whose elements belong to GF(q) are treated as elements of GF(q^{k_1}) and encoded by some (n_2, k_2) code over this field. This second code is called the *outer code*. In other words, all columns of matrix $\mathbf{X}_{11}$ are treated as one column of elements from GF(q^{k_1}) and encoded as such.

The decoder of the inner code detects and corrects errors. Symbols with detected errors are passed to the decoder of the outer code as erasures, which corrects the erasures and and remaining errors. Usually, the RS code is used as the outer code.

The performance of concatenated codes can be evaluated similarly to that of product codes. However, for the SSM model it is possible to evaluate code performance characteristics in two steps: first for the inner code and then for the outer code. The inner encoder output is fully characterized by these three matrix probabilities: the probability of correct decoding $\mathbf{P}_0$, the probability of incorrect decoding $\mathbf{P}_e$, and the probability of an erasure which is equal to the probability of detected errors $\mathbf{P}_* = \mathbf{P}^{n_1} - \mathbf{P}_0 - \mathbf{P}_e$. (The methods of calculating these probabilities are described in previous sections.) Then the concatenated code performance can be evaluated by the methods described in Sec 4.3.7.

EXAMPLE 4.3.10: Let us calculate the performance characteristics of a concatenated code whose inner code is the simple parity-check $(k+1, k)$ code and whose outer code is (N, K) RS code where $N=2^k-1$.

We determine the characteristics of the inner code first. It is obvious that the probability that a block of $k+1$ bits does not have errors $\mathbf{P}_0 = \mathbf{P}^{k+1}(0)$. The probability of undetected errors has been found in Example 4.3.1:

$$\mathbf{P}_e = 2^{-1}\mathbf{P}^{k+1} + 2^{-1}[\mathbf{P}(0) - \mathbf{P}(1)]^{k+1} - \mathbf{P}^{k+1}(0)$$

And the probability of an erasure is given by

$$\mathbf{P}_* = \mathbf{P}^{k+1} - \mathbf{P}_e - \mathbf{P}_0$$

which after simplification becomes

$$\mathbf{P}_* = 2^{-1}\mathbf{P}^{k+1} - 2^{-1}[\mathbf{P}(0) - \mathbf{P}(1)]^{k+1}$$

We are ready now to evaluate outer code performance. Since the RS code minimal distance is equal to $D=N-K+1$, this code corrects all the combinations of t errors and s erasures that satisfy the condition $2t+s \le N-K$ and only them. The probability of correct decoding is equal to the sum of the coefficients of z^m for $m=0,1,\ldots,N-K$ in the expansion of generating function (4.3.60) and is given by equations (4.3.61).

■

4.4 CONVOLUTIONAL CODE PERFORMANCE

Let us consider now methods of evaluating the performance of convolutional codes. Because of the nature of convolutional coding, its performance is usually characterized by some parameters of error sequences on a convolutional decoder output and, in particular, by decoder symbol-error probability.

4.4.1 Viterbi Algorithm Performance. It is difficult to find an exact equation for the probability of incorrect decoding when the Viterbi algorithm is used. However, for the SSM channel model the union bound on this probability can be expressed in the closed form. In Appendix 4.3 we derive the closed-form expression for the union bound on Viterbi algorithm performance for a general discrete-time system. According to equation (4.2.24), convolutional codes represent a particular case of discrete-time system (A.4.9). Therefore, we can apply the results of Appendix 4.3 to convolutional codes.

Suppose that we want to find the bit-error probability on the output of a decoder that uses the Hamming distance between the received and transmitted symbols as a branch metric

$$m(\mathbf{z}_k,\mathbf{w}_k) = |\,\mathbf{z}_k-\mathbf{y}_k\,| = |\,\mathbf{e}_k\,| \tag{4.4.1}$$

where $\mathbf{x}_k$ is the encoder input symbol, $\mathbf{y}_k$ is the encoder output symbol, $\mathbf{s}_k = (\mathbf{x}_{k-1},\mathbf{x}_{k-2},\ldots,\mathbf{x}_{k-m+1})$ is the encoder state, $\mathbf{w}_k = (\mathbf{x}_k,\mathbf{s}_k)$, and $\mathbf{z}_k$ is the decoder input symbol (see Sec 4.2.6 and Appendix 4.3). In this case the branch metric generating function (A.4.12) is

$$\mathbf{D}(v;\mathbf{w}_k,\hat{\mathbf{w}}_k) = \sum_{\mathbf{e}_k}\mathbf{P}(\mathbf{e}_k)v^{\,|\mathbf{e}_k|-|\hat{\mathbf{e}}_k|} \tag{4.4.2}$$

where $\hat{\mathbf{e}}_k = \mathbf{z}_k-\hat{\mathbf{y}}_k$, $\hat{\mathbf{y}}_k$ is the encoder output for $\hat{\mathbf{w}}_k$.

The distortion measure is the decoder output bit error

$$d(\mathbf{w}_k,\hat{\mathbf{w}}_k) = \begin{cases} 0 & \text{if } \mathbf{x}_k = \hat{\mathbf{x}}_k \\ 1 & \text{if } \mathbf{x}_k \neq \hat{\mathbf{x}}_k \end{cases}$$

This may be rewritten as

$$d(\mathbf{w}_k,\hat{\mathbf{w}}_k) = \delta(\mathbf{x}_k - \hat{\mathbf{x}}_k) \tag{4.4.3}$$

where $\delta(x) = 0.5\,[1-(-1)^{|x|}]$.

Generating function (A.4.17) in our case depends only on channel errors. If we assume that the encoder input symbols are independent and equiprobable, then we need not average in equation (A.4.16) and can assume that the all-zero sequence $\{\mathbf{y}_k = 0\}$ was transmitted.

The system's symmetry allows us to decrease the number of states as compared to the general case presented in Appendix 4.3: Instead of using the product states and considering all possible correct paths, we can assume that there is only one correct path—the all-zero path $\{\mathbf{w}_k = 0\}$. The symbol-error probability union bound (A.4.15) in this case takes the form

$$p_{sb} \le \frac{1}{2\pi j}\frac{d}{du}\{\oint_C (\frac{1}{v-1} - \frac{1}{2v})\mathbf{p}\mathbf{G}(u,v)\mathbf{1}dv\}\ |_{u=1} \tag{4.4.4}$$

where

$$\mathbf{G}(u,v) = \sum_{L=1}^{\infty}\sum_{\hat{\sigma}_L}\prod_{k=0}^{L-1}\mathbf{D}(v;0,\hat{\mathbf{w}}_k)u^{\delta(\hat{\mathbf{x}}_k)} \tag{4.4.5}$$

$C \in \{v : R_G \cap |v| > 1\}$, R_G is the region of convergence of (4.4.5).

Let us assume now that the symbols $\mathbf{y}_k = (y_{1k}, y_{2k}, \ldots, y_{bk})$ are transmitted serially bit-by-bit over an SSM channel. In this case, according to the SSM model,

$$\mathbf{P}(\mathbf{e}_k) = \prod_{i=1}^{b}\mathbf{P}(e_{ik}) \tag{4.4.6}$$

where e_{ik} denotes an i-th bit error and $\mathbf{e}_k = (e_{1k}, e_{2k}, \ldots, e_{bk})$. Also,

$$v^{|\mathbf{e}_k| - |\hat{\mathbf{e}}_k|} = \prod_{i=1}^{b} v^{\mu_{ik}}$$

where $\mu_{ik} = -(-1)^{e_{ik}}\hat{y}_{ik}$, so that equation (4.4.2) can be rewritten as

$$\mathbf{D}(v;0,\hat{\mathbf{w}}_k) = \prod_{i=1}^{b}[\mathbf{P}(0)v^{-\hat{y}_{ik}} + \mathbf{P}(1)v^{\hat{y}_{ik}}] \tag{4.4.7}$$

The generating function (4.4.5) can be found from the code state transition graph with branch weights

$$\mathbf{D}(v;0,\hat{\mathbf{w}}_k)u^{\delta(\hat{\mathbf{x}}_k)} = \prod_{i=1}^{b}[\mathbf{P}(0)v^{-\hat{y}_{ik}} + \mathbf{P}(1)v^{\hat{y}_{ik}}]u^{\delta(\hat{\mathbf{x}}_k)} \tag{4.4.8}$$

These weights can be composed from the encoder state diagram by using branch input/output symbols $(\hat{\mathbf{x}} / \hat{\mathbf{y}})$.

EXAMPLE 4.4.1: Let us illustrate the bit error probability union bound calculation for the code of Example 4.2.6.

The encoder state transition graph was shown in Fig. 4.4. To calculate the generating function (4.4.5) we break the loop at the state (0,0), which corresponds to correct decoding. Next we replace branch I/O symbols with weights (4.4.8). For instance, the branch symbol 1/01, which corresponds to the transition $(1,0) \to (1,1)$ in Fig. 4.4, is replaced with

$$\mathbf{D}(v;0,\hat{\mathbf{w}})u^{\delta(\hat{\mathbf{x}})} = \mathbf{PD}u$$

where $\mathbf{P} = \mathbf{P}(0)+\mathbf{P}(1)$, $\mathbf{D} = \mathbf{P}(0)v^{-1}+\mathbf{P}(1)v$. The generating function $\mathbf{G}(u,v)$ represents the transition weight from the state $\mathbf{0} = (0,0)$ back to it and can be calculated by absorption of the rest of the states. However, it is more convenient to duplicate the state $\mathbf{0}$ so that the edges, which ended at the state $\mathbf{0}$, go instead to its duplicate $\mathbf{0}'$, as shown in Fig. 4.11.

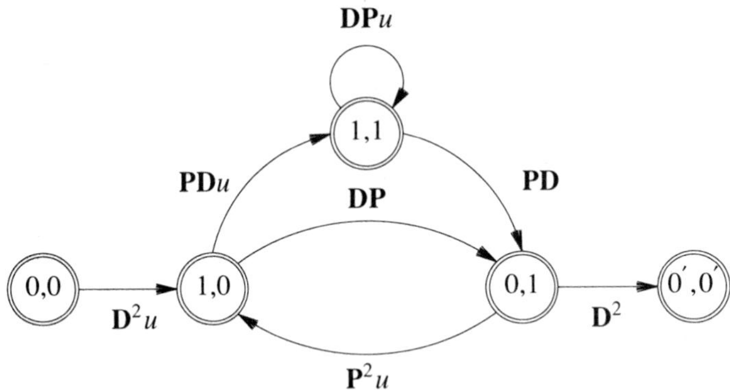

Figure 4.11. Signal flow graph for calculating bit-error probability.

After absorption of all the nodes (see Appendix 2.3) except for $\mathbf{0}$ and $\mathbf{0}'$, we obtain

$$\mathbf{G}(u,v) = u\mathbf{G}_1(u,v)[\mathbf{I} - u\mathbf{P}^2\mathbf{G}_1(u,v)]^{-1}\mathbf{D}^2$$

where

$$\mathbf{G}_1(u,v) = \mathbf{DP} + u\mathbf{PD}[\mathbf{I} - u\mathbf{DP}]^{-1}\mathbf{PD}$$

The bit-error probability union bound is obtained from (4.4.4). The contour integral in (4.4.4) can be found using the residue theorem. Indeed,

$$\mathbf{pG}(u,v)\mathbf{1} = a(u,v)/b(u,v)$$

is a rational function, and its derivative with respect to u at $u=1$

$$f(v) = [a'(1,v)b(1,v) - a(1,v)b'(1,v)]/b^2(1,v)$$

is also a rational function. Its poles are the roots of the polynomial $b(1,v)$, which can be written as

$$b(1,v) = \det[\Delta(v)\mathbf{I} - \mathbf{R}(v)]$$

where

$$\Delta(v) = \det[\mathbf{I} - \mathbf{DP}] \quad \mathbf{R}(v) = \mathbf{DP}+\mathbf{PDB}(v)\mathbf{PD}$$

and $\mathbf{B}(v)$ is the adjoint matrix of $\mathbf{I}-\mathbf{DP}$.

In the particular case of BSC with independent errors and the bit-error probability p_1, the calculations are much simpler. In this case $\mathbf{P}=1$, $\mathbf{D} = p_1 v + p_0 v^{-1}$,

$$\mathbf{G}(u,v) = u\mathbf{D}^5/(1 - 2u\mathbf{D})$$

and

$$f(v) = \mathbf{D}^5/(1 - 2\mathbf{D})^2$$

Union bound (4.4.4) can be rewritten as

$$p_{sb} \le \frac{1}{2\pi j} \oint_C \left(\frac{1}{v-1} - \frac{1}{2v}\right) \frac{(p_1 v + p_0 v^{-1})^5}{[1 - 2(p_1 v + p_0 v^{-1})]^2} dv$$

where $C \in \{ v : |2(p_1 v + p_0 v^{-1})| < 1 \cap |v| > 1 \}$. Applying the residue theorem, we obtain [108,109]

$$p_b \le 2^{-5}[\,5 - g(5 - 96p_0p_1)(1 - 16p_0p_1)^{-1.5} + 14p_1 + 12p_1^2 - 8p_1^3\,]$$

where $g = 1 - 2p_1$. As reported in Ref. [109], for $p_1 = 2^{-9}$ the measured error rate equals $\hat{p}_b = 3.3 \cdot 10^{-7}$, the above upper bound gives $3.9 \cdot 10^{-7}$ and the Viterbi upper bound [95] gives $7.9 \cdot 10^{-6}$.

■

4.4.2 Syndrome Decoding Performance. Consider now performance of the syndrome decoding discussed in Sec 4.2.6.3. In the case of definite decoding, error estimate $\hat{\mathbf{e}}_{lj}$ is a function of the syndromes that check this error:

$$\hat{\mathbf{e}}_{Ij} = \mathbf{f}(\mathbf{s}_j, \mathbf{s}_{j+1}, \ldots, \mathbf{s}_{j+m})$$

The received symbol $\mathbf{r}_{Ij}$ is decoded incorrectly if this estimate is not equal to the actual error $\mathbf{e}_{Ij}$. Thus, the probability of incorrect decoding is

$$p_u = Pr(\hat{\mathbf{e}}_{Ij} \neq \mathbf{e}_{Ij}) \qquad (4.4.9)$$

Since according to (4.2.38) a syndrome is a deterministic function of the finite number of errors, it can be expressed as a Markov function if we assume that the error source is described by an SSM. Therefore, we can calculate the probability of incorrect decoding by using matrix probabilities and generating functions.

Suppose that $j=0$ and consider the generating function

$$\mathbf{F}(\mathbf{t};\mathbf{z}) = \sum_{\mathbf{e}} \prod_{j=-m}^{m} \mathbf{P}(\mathbf{e}_j)\mathbf{z}^{\mathbf{S}}\mathbf{t}^{\mathbf{e}_{I0}}$$

where $\mathbf{e}_j = (\mathbf{e}_{Ij}, \mathbf{e}_{Pj})$, $\mathbf{z} = (\mathbf{z}_1, \mathbf{z}_2, \ldots, \mathbf{z}_m)$, and

$$\mathbf{z}^{\mathbf{S}} = \prod_{j=1}^{m} \mathbf{z}^{\mathbf{s}_j}$$

Since syndromes are linear functions of error values, this generating function can be calculated quite analogously to that of (4.3.9):

$$\mathbf{F}(\mathbf{t};\mathbf{z}) = \prod_{j=-m}^{m} \sum_{\mathbf{e}_j} \mathbf{P}(\mathbf{e}_j)\mathbf{z}^{\mathbf{e}_j\mathbf{H}_j}\mathbf{t}^{\delta(j)} \qquad (4.4.10)$$

where $\boldsymbol{\delta}(0) = \mathbf{e}_{I0}$ and $\boldsymbol{\delta}(j)=0$ for $j \neq 0$. The probability of incorrect decoding is equal to the sum of the coefficients of this function expansion into power series for the powers satisfying the condition

$$\mathbf{e}_{I0} \neq \mathbf{f}(\mathbf{s}_0, \mathbf{s}_1, \ldots, \mathbf{s}_m)$$

The inverse Fourier transform can be used to calculate the probability of the set of errors that satisfy the previous inequality.

Sometimes it is convenient to use conditional generating functions

$$\mathbf{F}_{\mathbf{e}_{I0}}(\mathbf{z}) = \prod_{j=-m}^{m} \sum_{\mathbf{e}_j} \mathbf{P}(\mathbf{e}_j)\mathbf{z}^{\mathbf{e}_j\mathbf{H}_j} \qquad (4.4.11)$$

where $\mathbf{e}_{I0}$ is fixed. For each of these generating functions we calculate the corresponding conditional matrix probabilities of incorrect decoding and then add them to obtain the matrix probability $\mathbf{P}_u$ of incorrect decoding.

EXAMPLE 4.4.2: To illustrate the probability of incorrect decoding calculation, consider a CSOC with the generator polynomial[110]

$$\mathbf{G}(D) = [1 \quad 1+D]$$

The equations (4.2.38) that check e_{Ij} are

$$\begin{aligned} s_j &= e_{I(j-1)} + e_{Ij} \qquad + e_{Pj} \\ s_{j+1} &= e_{Ij} \qquad + e_{I(j+1)} + e_{P(j+1)} \end{aligned} \tag{4.4.12}$$

Suppose that we use majority logic for decoding: $\hat{e}_{Ij} = s_j s_{j+1}$. Then the j-th symbol is incorrectly decoded if and only if the estimated error is not equal to the actual error: $e_{Ij} \neq s_j s_{j+1}$. The matrix probability of incorrect decoding is

$$\mathbf{P}_u = \sum_{e_{Ij} \neq s_j s_{j+1}} \mathbf{P}(e_{I(j-1)})\mathbf{P}\mathbf{P}(e_{Ij})\mathbf{P}(e_{Pj})\mathbf{P}(e_{I(j+1)})\mathbf{P}(e_{P(j+1)})$$

This probability can also be found by using generating function (4.4.10), which in our case takes the form

$$\mathbf{F}(t;z_1,z_2) = [\mathbf{P}(0)+z_1\mathbf{P}(1)]\mathbf{P}[\mathbf{P}(0)+tz_1z_2\mathbf{P}(1)][\mathbf{P}(0)+z_1\mathbf{P}(1)][\mathbf{P}(0)+z_2\mathbf{P}(1)]^2$$

Conditional generating functions (4.4.11) have the form

$$\mathbf{F}_0(z_1,z_2) = [\mathbf{P}(0)+z_1\mathbf{P}(1)]\mathbf{P}\mathbf{P}(0)[\mathbf{P}(0)+z_1\mathbf{P}(1)][\mathbf{P}(0)+z_2\mathbf{P}(1)]^2$$

$$\mathbf{F}_1(z_1,z_2) = [\mathbf{P}(0)+z_1\mathbf{P}(1)]\mathbf{P}\mathbf{P}(1)z_1z_2[\mathbf{P}(0)+z_1\mathbf{P}(1)][\mathbf{P}(0)+z_2\mathbf{P}(1)]^2$$

An error e_j is estimated incorrectly if $e_{Ij} = 0$, but $s_j s_{j+1} = 1$. In this case s_j=1 and s_{j+1}=1 so that the matrix probability of this event can be computed using [as in equation (4.3.12)] the inverse Fourier transform

$$\mathbf{P}_u^{(0)} = 2^{-2}[\mathbf{F}_0(1,1)-\mathbf{F}_0(-1,1)-\mathbf{F}_0(1,-1)+\mathbf{F}_0(-1,-1)]$$

The error is also estimated incorrectly if $e_{Ij} = 1$ but $s_j s_{j+1} = 0$. The matrix probability of the complementary event $s_j s_{j+1}$ is given by the previous equation, in which $\mathbf{F}_0(\cdot,\cdot)$ is replaced with $\mathbf{F}_1(\cdot,\cdot)$. Therefore, the matrix probability of $s_j s_{j+1}$ can be expressed as

$$\mathbf{P}_u^{(1)} = \mathbf{P}^2\mathbf{P}(1)\mathbf{P}^2 - 2^{-2}[\mathbf{F}_1(1,1) - \mathbf{F}_1(-1,1) - \mathbf{F}_1(1,-1) + \mathbf{F}_1(-1,-1)]$$

Finally, the matrix probability of incorrect decoding is equal to the sum of the conditional probabilities:

$$\mathbf{P}_u = \mathbf{P}_u^{(0)} + \mathbf{P}_u^{(1)}$$

In particular, for the BSC with independent errors and $\mathbf{P}(0) = p_0$ and $\mathbf{P}(1) = p_1$ this equation takes the form

$$\mathbf{P}_u = p_u = p_1 - 4p_1^3 p_0^2 + 4p_1^2 p_0^3$$

which coincides with the result of Ref. [110].

■

Consider now the performance of feedback decoding. The state of syndrome registers of the feedback decoder is given by equation (4.2.42)

$$\mathbf{u}_{j+1} = \mathbf{u}_j\mathbf{A} + \mathbf{s}_{j+m}\mathbf{B} - \hat{\mathbf{e}}_j\mathbf{G}_P \tag{4.4.13}$$

and decision function (4.2.52) can be rewritten as

$$\hat{\mathbf{e}}_{Ij} = \mathbf{f}(\mathbf{u}_j, \mathbf{s}_{j+m}) \tag{4.4.14}$$

which depends not only on actual channel errors, but also on the previously estimated errors. These equations define an SSM whose output $\hat{\mathbf{e}}_{Ij}$ depends only on its input $\mathbf{s}_{j+m}$ and state $\mathbf{u}_j$. However, this description does not let us calculate the probability of incorrect decoding, because the state $\mathbf{u}_j$ and the syndrome $\mathbf{s}_{j+m}$ do not uniquely identify the error $\mathbf{e}_{Ij}$. To find the probability of incorrect decoding (4.4.9), we introduce the decoder extended state

$$\mathbf{v}_j = (\mathbf{e}_{I(j+m-1)}, \mathbf{e}_{I(j+m-2)}, \ldots, \mathbf{e}_{I(j-1)}, \mathbf{u}_j)$$

The matrix probability of incorrect decoding can be found using several methods. The first method [110] uses the total probability theorem: The probability of incorrect decoding is presented as a sum of the stationary probabilities of the extended states times the conditional probabilities of incorrect decoding at these states.

We can also find the probability of incorrect decoding by introducing states $\mathbf{w}_j = (\mathbf{v}_j, \mathbf{s}_{j+m})$. Decision rule (4.4.14) partitions these states into two subsets: "good" (G) and "bad" (B). A state $\mathbf{w}_j$ belongs to G if and only if $\mathbf{e}_{Ij} = \mathbf{f}(\mathbf{u}_j, \mathbf{s}_{j+m})$; otherwise it belongs to B. The probability of incorrect decoding is equal to the sum of the stationary probabilities of the states that belong to subset B:

$$p_u = \sum_{\mathbf{w} \in B} \boldsymbol{\pi}_{\mathbf{w}}$$

EXAMPLE 4.4.3: Consider the performance of the code in Example 4.4.2 when feedback decoding is used.

Equations (4.4.13) and (4.4.14) take the form

$$u_{j+1} = s_{j+1} + \hat{e}_j \qquad \hat{e}_j = s_{j+1} u_j \tag{4.4.15}$$

and the syndrome s_{j+1} is defined by equation (4.4.12). The decoder extended state $\mathbf{v}_j = (e_{Ij}, u_j)$ represents an SSM with transition matrix probabilities

$$\mathbf{P}(\mathbf{v}_{j+1} \mid \mathbf{v}_j) = \mathbf{P}(e_{I(j+1)}) \mathbf{P}(u_{j+1} \mid e_{Ij}, e_{I(j+1)}, u_j)$$

which after substitutions (4.4.12) and (4.4.15) take the following form

$$\mathbf{P}(\mathbf{v}_{j+1} \mid \mathbf{v}_j) = \mathbf{P}(e_{I(j+1)}) \mathbf{Pr}\{u_{j+1} = [e_{Ij} + e_{I(j+1)} + e_{P(j+1)}](u_j + 1)\}$$

Using this formula, we obtain the SSM transition probability matrix

$$\begin{bmatrix} \mathbf{P}^2(0) & \mathbf{P}(0)\mathbf{P}(1) & \mathbf{P}^2(1) & \mathbf{P}(1)\mathbf{P}(0) \\ \mathbf{P}(0)\mathbf{P} & 0 & \mathbf{P}(1)\mathbf{P} & 0 \\ \mathbf{P}(0)\mathbf{P}(1) & \mathbf{P}^2(0) & \mathbf{P}(1)\mathbf{P}(0) & \mathbf{P}^2(1) \\ \mathbf{P}(0)\mathbf{P} & 0 & \mathbf{P}(1)\mathbf{P} & 0 \end{bmatrix}$$

The SSM state stationary distribution can be found from system (1.1.8), which in our case is given by

$$\begin{aligned} \mathbf{v}_{00} &= \mathbf{v}_{00}\mathbf{P}^2(0) + \mathbf{v}_{01}\mathbf{P}(0)\mathbf{P} + \mathbf{v}_{10}\mathbf{P}(0)\mathbf{P}(1)\mathbf{v}_{11}\mathbf{P}(0)\mathbf{P} \\ \mathbf{v}_{01} &= \mathbf{v}_{00}\mathbf{P}(0)\mathbf{P}(1) \quad + \quad \mathbf{v}_{10}\mathbf{P}^2(0) \\ \mathbf{v}_{10} &= \mathbf{v}_{00}\mathbf{P}^2(1) + \mathbf{v}_{01}\mathbf{P}(1)\mathbf{P} + \mathbf{v}_{10}\mathbf{P}(1)\mathbf{P}(0) + \mathbf{v}_{11}\mathbf{P}(1)\mathbf{P} \\ \mathbf{v}_{11} &= \mathbf{v}_{00}\mathbf{P}(1)\mathbf{P}(0) \quad + \quad \mathbf{v}_{10}\mathbf{P}^2(1) \end{aligned}$$

$$(\mathbf{v}_{00} + \mathbf{v}_{01} + \mathbf{v}_{10} + \mathbf{v}_{11})\mathbf{1} = 1$$

The solution of this system can be written as

$$\mathbf{v}_{00} = \boldsymbol{\pi}[\ \mathbf{P}(0)\mathbf{P}-\mathbf{P}(1)\mathbf{P}\mathbf{X}\mathbf{P}^2(0)\ ]\mathbf{Y} \quad \mathbf{v}_{01} = \boldsymbol{\pi}\mathbf{P}(0)\mathbf{P}-\mathbf{v}_{00}$$
$$\mathbf{v}_{10} = [\ \boldsymbol{\pi}\mathbf{P}(1)\mathbf{P}-\mathbf{v}_{00}\mathbf{P}(1)\mathbf{P}(0)\ ]\mathbf{X} \quad \mathbf{v}_{11} = \boldsymbol{\pi}\mathbf{P}(1)\mathbf{P}-\mathbf{v}_{10}$$

where

$$\mathbf{X} = [\ \mathbf{I} + \mathbf{P}^2(1)\ ]^{-1} \quad \mathbf{Y} = [\ \mathbf{I} + \mathbf{P}(0)\mathbf{P}(1) - \mathbf{P}(1)\mathbf{P}(0)\mathbf{X}\mathbf{P}^2(0)\ ]^{-1}$$

Here $\boldsymbol{\pi}$ is the stationary distribution of the error source states ($\boldsymbol{\pi}\mathbf{P} = \boldsymbol{\pi}$, $\boldsymbol{\pi}\mathbf{1} = 1$).

The probability of incorrect decoding can be found using the total probability theorem

$$p_u = \sum_{i=0}^{1}\sum_{j=0}^{1}\mathbf{v}_{st}\mathbf{P}_u^{(et)}\mathbf{1}$$

where $\mathbf{P}_u^{(et)}$ is the matrix probability of incorrect decoding, given that the system is in the state (e,t), or in other words, the probability that the estimated error $\hat{e}_{Ij} = s_{j+1}t$ is not equal to the actual error e:

$$\mathbf{P}_u^{(et)} = \mathbf{Pr}\{t(e_{I(j+1)} + e_{P(j+1)}) = e\,(t+1)+1\}$$

This equation gives

$$\mathbf{P}_u^{(00)} = 0 \quad \mathbf{P}_u^{(10)} = \mathbf{P}^2 \quad \mathbf{P}_u^{(01)} = \mathbf{P}_u^{(11)} = \mathbf{P}(0)\mathbf{P}(1) + \mathbf{P}(1)\mathbf{P}(0)$$

In the particular case of the BSC with independent errors and $\mathbf{P}(0) = p_0$ and $\mathbf{P}(1) = p_1$, we find

$$\mathbf{v}_{00} = (1-2p_1+3p_1^2-2p_1^3)R \quad \mathbf{v}_{01} = (p_1-3p_1^3+2p_1^4)R$$
$$\mathbf{v}_{11} = (p_1-3p_1^2+5p_1^3-2p_1^4)R \quad \mathbf{v}_{10} = (3p_1^2-2p_1^3)R$$

where $R = 1\,/\,(1+3p_1^2-2p_1^3)$. Thus,

$$p_u = (7p_1^2-12p_1^3+10p_1^4-4p_1^5)R$$

which coincides with the result reported in Ref. [110]. ■

4.5 COMPUTER SIMULATION

Computer simulation of a code operation is a powerful tool for analyzing code performance. As we mentioned in Sec 2.5, a direct bit-by-bit error simulation and decoder operation would be inefficient; the simulation of intervals between errors dramatically increases simulation speed.

Suppose that we want to evaluate a block code performance characteristic. If this code corrects up to t errors and detects up to $d-1$ errors, we do not have to simulate decoding of each block. The blocks that contain less than t errors can be identified immediately as the blocks that decoded correctly. Our SSM error source simulator produces intervals between consecutive errors, as shown in Fig. 4.12.

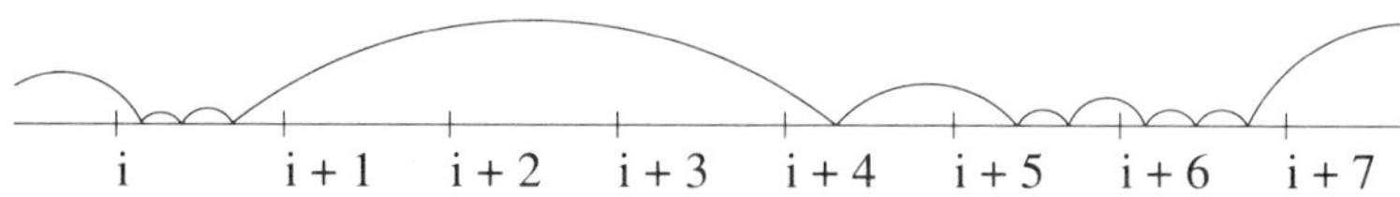

Figure 4.12. Example of an error sequence.

If the code corrects a single error and detects up to two errors, we need to apply the decoding algorithm only to i-th and i+6-th blocks. Suppose that the i-th block was incorrectly decoded and that the decoder detected errors in the i+6-th block. Then, for the situation depicted in Fig. 4.12, the counter of correct blocks on the decoder output is incremented by 4, the counter of blocks with detected (but not corrected) errors is incremented by 2, and the counter of incorrectly decoded blocks is incremented by one.

During the simulation two possibilities exist: either to memorize the error configuration while errors are occurring in the same block and start calculating error syndromes when the number of errors in the block exceeds $d-1$, or to start calculating syndromes with the first error in the block. It is the usual performance and memory trade-off: In the majority of cases the number of errors in the block is less than $d-1$, so there is no need for calculating syndromes.

Similar methods apply to simulation of a convolutional code with the decoder shift registers playing the role of a block. If the number of errors inside the shift registers does not exceed the number of errors that the code can correct, and if definite decoding is used, we can avoid calculating error syndromes. In the case of feedback decoding our approach is different: As long as the number of errors does not exceed the code error-correcting capability, the simulation proceeds the same way as in definite decoding, but when the number of errors exceeds this limit the syndrome registers' contents must be evaluated and the decoding algorithm applied until all the errors are flushed out.

Consider, for example, simulating (8,7,47) CSOC with the generator polynomial matrix[100]

$$G_P(D) = \begin{bmatrix} 1+D^3+D^{19}+D^{42} \\ 1+D^{21}+D^{34}+D^{43} \\ 1+D^{29}+D^{33}+D^{47} \\ 1+D^{25}+D^{36}+D^{37} \\ 1+D^{15}+D^{20}+D^{46} \\ 1+D^2+D^8+D^{32} \\ 1+D^7+D^{17}+D^{45} \end{bmatrix}$$

This code is popular in satellite communications. [100,111] The code is majority-logic decodable and is able to correct up to two errors.

To analyze the effect of feedback decoding on error-burst distribution on the decoder output, we simulated the decoder operation in a BSC without memory. The error sequence on the decoder output may be described as an SSM [110] whose states are defined by the contents of the shift registers. Since the number of states is large, direct SSM simulation is not feasible. However, if we simulate the decoder operation only when shift registers contain errors, the simulation may be performed quite successfully.

Table 4.1

Convolutional Decoder Output Error Burst Weight Distribution

w	1	2	3	4	5	6	>6
$P(w)$	.736	.01	.15	.075	.021	.004	.004

The simulation outcome almost precisely matches (to well within the statistical margin of error) the hardware measurements. Table 4.1 contains the conditional distribution

$$p_n(m)/[1-p_n(0)]$$

of the number of errors in a block of 1024 bits on the decoder output. It shows a very mild increase of burstiness as compared to other types of decoders, which makes CSOCs attractive candidates for concatenation.[100]

4.6 ZERO-REDUNDANCY CODES

4.6.1 Interleaving. As noted above, the majority of codes that are actually implemented are not quite effective in an actual channel, due to the grouping of errors. Special coding procedures are designed to fight against bursty errors, *interleaving* being one them. This method consists of reordering the symbols before transmitting them over a channel so that the symbols of the same code word are not hit by the same burst. The receiver performs the inverse operation, called *deinterleaving*. If the interleaving depth is large, we may treat errors on the deinterleaver output as independent.

The simplest interleaver separates symbols of the same code word by a fixed number of positions. It can be implemented as a memory array: The encoded symbols are read into the array by columns and transmitted by rows. If the array has j columns, then the adjacent symbols of the same code word (column) are separated by $j-1$ symbols on the interleaver output. When the interleaving depth j grows, the SSM model approaches the memoryless error source model. Indeed, in this case the probability of having some combination $\mathbf{e}_1, \mathbf{e}_2, ..., \mathbf{e}_n$ of errors in the block with interleaved positions is equal to

$$p_j(\mathbf{e}_1, \mathbf{e}_2, ..., \mathbf{e}_n) = \boldsymbol{\pi} \prod_{i=1}^{n} \mathbf{P}(\mathbf{e}_i)\mathbf{P}^j \mathbf{1}. \tag{4.6.1}$$

Due to Markov chain regularity (see Appendix 6)

$$\lim_{j \to \infty} \mathbf{P}^j = \mathbf{1}\boldsymbol{\pi}$$

Therefore $\mathbf{P}^j$ is approximately equal to $\mathbf{1}\boldsymbol{\pi}$ for large values of j. Asymptotically, relationship (4.6.1) will have the form

$$p_{\infty}(\mathbf{e}_1, \mathbf{e}_2, ..., \mathbf{e}_n) = \prod_{i=1}^{n} \boldsymbol{\pi}\mathbf{P}(\mathbf{e}_i)\mathbf{1} \tag{4.6.2}$$

Since

$$p(\mathbf{e}_i) = \boldsymbol{\pi}\mathbf{P}(\mathbf{e}_i)\mathbf{1}$$

is a number, equation (4.6.2) can be rewritten as

$$p_{\infty}(\mathbf{e}_1, \mathbf{e}_2, ..., \mathbf{e}_n) = \prod_{i=1}^{n} p(\mathbf{e}_j)$$

This proves that asymptotically, when the interleaving depth tends to infinity, the interleaved errors are statistically independent. From this immediately follows that for large j the effectiveness of codes with interleaving is close to that of codes in the channel without memory. However, in order to achieve better performance, the interleaving depth might be large, thus requiring it to have a large capacity of transmitter and receiver memory as well as a large delay in delivering information to the customer. Using some sophisticated interleaving techniques,[112] it is possible to improve interleaving efficiency.

It follows from the previously derived equations that we can use all the methods of code performance evaluation by replacing $\mathbf{P}(\mathbf{e})$ with $\mathbf{P}_j(\mathbf{e}) = \mathbf{P}(\mathbf{e})\mathbf{P}^j$ when dealing with interleaving. To evaluate an interleaver performance one may use the interleaving effectiveness coefficient

$$\theta_j = p_u^{(j)}/p_u$$

This coefficient is equal to the ratio of the probability of incorrect decoding of an interleaved code word to that for the code without interleaving.

The interleaving may be treated as a particular case of coding with zero redundancy. For instance, the product code of the (n_1,n_1) code and (n_2,k_2) code represents the (n_2,k_2) code with depth n_1 canonical interleaving. There are some other zero-redundancy codes that have numerous applications, which are considered in the next section.

4.6.2 Encryptors and Scramblers. Usually, scramblers are implemented as rate 1 convolutional codes. They are used to randomize the transmitted bit sequence, thereby improving receiver clock recovery. They are also used to eliminate periodic components from a transmitted signal so that it spans more evenly the frequency band allocated to it. Sometimes they are used as inexpensive encryptors to prevent unauthorized reception of the transmitted information.

Encryptors, on the other hand, are viewed as more complex and general transformers than scramblers. They usually represent nonlinear data transformation, which is very difficult to invert unless some additional information (called *key information*) is provided. Block encryption is similar to a block code: The information is transmitted in blocks, which are decrypted independently. Stream encryption is similar to convolutional coding.

Encryptors and scramblers degrade the quality of data communication, and their impact can be evaluated similarly to that of general codes. Usually, they cause error multiplication, and therefore special analysis should be done before they are added to a network. For instance, if a descrambler with three taps precedes a single error correction decoder in BSC with independent errors, then the majority of errors that enter the decoder will not be corrected. For the same reason, an encryptor should be positioned before an encoder, so that the decoder does not have to correct bursts of errors caused by the decryptor.

4.7 CONCLUSION

Using methods developed in Chapter 2, we found basic performance characteristics of error correction and error detection systems. Transform methods proved to be most efficient in calculating matrix probabilities of incorrect decoding, undetected errors, and postdecoding errors. We used characters of GF(q) to calculate these probabilities. In the case of a channel with independent errors some of the results are well known. The structure of the generating functions developed in this chapter allows us to use fast algorithms for calculating matrix probabilities.

The major part of Sec 4.2 represents an introduction to error control coding and is included to simplify understanding of the coding performance evaluation. In this section, we also propose a modified Viterbi algorithm for channels with memory.

Appendix 4 contains results related to this chapter. In Appendix 4.2 we derive closed-form expression for the code-weight generating function of the maximum-distance-separable code. Closed-form expression for the error-event probability of the Viterbi algorithm applied to a general discrete-time communication system with an SSM channel is given in Appendix 4.3.

CHAPTER 5

PERFORMANCE ANALYSIS OF COMMUNICATION PROTOCOLS

5.1 BASIC CHARACTERISTICS OF TWO-WAY SYSTEMS

In two-way systems the information between two stations is transmitted in both directions. The set of rules that governs the communication between the stations is usually called the *link control protocol*. These systems can be divided into three categories: systems with repeat-request, systems with comparison, and those combined.

In systems with automatic repeat-request (ARQ), a receiver instructs the transmitter by sending a negative acknowledgement (NAK) through the return channel to retransmit the same message if it detects noncorrectable errors. Else the receiver acknowledges receipt of the message by sending a positive acknowledgement (ACK). Thus, in these systems the receiver controls the transmission.

In systems using comparison, a transmitter controls data flow. In the simplest case the receiver echoes every message it receives back through the return channel and the transmitter decides whether the message is correct. This method is often used in human-oriented protocols, in which a computer echoes every character it receives and an operator decides whether it is correct. To delete an incorrect character, the operator sends a special erase character.

In combined systems the decision is made by both receiver and transmitter. For example, if an operator misspelled a command and did not notice the error, the computer might ask the operator to retransmit the command.

We consider here only some aspects of link control protocols, as a detailed description of any of them would require its own book and may be found in the corresponding technical documentation. The protocols can be described using sequential machines.[113] Let $\{\mathbf{u}_k\}$ be the source sequence. The transmitter sends a message $\mathbf{x}_k$, which depends on the transmitter state $\mathbf{s}_k$ and the acknowledgement $\boldsymbol{\rho}_{k-1}$ of the previous message received over the return channel. Then the transmitter changes its state to $\mathbf{s}_{k+1}$:

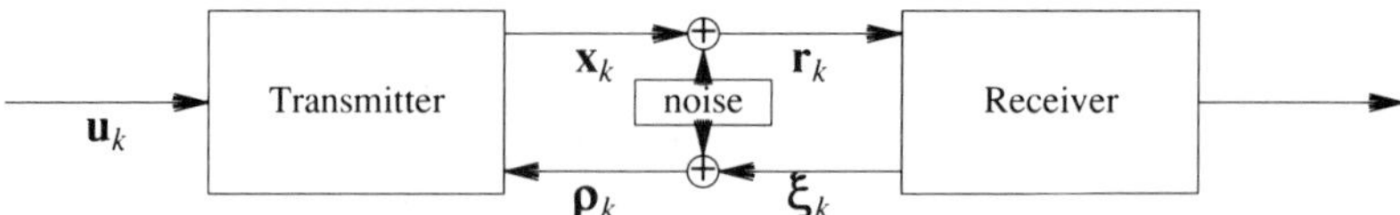

Figure 5.1. Two-way system.

$$\begin{aligned}\mathbf{x}_k &= \mathbf{f}_T(\mathbf{u}_k,\mathbf{s}_k,\boldsymbol{\rho}_{k-1})\\ \mathbf{s}_{k+1} &= \mathbf{g}_T(\mathbf{u}_k,\mathbf{s}_k,\boldsymbol{\rho}_{k-1})\end{aligned}$$

The receiver sends an acknowledgement that depends on the received message $\mathbf{r}_k$ and the receiver state $\boldsymbol{\sigma}_k$:

$$\begin{aligned}\boldsymbol{\xi}_k &= \mathbf{f}_R(\boldsymbol{\sigma}_k,\mathbf{r}_k)\\ \boldsymbol{\sigma}_{k+1} &= \mathbf{g}_R(\boldsymbol{\sigma}_k,\mathbf{r}_k)\end{aligned}$$

In this chapter we analyze two-way system performance on the basis of an SSM error source model. We assume that the matrix probability of receiving

$$\mathbf{r}_k = \mathbf{x}_k + \mathbf{e}_k^{(d)} \qquad \boldsymbol{\rho}_k = \boldsymbol{\xi}_k + \mathbf{e}_k^{(r)}$$

can be expressed as

$$\mathbf{P}(\mathbf{e}_k^{(d)},\mathbf{e}_k^{(r)}) = \mathbf{P}(\mathbf{e}_k^{(d)})\mathbf{P}(e_k^{(r)})$$

This assumption is valid when the same channel is used for both messages and responses (half-duplex systems). It is also valid when the direct and return channels are statistically independent:

$$\mathbf{P}(\mathbf{e}_k^{(d)},\mathbf{e}_k^{(r)}) = \mathbf{P}_d(\mathbf{e}_k^{(d)})\otimes \mathbf{P}_r(\mathbf{e}_k^{(r)}) = \mathbf{P}(\mathbf{e}_k^{(d)})\mathbf{P}(e_k^{(r)})$$

where $\mathbf{P}(\mathbf{e}_k^{(d)}) = \mathbf{P}_d(\mathbf{e}_k^{(d)})\otimes \mathbf{I}$, $\mathbf{P}(\mathbf{e}_k^{(r)}) = \mathbf{I}\otimes \mathbf{P}_r(\mathbf{e}_k^{(r)})$, and the subscripts d and r correspond to direct and return-channel models. If errors in the channels are dependent, it is possible to obtain the same result by increasing the number of model states.

We compare different systems and protocols using the following performance characteristics:

- The probability of correct message reception (p_c)
- The probability of incorrect message reception (p_u)
- The probability $p(\gamma)$ that a message was retransmitted γ times
- The probability of receiving a duplicate message (insertion)
- The probability of losing a message
- The maximum possible link utilization (throughput)

One of the basic goals in designing a communication system is to achieve the information throughput that satisfies customer's needs. Sometimes it is assumed that the information flow rate is the same as that defined by the channel baud rate. However, in noisy channels the same message might be retransmitted several times, so that the information throughput there would be lower than that in a noiseless channel.

In order to design a system with sufficient throughput it is important to know the information source statistics. It is often assumed that the source contains an infinite number of messages and that the system is always busy transmitting them. Another popular assumption is that messages or their components arrive randomly with constant probability (binomial or Poisson data flow). Sometimes it is assumed that messages are generated periodically, as in some telemetry systems, in which control messages and status reports are sent regularly.

The other important system parameter that must be taken into account is the maximum tolerable delay in message transmission. This limitation could be "hard" (each message must not stay in the system longer than the specified time period) or "soft" (the limit is imposed on the message delay probability).

The interaction between messages received over the direct channel and the corresponding control messages sent over the return channel is the most important part of a link control protocol. The direct-channel messages are used for information transmission and are similar to those of the one-way systems we discussed before. Let us now consider return-channel messages.

5.2 RETURN-CHANNEL MESSAGES

As noted above, return-channel noise can cause loss of synchronization and thereby may lead to incorrect reception of long sequences of messages. Therefore, return-channel messages should be encoded in such a way that the probability of their mutual transformation is small. Let us consider various methods of coding the acknowledgement.

We consider first the case in which the return channel is used solely for transmitting the acknowledgement messages. Suppose that $\mathbf{f}_0=(f_{01},f_{02},...,f_{0n})$ represents an ACK and $\mathbf{f}_1=(f_{11},f_{12},...,f_{1n})$ represents an NAK. The ACK is decoded if a received message $\mathbf{f}=(f_1,f_2,...,f_n)$ belongs to a vector set V_0, and the NAK is decoded if $\mathbf{f}\in V_1$. If the ACK and NAK are the only control signals that are sent over the return channel, then V_0 is the complement of V_1.

The probability of correctly decoding these messages is equal to

$$r_{uu} = Pr\{\mathbf{f}_u + \mathbf{e} \in V_u\} \quad u = 0,1 \tag{5.2.1}$$

and the probability of their mutual transformation equals

$$r_{uv} = Pr\{\mathbf{f}_u + \mathbf{e} \in V_v\} \quad u \neq v \tag{5.2.2}$$

where $\mathbf{e} = (e_1, e_2, ..., e_n)$ is the return-channel error.

Equations (5.2.1) and (5.2.2) can be rewritten as

$$r_{uv} = Pr\{\mathbf{e} \in W_{uv}\}$$

where

$$W_{uv} = \{\mathbf{g} : \mathbf{g} = \mathbf{f} - \mathbf{f}_v \quad \mathbf{f} \in V_u\}$$

is the set of differences between vectors of V_u and f_v. The corresponding matrix probability of decoding $\mathbf{f}_v$ when $\mathbf{f}_u$ was sent is

$$\mathbf{R}_{uv} = \sum_{\mathbf{e} \in W_{uv}} \prod_{i=1}^{n} \mathbf{P}(e_i) \tag{5.2.3}$$

Let us analyze several methods of encoding the acknowledgement. If the minimum Hamming distance decoding is used, then $\mathbf{f}_0$ should be the bitwise complement of $\mathbf{f}_1$. (For instance, $\mathbf{f}_0 = (0,0,...,0)$ and $\mathbf{f}_1 = (1,1,...,1)$.) It is natural to define V_0 as the set of all vectors whose Hamming weight is not greater than h and V_1 as its complement ($V_1 = \bar{V}_0$). According to the results of Sec. 2.3.2,

$$\mathbf{R}_{00} = \mathbf{P}_n(t \le h) \quad \mathbf{R}_{01} = \mathbf{P}^n - \mathbf{R}_{00}$$

$$\mathbf{R}_{11} = \mathbf{P}_n(t \le n-h) \quad \mathbf{R}_{10} = \mathbf{P}^n - \mathbf{R}_{11}$$

In some full-duplex systems the acknowledgement messages are encoded together with some other information that is sent over the return channel. The acknowledgement message occupies k_1 symbols out of total k information symbols protected by the (n,k) code. The return-channel decoder corrects up to t_1 and detects up to $d-1$ errors in the block. If the decoder detects non-correctable errors or the designated k_1 symbols do not represent an ACK, the message is decoded as an NAK; otherwise it is decoded as an ACK.

To calculate the matrix probabilities $\mathbf{R}_{uv}$, let us consider all possible message decoding outcomes:

A The message is decoded correctly

B The decoder detects errors in the message, but does not correct them

C The message is decoded incorrectly

U The acknowledgement portion of the message is correct

V The acknowledgement portion of the message is incorrect

The above-defined error sets W_{uv} may be expressed as

$$W_{00} = A \cup (C \cap U) \quad W_{01} = B \cup (C \cap V)$$
$$W_{10} = C \cap V \quad W_{01} = A \cup B \cup (C \cap V)$$

The matrix probabilities $\mathbf{R}_{uv}$ are defined by equation (5.2.3) and can be found using methods developed in Sec. 4.3. Usually it is possible to neglect the probabilities of events $C \cap U$ and $C \cap V$ as compared to the probabilities of A and B. In this case,

$$\begin{aligned} \mathbf{R}_{00} &\approx \mathbf{P}_n(t \le t_1) \quad \mathbf{R}_{01} = \mathbf{P}^n - \mathbf{R}_{00} \\ \mathbf{R}_{10} &= \mathbf{P}_{cv} \quad \mathbf{R}_{11} = \mathbf{P}^n - \mathbf{R}_{10} \end{aligned} \tag{5.2.4}$$

where $\mathbf{P}_{cv}$ is the matrix probability of not detecting errors in the block and of misinterpreting the control sub-block for the ACK when the NAK was transmitted. This probability can be calculated using the modified equation (4.3.27) in which only the errors that cause the NAK $\to$ ACK transformation are considered. Suppose that the first k_1 bits of a code word are used to encode the acknowledgement information

$$\text{ACK} \sim \mathbf{f}_0 = (0,0,\ldots,0) \text{ and NAK} \sim \mathbf{f}_1 = (1,1,\ldots,1)$$

Then we obtain from modified equation (4.3.27)

$$\mathbf{R}_{00} = 2^{-r} \sum_{|\mathbf{e}| \le t} \sum_{\mathbf{v}} (-1)^{-(\mathbf{eH}')\mathbf{v}} \prod_{j=1}^{n} [\mathbf{P}(0) + \delta_j \mathbf{P}(1)(-1)^{\mathbf{vh}_j}] \tag{5.2.5}$$

$$\mathbf{R}_{10} = 2^{-r} \sum_{|\mathbf{e}| \le t} \sum_{\mathbf{v}} (-1)^{-(\mathbf{eH}')\mathbf{v}} \prod_{j=1}^{n} [\delta_j \mathbf{P}(0) + \mathbf{P}(1)(-1)^{\mathbf{vh}_j}] \tag{5.2.6}$$

where $\delta_j = 0$ for $j \le k_1$ and $\delta_j = 1$ otherwise.

EXAMPLE 5.2.1: Suppose that the (7,4) Hamming code is used for error detection in the return channel and the first bit of each code word is equal to 0 if it is an ACK, otherwise this bit is set to 1. Let us calculate the return-channel characteristics $\mathbf{R}_{uv}$.

Equation (5.2.5) gives (see Example 4.3.2)

$$\mathbf{R}_{00} = 2^{-3}\mathbf{P}(0) \sum_{i,j,k=0}^{1} \mathbf{F}_1((-1)^i,(-1)^j,(-1)^k)$$

where

$$\mathbf{F}_1(z_1,z_2,z_3) = [\mathbf{P}(0)+\mathbf{P}(1)z_1z_2][\mathbf{P}(0)+\mathbf{P}(1)z_1z_2z_3]\times$$
$$[\mathbf{P}(0)+\mathbf{P}(1)z_2z_3][\mathbf{P}(0)+\mathbf{P}(1)z_1z_3][\mathbf{P}(0)+\mathbf{P}(1)z_2][\mathbf{P}(0)+\mathbf{P}(1)z_3]$$

and equation (5.2.6) becomes

$$\mathbf{R}_{10} = 2^{-3}\mathbf{P}(1) \sum_{i,j,k=0}^{1} (-1)^i\mathbf{F}_1((-1)^i,(-1)^j,(-1)^k)$$

From (5.2.4) we obtain the rest of the matrix probabilities: $\mathbf{R}_{01} = \mathbf{P}^7-\mathbf{R}_{00}$, $\mathbf{R}_{11} = \mathbf{P}^7-\mathbf{R}_{10}$.

In the case of a BSC with $\mathbf{P}(0) = q$ and $\mathbf{P}(1) = p$ these equations are simplified:

$$r_{00} = 2^{-3}q[1+(q-p)^3(7-8p)] \qquad r_{01} = 1 - r_{00}$$
$$r_{10} = 2^{-3}p[1+(q-p)^3(1-8p)] \qquad r_{11} = 1 - r_{10}$$

■

Consider now the case when an ACK is encoded by any code vector, while an NAK is represented by a certain noncode vector. The ACK is decoded correctly if the decoder does not detect errors, that is, when a syndrome $\mathbf{S}$ belongs to the set C of correctable errors. As in (4.3.27), we obtain the matrix probability of decoding the ACK correctly:

$$\mathbf{R}_{00} = 2^{-r} \sum_{|\mathbf{e}|\le t} \sum_{\mathbf{v}} (-1)^{-(\mathbf{eH}')\mathbf{v}} \prod_{k=1}^{n} [\mathbf{P}(0) + \mathbf{P}(1)(-1)^{\mathbf{vh}_k}]$$

The NAK can be encoded either by selecting a predefined noncode vector or by deliberately adding errors into transmitted blocks.

If a fixed non-code vector $\mathbf{f}$ represents the NAK, then the probability of its incorrect decoding is equal to the probability of receiving a word $\mathbf{f}+\mathbf{e}$ whose syndrome belongs to the set of correctable errors

$$\mathbf{R}_{10} = \mathbf{Pr}\{\mathbf{S}(\mathbf{f} + \mathbf{e}) \in C\}$$

Since a syndrome is a linear function, the previous equation may be written as

$$\mathbf{R}_{10} = \mathbf{Pr}\{\mathbf{S}(\mathbf{e}) \in C_{\Delta}\}$$

where $C_{\Delta} = \{\mathbf{h} : \mathbf{h} = \mathbf{S}-\mathbf{S}(\mathbf{f}),\ \mathbf{S} \in C\}$. Using equation (4.3.25) we can express the probability $\mathbf{R}_{10}$ through generating functions (see Sec. 4.2.2):

$$\mathbf{R}_{10} = 2^{-r} \sum_{\mathbf{S} \in C_{\Delta}} \sum_{\mathbf{v}} z^{-\mathbf{S}\mathbf{v}} \mathbf{F}(z^{\mathbf{v}})$$

Consider now the case when the NAK is formed by adding to a code vector $\mathbf{c}_i$ of a fixed error configuration $\boldsymbol{\psi}$, that is, the NAK is encoded as $\mathbf{c}_i+\boldsymbol{\psi}$. If all code words are equally probable, we can assume that the all-zero code word ($\mathbf{c}_i = 0$) was transmitted, so the probability $\mathbf{R}_{10}$ can be computed using the previous equation.

5.3 SYNCHRONIZATION

Synchronization is one of the most important tasks in digital communications. The receiver is synchronized with the transmitter if it recognizes the beginning and the end of each block of data. Depending on the size of the blocks, several synchronization levels exist (bit, byte, message, etc). For each level there are procedures of restoring and maintaining the synchronization that are usually based on block framing: Some redundant symbols are sent before (*preamble*) and after (*postamble*) the block. The framed block of data is called a *frame*.

Synchronization recovery can be done most effectively together with message decoding. Special codes enable fast synchronization recovery.[88] However, in the majority of systems only framing symbols are used, and synchronization recovery is performed independently of information decoding. In systems with fixed block size, framing information is transmitted periodically, so that a synchronization algorithm can correlate the framing symbols of several blocks. If messages have variable lengths, then each block is usually synchronized independently of the others, using its own framing symbols.

The simplest technique of synchronization recovery consists of using a predefined sequence of framing symbols. If this sequence does not match the received candidate sequence, the next symbol is tried as a possible framing position, and so on until a satisfactory agreement is reached. In *asynchronous* communication each character (usually eight bits) is framed by so-called start and stop bits, and synchronization is achieved at the beginning of each character. In *synchronous* communication the synchronization is maintained continuously.

In block coding it is important to synchronize at the beginning of each block. Suppose that every code word is framed with a preamble $\mathbf{h} = (h_1, h_2, \ldots, h_u)$ and a postamble $\mathbf{t} = (t_1, t_2, \ldots, t_v)$. Except for the first message, we can treat those two sequences as one [$\mathbf{s} = (\mathbf{t},\ \mathbf{h})$] and call it a synchronization block (SYNC).

In calculating a one-sided system's performance characteristics we assumed that code words were perfectly synchronized. Without this assumption the calculations would have been more complex. We must add the condition of synchronization to every calculation. For instance, if each code word is synchronized independently of the others, then the matrix probability of correct decoding can be expressed as

$$\mathbf{P}_{\text{syn}}\mathbf{P}_c$$

where $\mathbf{P}_{\text{syn}}$ is the matrix probability that the block is synchronized. For example, if the receiver is decoding SYNC that contains less than b errors, then the synchronization matrix probability is given by the matrix binomial probability (defined in Sec 2.3.3):

$$\mathbf{P}_{\text{syn}} = \mathbf{P}_{u+v}(m<b)$$

In the particular case when the receiver is looking for a perfect match

$$\mathbf{P}_{\text{syn}} = \mathbf{P}^{u+v}(0)$$

The matrix probabilities of other events that characterize block code performance can be found in a similar manner.

If the receiver is not able to identify a message start, the whole message and several following messages might be lost or incorrectly decoded. However, the probability of incorrectly decoding a misframed message is usually much smaller than the probability of losing the message, and it therefore can be neglected. The assumption that the synchronization may be lost introduces an additional event—loss of the message—to the set of possible outcomes of the message transmission. The matrix probability of a frame loss is given by

$$\mathbf{P}_{\Lambda} = \mathbf{P}^{u+v} - \mathbf{P}_{\text{syn}}$$

EXAMPLE 5.3.1: Binary Synchronous Communication Control (BISYNC) protocol[114] uses two eight-bit characters (SYN SYN) for data-block framing. The frame loss probability is given by

$$p_{\Lambda} = 1 - \boldsymbol{\pi}\mathbf{P}^{16}(0)\mathbf{1}$$

if we assume that an exact match is required.

■

In synchronous communications the receiver has usually two modes of operation: an in-frame mode and a framing search mode. In contrast with asynchronous communications, there is no need to resynchronize each frame independently from the previous frames. If the previous frames were synchronized and the framing pattern of the current frame is slightly different from that expected, the receiver may decide that this difference is caused by channel errors and does not initiate a framing search. The out-of-frame detection algorithm must be robust enough to avoid the interruption of transmission due to unnecessary framing searches. On the other hand, the algorithm must be sensitive enough to avoid the reception of long sequences of misframed information.

EXAMPLE 5.3.2: Let us calculate the probability distribution of the interval between false out-of-frame detection (misframing) by the algorithm used in the T1 channel format. A T1 frame consists of 192 information bits and one framing bit. The T1 framing algorithm declares an out-of-frame condition when two errors in four consecutive framing bits are detected.

This algorithm can be implemented in a four-bit shift register. Each framing bit is shifted into this register. The receiver compares the register contents with the locally generated framing bit pattern. If they disagree in more than one position, the algorithm declares the out-of-frame condition.

Suppose that $\boldsymbol{\alpha}$ is the probability distribution of SSM model states after synchronization recovery and that the shift register contains no errors; this state of the shift register we denote as $\mathbf{s}_0 = [0\ 0\ 0\ 0]$. If this register contains only one error (that is, $\mathbf{s}_1 = [0\ 0\ 0\ 1]$, or $\mathbf{s}_2 = [0\ 0\ 1\ 0]$, or $\mathbf{s}_3 = [0\ 1\ 0\ 0]$, or $\mathbf{s}_4 = [1\ 0\ 0\ 0]$), there is no misframing. The receiver erroneously declares an out-of-frame condition ($\mathbf{s}_5$) when the shift register contains at least two errors. As we pointed out in the previous chapter, shift register behavior can be described using Markov functions. In our case, the transition probability matrix has the form

$$\begin{bmatrix} \mathbf{Q}(0) & \mathbf{Q}(1) & 0 & 0 & 0 & 0 \\ 0 & 0 & \mathbf{Q}(0) & 0 & 0 & \mathbf{Q}(1) \\ 0 & 0 & 0 & \mathbf{Q}(0) & 0 & \mathbf{Q}(1) \\ 0 & 0 & 0 & 0 & \mathbf{Q}(0) & \mathbf{Q}(1) \\ \mathbf{Q}(0) & 0 & 0 & 0 & 0 & \mathbf{Q}(1) \\ 0 & 0 & 0 & 0 & 0 & \mathbf{I} \end{bmatrix}$$

where $\mathbf{Q}(0) = \mathbf{P}(0)\mathbf{P}^{192}$, $\mathbf{Q}(1) = \mathbf{P}(1)\mathbf{P}^{192}$. According to Sec. 2.4.3, the interval between misframes has the matrix-geometric distribution

$$Pr(l) = \mathbf{p}\mathbf{P}^{l-2}(A_1 | A_1)\mathbf{q}$$

where $\mathbf{p} = [\ \boldsymbol{\alpha}\ \ 0\ \ 0\ \ 0\ \ 0\ ]$, $\mathbf{q} = [\ 0\ \ \mathbf{Q}(1)\ \ \mathbf{Q}(1)\ \ \mathbf{Q}(1)\ \ \mathbf{Q}(1)\ ]'$ is the last column of the previous matrix, and

$$\mathbf{P}(A_1 | A_1) = \begin{bmatrix} \mathbf{Q}(0) & \mathbf{Q}(1) & 0 & 0 & 0 \\ 0 & 0 & \mathbf{Q}(0) & 0 & 0 \\ 0 & 0 & 0 & \mathbf{Q}(0) & 0 \\ 0 & 0 & 0 & 0 & \mathbf{Q}(0) \\ \mathbf{Q}(0) & 0 & 0 & 0 & 0 \end{bmatrix}$$

Let us now calculate the mean time between misframes. The generating function of the misframe interval distribution (see Sec. 2.4.3) is equal to

$$\theta_1(z) = z\mathbf{p}[\mathbf{I} - z\mathbf{P}(A_1 | a_1)]^{-1}\mathbf{q}$$

The mean time between misframes can be expressed as a derivative of this generating function:

$$\bar{l} = \theta'_1(1) = \mathbf{p}[\mathbf{I} - \mathbf{P}(A_1 | A_1)]^{-1}\mathbf{q} + \mathbf{p}[\mathbf{I} - \mathbf{P}(A_1 | A_1)]^{-1}\mathbf{P}(A_1 | A_1)\mathbf{q}$$

which after applying the formula

$$[\mathbf{I} - \mathbf{P}(A_1 | A_1)]^{-1}\mathbf{q} = \mathbf{1}$$

becomes

$$\bar{l} = \theta'_1(1) = \mathbf{p}[\mathbf{I} - \mathbf{P}(A_1 | A_1)]^{-1}\mathbf{1}$$

After calculating the inverse matrix we obtain

$$\bar{l} = 1 + \boldsymbol{\alpha}\mathbf{X}\mathbf{Q}(1)[\mathbf{I} - \mathbf{Q}(0)]^{-1}\mathbf{1}$$

where

$$\mathbf{X} = [\mathbf{I} - \mathbf{Q}(0) - \mathbf{Q}(1)\mathbf{Q}^4(0)]^{-1}$$

In particular, for a memoryless channel with bit-error probability p, the previous expression takes the form

$$\bar{l} = \frac{1 + p + pq + pq^2 + pq^2}{p^2(1 + q + q^2 + q^3)} \qquad \blacksquare$$

In two-sided systems the problem of synchronization becomes even more important, because we also have to maintain correspondence between the messages that are sent over the channels in both directions. The synchronization between the messages that are sent over the direct channel and the responses received from the return channel we call an MR-sync. Any insertion or loss in the absence of supplementary redundancy may lead to incorrect reception of a group of messages as a result of synchronization loss between the messages and their acknowledgements. Therefore, the ARQ system should have a provision to maintain and recover correspondence between the messages and their acknowledgements.

One method of maintaining synchronization consists of sending each message number as a part of the message and of its acknowledgement. If the same message is received more than once, the receiver recognizes it by its number and the message is discarded. If a message is lost, the receiving station sends a request for its retransmission. In a case when no acknowledgement is received, the station retransmits all the nonacknowledged messages after waiting for an acknowledgement for a certain period of time (timeout). The details of this technique are considered in the following sections.

5.4 ARQ PERFORMANCE CHARACTERISTICS

The matrix probabilities $\mathbf{R}_{uv}$ may be used to determine the probabilities of control message decoding outcomes. However, they alone cannot be used as ARQ system performance characteristics. We must describe the system operation (including the direct channel) and find the probabilities of the system states.

For a frame transmitted over a direct channel, we define the following events, which are related to the results of its decoding:

A The frame is decoded correctly

B The decoder detects, but does not correct, errors in the frame

C The frame is decoded incorrectly

We also denote $F = A \cup B$ and $Q = B \cup C$. The events related to decoding of the return-channel control messages we defined as W_{uv} in Sec. 5.2, where

$u=0$ If an ACK was transmitted

$u=1$ If an NAK was transmitted

$v=0$ If the ACK was decoded

$v=1$ If the NAK was decoded

As mentioned before, we assume that the matrix probabilities of the sequence of events related to both channels can be expressed as products of the matrix probabilities of individual events. For instance, the matrix probability that a frame has been decoded correctly (A) and the corresponding ACK was also decoded correctly (W_{00}) can be expressed as

$$\mathbf{P}_{A0} = \mathbf{Pr}\{A \cap W_{00}\} = \mathbf{P}_A \mathbf{R}_{00}$$

Quite analogously, we define the rest of the matrix probabilities that are needed for describing the system operation:

$$\mathbf{P}_{Av} = \mathbf{Pr}\{A \cap W_{0v}\} = \mathbf{P}_A \mathbf{R}_{0v} \quad \mathbf{P}_{Cv} = \mathbf{Pr}\{C \cap W_{0v}\} = \mathbf{P}_C \mathbf{R}_{0v}$$
$$\mathbf{P}_{Bv} = \mathbf{Pr}\{B \cap W_{1v}\} = \mathbf{P}_B \mathbf{R}_{1v} \quad v=0,1$$

5.4.1 Stop-and-Wait ARQ. Stop-and-wait is one of the simplest and most broadly used ARQ link controls.[114—117] With this scheme, after a communication link has been established the transmitter sends a frame, then stops and waits for an acknowledgement.

Let us consider first the simplest version of the system, which assumes that frames are perfectly synchronized in each channel and that only an MR-sync loss may happen. There is no frame numbering; every direct-channel frame is acknowledged with an ACK or NAK, depending on the result of decoding the frame. We assume that each message occupies a single frame, so that the frame loss affects only a single message and does not lead to discarding adjacent frames. The system operation is shown in Fig. 5.2.

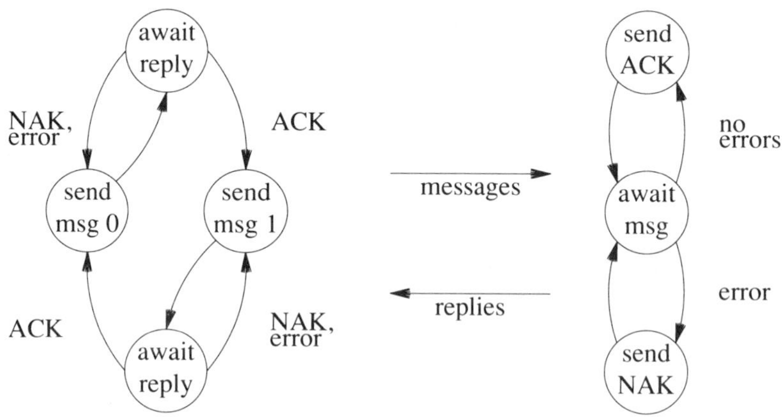

Figure 5.2. Simple ARQ system.

In order to evaluate system performance, let us define all possible outcomes of transmitting a message.

- G The message is received correctly the first time. This outcome occurs if the previous message has been positively acknowledged and the current message has been decoded correctly; the matrix probability of the transition $G \rightarrow G$ is equal to $\mathbf{P}_{GG} = \mathbf{R}_{00}\mathbf{P}_A$.
- U The message is received incorrectly the first time. This outcome occurs if the previous message has been positively acknowledged and the current

message has been decoded incorrectly; the matrix probability of the transition $G \rightarrow U$ can be expressed as $\mathbf{P}_{GU} = \mathbf{R}_{00}\mathbf{P}_C$.

I The message is received without detected errors, but not the first time. This outcome occurs if the returning ACK has been improperly interpreted as an NAK and no errors in the direct-channel message have been detected; the matrix probability of the transition $G \rightarrow I$ is given by $\mathbf{P}_{GI} = \mathbf{R}_{01}\mathbf{P}_F$, where $\mathbf{P}_F = \mathbf{P}_A + \mathbf{P}_C$.

S The message is received, but not the first time, and has detectable errors. This outcome occurs if the returning ACK has been improperly interpreted as an NAK and errors in the direct-channel message have been detected; the matrix probability of the transition $G \rightarrow S$ is given by $\mathbf{P}_{GS} = \mathbf{R}_{01}\mathbf{P}_B$.

D The message will be retransmitted next time. This outcome occurs if the previous ACK was decoded correctly, the transmitted message contains detected errors, and its NAK is decoded correctly; the matrix probability of the transition $G \rightarrow D$ is equal to $\mathbf{P}_{GD} = \mathbf{R}_{00}\mathbf{P}_B\mathbf{R}_{11}$.

L The message is lost. This outcome occurs if the previous ACK was decoded correctly, the transmitted message contains detected errors, and its NAK is decoded incorrectly; the matrix probability of the transition $G \rightarrow L$ is equal to $\mathbf{P}_{GL} = \mathbf{R}_{00}\mathbf{P}_B\mathbf{R}_{10}$.

These events make it possible to describe the process on the receiver output as an SSM. For instance, let

$$A_1, A_2, A_2, C_2, B_2, A_3, A_4, B_5, B_5, A_6,$$
$$B_6, A_6, B_7, B_8, B_9, B_9, A_9, C_{10}, A_{10}$$

be a sequence of received messages where capital letters denote the previously defined results of message decoding and indices denote the transmitter output message sequence number. Thus, A_1 means that the first message was decoded correctly, and B_9 means that the decoder detected errors in the ninth source message. Using the above-defined message transmission outcomes, the previous sequence can be transformed into

$$G, G, I, I, S, G, G, D, L, G, S, I, L, L, D, D, G, U, I \tag{5.4.1}$$

The transformed sequence of events allows us to determine all basic performance characteristics. The probability of receiving a correct message is the probability of *G* in the sequence of the events, the probability of inserting a message is the probability of *I*, and so on. The transformed sequence can be described as an SSM, because the outcomes of message transition depend deterministically on the sequence of errors in both channels. The transition probability matrix of this SSM has the form

$$\mathbf{T} = [\mathbf{B}_G \quad \mathbf{B}_G \quad \mathbf{B}_G \quad \mathbf{B}_S \quad \mathbf{B}_M \quad \mathbf{B}_M]' \tag{5.4.2}$$

where

$$\begin{aligned}
\mathbf{B}_G &= [\,\mathbf{P}_{GG} \quad \mathbf{P}_{GU} \quad \mathbf{P}_{GI} \quad \mathbf{P}_{GS} \quad \mathbf{P}_{GD} \quad \mathbf{P}_{GL}\,] \\
\mathbf{B}_S &= [\,\mathbf{R}_{10}\mathbf{P}_A \quad \mathbf{R}_{10}\mathbf{P}_C \quad \mathbf{R}_{11}\mathbf{P}_F \quad \mathbf{R}_{11}\mathbf{P}_B \quad \mathbf{R}_{10}\mathbf{P}_B\mathbf{R}_{11} \quad \mathbf{R}_{10}\mathbf{P}_B\mathbf{R}_{11}\,] \\
\mathbf{M}_S &= [\,\mathbf{P}_A \quad \mathbf{P}_C \quad 0 \quad 0 \quad \mathbf{P}_B\mathbf{R}_{11} \quad \mathbf{P}_B\mathbf{R}_{10}\,]
\end{aligned}$$

Calculation of the first row $\mathbf{B}_G$ of matrix (5.4.2) was explained during the message outcome definition; the remaining transition probabilities are calculated quite analogously.

The ARQ system performance characteristics are expressed as the stationary probabilities

$$\mathbf{p} = [\ \mathbf{p}(G) \ \ \mathbf{p}(U) \ \ \mathbf{p}(I) \ \ \mathbf{p}(S) \ \ \mathbf{p}(D) \ \ \mathbf{p}(L)\ ] \tag{5.4.3}$$

of matrix (5.4.2), assuming that this matrix is regular. The stationary probabilities are found from the system

$$\mathbf{pT} = \mathbf{p} \qquad \mathbf{p1} = 1$$

which can be simplified, because the matrix has several identical rows. If we denote

$$\mathbf{X} = \mathbf{p}(G)+\mathbf{p}(U)+\mathbf{p}(I) \quad \mathbf{Y} = \mathbf{p}(S) \quad \mathbf{Z} = \mathbf{p}(D)+\mathbf{p}(L)$$

$$\mathbf{c}_1 = \mathbf{XR}_{00} + \mathbf{YR}_{10} + \mathbf{Z} \quad \mathbf{c}_2 = \mathbf{XR}_{01} + \mathbf{YR}_{11}$$

then the previous system can be written as

$$\begin{aligned}
&\mathbf{p}(G) = \mathbf{c}_1\mathbf{P}_A \quad \mathbf{p}(U) = \mathbf{c}_1\mathbf{P}_C \quad \mathbf{p}(I) = \mathbf{c}_2\mathbf{P}_F \\
&\mathbf{p}(S) = \mathbf{c}_2\mathbf{P}_B \quad \mathbf{p}(D) = \mathbf{c}_2\mathbf{P}_B\mathbf{R}_{11} \quad \mathbf{p}(L) = \mathbf{c}_2\mathbf{P}_B\mathbf{R}_{10}
\end{aligned}$$

From this system it follows that

$$\begin{aligned}
\mathbf{X} &= [\ (\mathbf{X} + \mathbf{Y})\mathbf{S} + \mathbf{Z}\ ]\mathbf{P}_F \\
\mathbf{Y} &= (\mathbf{XR}_{01} + \mathbf{YR}_{11})\mathbf{P}_B \\
\mathbf{Z} &= (\mathbf{XR}_{00} + \mathbf{YR}_{10} + \mathbf{Z})\mathbf{P}_B\mathbf{S}
\end{aligned}$$

and

$$(\mathbf{X} + \mathbf{Y})\mathbf{S} + \mathbf{Z} = \boldsymbol{\pi}$$

where $\mathbf{S} = \mathbf{P}_F + \mathbf{P}_B$ and $\boldsymbol{\pi}$ is the stationary distribution of the Markov chain with the matrix $\mathbf{S}$: $\boldsymbol{\pi}\mathbf{S} = \boldsymbol{\pi}$, $\boldsymbol{\pi}\mathbf{1} = 1$. Solving this system, we obtain

$$\mathbf{p}(G) = \mathbf{c}_1\mathbf{P}_A \quad \mathbf{p}(U) = \mathbf{c}_1\mathbf{P}_C \quad \mathbf{p}(I) = \mathbf{c}_2\mathbf{P}_F$$

$$\mathbf{p}(S) = \mathbf{c}_2\mathbf{P}_B \quad \mathbf{p}(D) = \mathbf{c}_2\mathbf{P}_B\mathbf{R}_{11} \quad \mathbf{p}(L) = \mathbf{c}_2\mathbf{P}_B\mathbf{R}_{10}$$

where

$$\mathbf{c}_1 = \boldsymbol{\pi}\mathbf{H}_0(\mathbf{I} - \mathbf{P}_B\mathbf{R}_{11})^{-1} \qquad \mathbf{H}_0 = \mathbf{P}_F\mathbf{R}_{00} + \mathbf{P}_B\mathbf{R}_{10}$$

$$\mathbf{c}_2 = \boldsymbol{\pi}\mathbf{P}_F\mathbf{R}_{01}(\mathbf{I} - \mathbf{P}_B\mathbf{R}_{11})^{-1}$$

Using stationary distribution (5.4.3), we can find the ARQ system performance characteristics:

$\mathbf{c}_1\mathbf{P}_A\mathbf{1}$	The probability that the message is received correctly (for the first time)
$\mathbf{c}_1\mathbf{P}_C\mathbf{1}$	The probability that the message is received incorrectly (for the first time)
$\mathbf{c}_2\mathbf{P}_F\mathbf{1}$	The probability that the message is an insertion of the previous message
$\mathbf{c}_1\mathbf{P}_B\mathbf{R}_{10}\mathbf{1}$	The probability that the message is lost

To calculate the system throughput, we denote $n+\Delta$ the minimal amount of time between transmissions of two adjacent messages (measured in the number of symbols that the transmitter could have sent during this period). If there were no retransmissions, insertions, or losses, the throughput would have been $k/(n+\Delta)$. However, only the messages of the type $G \cup U$ should be taken into account, so that the throughput is

$$\tau = [\mathbf{p}(G) + \mathbf{p}(U)]\mathbf{1}k/(n+\Delta)$$

Let us now find interval distributions that characterize sequences of messages on decoder output.

5.4.2 Message Delay Distribution. Let us calculate the probability $Pr(\gamma)$ that the number of message retransmissions until the message is received (correctly or incorrectly) for the first time is equal to γ. This probability can be found as the conditional probability that the previous message has been positively acknowledged and the current message remains in one of the states D for the period γ until it reaches $G \cup U$ for the first time. In other words, $Pr(\gamma)$ is the probability of a series of γ events D in sequence (5.4.1) that ends with G or U.

Using methods described in Sec. 2.4.3, we obtain

$$Pr(\gamma) = \mathbf{c}_1(\mathbf{I} - \mathbf{P}_B\mathbf{R}_{11})(\mathbf{P}_B\mathbf{R}_{11})^{\gamma}\mathbf{P}_F\mathbf{1} \,/\, \mathbf{c}_1\mathbf{P}_F\mathbf{1} \tag{5.4.4}$$

The generating function of this distribution has the form

$$\phi(z) = \sum_{\gamma=0}^{\infty} Pr(\gamma)z^{\gamma} = \mathbf{K}(\mathbf{I} - \mathbf{P}_B\mathbf{R}_{11})(\mathbf{I} - \mathbf{P}_B\mathbf{R}_{11}z)^{-1}\mathbf{P}_F\mathbf{1}$$

where $\mathbf{K} = \mathbf{c}_1/\mathbf{c}_1\mathbf{P}_F\mathbf{1}$. This generating function can be used to calculate the distribution moments. The average delay is equal to

$$\bar{\gamma} = \phi'(1) = \mathbf{K}(\mathbf{I} - \mathbf{P}_B\mathbf{R}_{11})^{-1}\mathbf{P}_B\mathbf{R}_{11}\mathbf{P}_F\mathbf{1}, \tag{5.4.5}$$

and the variance of the delay is

$$\sigma_{\gamma}^2 = \phi''(1) - \bar{\gamma} - \bar{\gamma}^2$$

where

$$\phi''(1) = 2\mathbf{K}(\mathbf{I} - \mathbf{P}_B\mathbf{R}_{11})^{-2}(\mathbf{P}_B\mathbf{R}_{11})^2\mathbf{P}_F\mathbf{1}$$

The message-delay distribution differs from the period of time that the message occupies the system, because it might be received several times (in the case of MR-sync loss). Obviously, the period of time that the system is busy transmitting the message is equal to the length of the series of NAKs received over the return channel. Similarly to (5.4.4), we obtain

$$Pr(\beta) = \boldsymbol{\pi}\mathbf{H}_0\mathbf{H}_1^{\beta-1}\mathbf{H}_0\mathbf{1} \,/\, \boldsymbol{\pi}\mathbf{H}_0\mathbf{1}$$

where $\mathbf{H}_1 = \mathbf{P}_F\mathbf{R}_{01} + \mathbf{P}_B\mathbf{R}_{11}$. The average busy period is equal to

$$\bar{\beta} = 1/\boldsymbol{\pi}\mathbf{H}_0\mathbf{1}$$

and can be found quite analogously to (5.4.5).

5.4.3 Insertion and Loss Interval Distribution. It is important to know the distribution of synchronization loss time. According to our assumptions, only MR-sync loss can happen as a result of incorrectly decoding the acknowledgements. Let us first find the distribution of the intervals of the positive synchronization loss (which is a series of events $I \cup S$). The length i of the series is equal to the time that the Markov chain with matrix (5.4.2) stays in the states $I \cup S$. The insertion length distribution can be found using the methods of Sec. 2.4.3. Denote

$$\mathbf{U}_{00} = \begin{bmatrix} \mathbf{R}_{01}\mathbf{P}_F & \mathbf{R}_{01}\mathbf{P}_B \\ \mathbf{R}_{11}\mathbf{P}_F & \mathbf{R}_{11}\mathbf{P}_B \end{bmatrix} = \begin{bmatrix} \mathbf{R}_{01} \\ \mathbf{R}_{11} \end{bmatrix} \begin{bmatrix} \mathbf{P}_F & \mathbf{P}_B \end{bmatrix} \tag{5.4.6}$$

the sub-block of matrix (5.4.2), which corresponds to transitions between states $I \cup S$. Then the probability distribution of the insertion series length can be expressed as

$$Pr(i) = \mathbf{q}(\mathbf{I}-\mathbf{U}_{00})\mathbf{U}_{00}^{i-1}(\mathbf{I}-\mathbf{U}_{00})\mathbf{1} \,/\, \mathbf{q}(\mathbf{I}-\mathbf{U}_{00})\mathbf{1}$$

where $\mathbf{q} = [\ \mathbf{p}(I)\ \ \mathbf{p}(S)\ ]$. Using factorization (5.4.6), we can simplify this expression:

$$Pr(i) = \mathbf{c}_2\mathbf{H}_1^{i-1}\mathbf{H}_0^2\mathbf{1} \,/\, \mathbf{c}_2\mathbf{H}_0\mathbf{1}$$

The average length of the series is equal to

$$\bar{i} = \mathbf{c}_2\mathbf{1} \,/\, \mathbf{c}_2\mathbf{H}_0\mathbf{1}$$

The probability distribution and the average of the series of consecutive frame losses can be found quite analogously:

$$Pr(l) = \mathbf{c}_3\mathbf{P}_{B0}^{l-1}(\mathbf{I}-\mathbf{P}_{B0})\mathbf{1} \,/\, \mathbf{c}_3\mathbf{1} \qquad \bar{l} = \mathbf{p}(L) \,/\, \mathbf{c}_3\mathbf{1}$$

where $\mathbf{c}_3 = \mathbf{p}(L)(\mathbf{I}-\mathbf{P}_{B0})$ and $\mathbf{P}_{B0} = \mathbf{P}_B\mathbf{R}_{10}$.

5.4.4 Accepted Messages. The SSM with matrix (5.4.2) describes the sequence of messages on decoder output. Some of the messages that contain detected errors have no effect on the messages that are ultimately accepted. The sequence of accepted messages is obtained by the type of D and S messages removed from the decoder output. For example, the sequence (5.4.1) after such removal becomes

$$G,\ G,\ I,\ I,\ G,\ G,\ L,\ G,\ I,\ L,\ L,\ G,\ U,\ I.$$

Obviously, this sequence represents an SSM. Its matrix is obtained from matrix (5.4.2) by eliminating the unobserved sets of states D and S (see Appendix 6.1.3). After performing the necessary transformations we obtain

$$\mathbf{T}_1 = \begin{bmatrix} \mathbf{W}_1\mathbf{P}_A & \mathbf{W}_1\mathbf{P}_C & \mathbf{W}_2\mathbf{P}_F & \mathbf{W}_1\mathbf{P}_B\mathbf{R}_{10} \\ \mathbf{W}_1 P_A & \mathbf{W}_1 P_C & \mathbf{W}_2 P_F & \mathbf{W}_1 P_B\mathbf{R}_{10} \\ \mathbf{W}_1 P_A & \mathbf{W}_1 P_C & \mathbf{W}_2 P_F & \mathbf{W}_1 P_B\mathbf{R}_{10} \\ \mathbf{W}_3 P_A & \mathbf{W}_3 P_C & 0 & \mathbf{W}_3 P_B\mathbf{R}_{10} \end{bmatrix} \tag{5.4.7}$$

where $\mathbf{W}_3 = (\mathbf{I}-\mathbf{P}_B\mathbf{R}_{11})^{-1}$, $\mathbf{W}_2 = \mathbf{R}_{01}\mathbf{W}_3$, and $\mathbf{W}_1 = (\mathbf{R}_{00}+\mathbf{R}_{01}\mathbf{W}_3\mathbf{P}_B\mathbf{R}_{01})\mathbf{W}_3$. The corresponding stationary distribution

$$[\ \mathbf{p}_1(G)\ \ \mathbf{p}_1(U)\ \ \mathbf{p}_1(I)\ \ \mathbf{p}_1(L)\ ] \tag{5.4.8}$$

may be obtained from stationary distribution (5.4.3) by the removal of $\mathbf{p}(D)$ and $\mathbf{p}(S)$, with the subsequent normalization:

$$\mathbf{p}_1(G) = \mu\mathbf{p}(G) \quad \mathbf{p}_1(U) = \mu\mathbf{p}(U)$$
$$\mathbf{p}_1(I) = \mu\mathbf{p}(I) \quad \mathbf{p}_1(L) = \mu\mathbf{p}(L)$$

where

$$\mu = [\mathbf{c}_1(\mathbf{P}_F+\mathbf{P}_B\mathbf{R}_{10})\mathbf{1}+\mathbf{c}_2\mathbf{P}_F\mathbf{1}]^{-1}$$

Matrices (5.4.7) and (5.4.8) characterize the sequence of accepted messages. Using these matrices we can obtain the system performance characteristics:

$\mu\mathbf{c}_1\mathbf{P}_A\mathbf{1}$ The probability that the message is received correctly (for the first time)

$\mu\mathbf{c}_1\mathbf{P}_C\mathbf{1}$ The probability that the message is received incorrectly (for the first time)

$\mu\mathbf{c}_2\mathbf{P}_F\mathbf{1}$ The probability that the message is an insertion of the previous message

$\mu\mathbf{c}_1\mathbf{P}_B\mathbf{R}_{10}\mathbf{1}$ The probability that the message is lost

5.4.5 Alternative Assumptions. In some control systems a message is considered to be accepted correctly if in a series of its insertions it appears at least once without errors. To investigate these systems' performance, we introduce the following events (in addition to those previously defined):

G_0 The first correct reception of the message in the series of insertions

U_0 The message in the series of insertions is decoded incorrectly

D_0 The message in the series of insertions contains detected errors

C_0 The message in the series of insertions appears after it has been received correctly at least once

The transition probability matrix for the events G, U, G_0, U_0, D_0, C_0, D, and L has the form

$$[\mathbf{M}_g \quad \mathbf{M}_g \quad \mathbf{M}_b \quad \mathbf{M}_g \quad \mathbf{M}_g \quad \mathbf{M}_b \quad \mathbf{M}_g \quad \mathbf{M}_g]'$$

where

$$\mathbf{M}_g = [\mathbf{P}_{A0} \quad \mathbf{P}_{C0} \quad \mathbf{P}_{A1} \quad \mathbf{P}_{C1} \quad 0 \quad 0 \quad \mathbf{P}_{B1} \quad \mathbf{P}_{B0}]$$
$$\mathbf{M}_b = [\mathbf{H}_0 \quad 0 \quad 0 \quad 0 \quad 0 \quad \mathbf{H}_1 \quad 0 \quad 0]$$

The stationary distribution of the Markov chain with this matrix satisfies the following equations:

$$\mathbf{p}(G) = \mathbf{a}_1\mathbf{P}_{A0} + \mathbf{a}_2\mathbf{H}_0 \quad \mathbf{p}(U) = \mathbf{a}_1\mathbf{P}_{C0} \quad \mathbf{p}(G_0) = \mathbf{a}_1\mathbf{P}_{A1} \quad \mathbf{p}(U_0) = \mathbf{a}_1\mathbf{P}_{C1}$$
$$\mathbf{p}(D_0) = \mathbf{a}_3\mathbf{P}_{B1} \quad \mathbf{p}(C_0) = \mathbf{a}_2\mathbf{H}_1 \quad \mathbf{p}(D) = (\mathbf{a}_1 - \mathbf{a}_3)\mathbf{P}_{B1} \quad \mathbf{p}(L) = \mathbf{a}_1\mathbf{P}_{B0}$$

where

$$\mathbf{a}_1 = \mathbf{p}(G) + \mathbf{p}(U) + \mathbf{p}(U_0) + \mathbf{p}(D_0) + \mathbf{p}(D) + \mathbf{p}(L)$$
$$\mathbf{a}_2 = \mathbf{p}(G_0) + \mathbf{p}(C_0) \quad \mathbf{a}_3 = \mathbf{p}(U_0) + \mathbf{p}(D_0)$$

Solving this system, we obtain

$$\mathbf{a}_1 = \boldsymbol{\pi}\mathbf{H}_0(\mathbf{I} - \mathbf{P}_{V1})^{-1} \quad \mathbf{a}_2 = \boldsymbol{\pi}\mathbf{P}_{A1}(\mathbf{I} - \mathbf{P}_{V1})^{-1} \quad \mathbf{a}_3 = \mathbf{a}_1\mathbf{P}_{C1}(\mathbf{I} - \mathbf{P}_{B1})^{-1}$$

where $\mathbf{P}_{V1} = \mathbf{P}_{B1} + \mathbf{P}_{C1}$.

For the sequence of accepted messages the basic probabilities that characterize the system's performance are:

$\lambda\mathbf{p}(G)\mathbf{1}$ The probability that the message is received correctly

$\lambda\mathbf{p}(U)\mathbf{1}$ The probability that the message is received incorrectly

$\lambda\mathbf{p}(I_0)\mathbf{1}$ The probability that the message is a duplicate of the previous message ($I_0 = G_0 \cup U_0 \cup C_0$)

$\lambda\mathbf{p}(L)\mathbf{1}$ The probability that the message is lost

The coefficient λ is obtained by excluding the states corresponding to D_0 and D

with the appropriate normalization:

$$\lambda = [1 - \mathbf{p}(D_0)\mathbf{1} - \mathbf{p}(D)\mathbf{1}]^{-1}$$

5.4.6 Modified Stop-and-Wait ARQ. The simple stop-and-wait ARQ considered in the previous section has several deficiencies that may cause a permanent synchronization loss or create a deadlock. If a message or a reply is lost in one of the channels and the loss goes undetected, the system may be deadlocked, with the receiver waiting for the message and the transmitter waiting for an acknowledgement.

The transmission protocol can be modified to avoid this problem. A timeout can be used to break a deadlock: The previous message is retransmitted if no reply has been received during the timeout period. If the direct-channel message has been accepted, but the corresponding ACK has been lost, the retransmitted message duplicates the previous one. To avoid inserting the message, the frames and acknowledgements are modulo 2 numbered (ACK0 and ACK1). The flow control protocol may be described as follows (see Fig. 5.3):

Transmitter:

- If the response was the ACK with the previous message number, the next message is sent.
- If the response was different from the ACK with the previous message number, or no response was received during the timeout period, the previous message is retransmitted.

Receiver:

- If the decoder does not detect errors and the received message number differs from the previous ACK number, the message is accepted and the ACK with this message number is sent to the transmitter.
- If the decoder detects errors in the received message, it sends the NAK.
- If the decoder does not detect errors and the received message number coincides with the previous ACK number, the message is discarded and the ACK with this message number is sent to the transmitter.

If each channel is perfectly synchronized, the message numbering decreases the probability of the MR-sync loss, because only the ACK with the transmitted message number serves as the positive acknowledgement. Now we have three different acknowledgements that can be sent over the return channel, so that we need to consider the probabilities of their mutual transformations W_{uv} where

$u=a_0$ If an ACK0 was transmitted

$u=a_1$ If an ACK1 was transmitted

$u=1$ If an NAK was transmitted

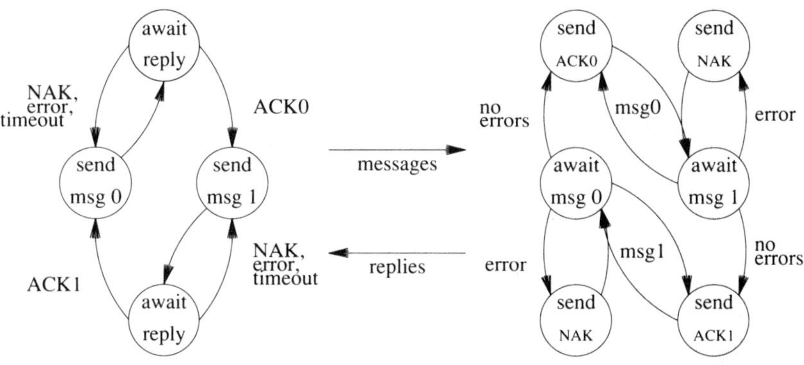

Figure 5.3. Modified ARQ system.

$v=a_0$ If the ACK0 was decoded

$v=a_1$ If the ACK1 was decoded

$v=1$ If the NAK was decoded

The process on the decoder output can be described using an SSM as for the simple stop-and-wait ARQ. However, the message numbering requires consideration of all possible outcomes of the transmission of even-numbered messages G_0, U_0, I_0, S_0, D_0, L_0 and odd-numbered messages $G_1, U_1, I_1, S_1, D_1, L_1$. The process transition probability matrix can be found analogously to (5.4.2). Since the matrix contains many zero blocks, it is convenient to describe it in terms of nonzero matrix transitional probabilities:

$$\mathbf{P}_{X_i,G_j} = \mathbf{R}_{a_i a_i}\mathbf{P}_A \quad \mathbf{P}_{X_i,U_j} = \mathbf{R}_{a_i a_i}\mathbf{P}_C \quad \mathbf{P}_{X_i,I_i} = \mathbf{R}_{a_i,1}\mathbf{P}_F$$
$$\mathbf{P}_{X_i,S_i} = \mathbf{R}_{a_i,1}\mathbf{P}_B \quad \mathbf{P}_{X_i,D_i} = \mathbf{R}_{a_i a_i}\mathbf{P}_B\mathbf{R}_{11}^{(i)} \quad \mathbf{P}_{X_i,L_i} = \mathbf{R}_{a_i a_i}\mathbf{P}_B\mathbf{R}_{1,a_i}$$

$$\mathbf{P}_{S_i,G_j} = \mathbf{R}_{1,a_i}\mathbf{P}_A \quad \mathbf{P}_{S_i,U_j} = \mathbf{R}_{1,a_i}\mathbf{P}_C \quad \mathbf{P}_{S_i,I_i} = (\mathbf{R}_{1,a_j}+\mathbf{R}_{11})\mathbf{P}_F$$
$$\mathbf{P}_{S_i,S_i} = (\mathbf{R}_{1,a_j}+\mathbf{R}_{11})\mathbf{P}_B \quad \mathbf{P}_{S_i,D_i} = \mathbf{R}_{1,a_i}\mathbf{P}_B\mathbf{R}_{11}^{(i)} \quad \mathbf{P}_{S_i,L_i} = \mathbf{R}_{1,a_i}\mathbf{P}_B\mathbf{R}_{1,a_i}$$

$$\mathbf{P}_{Y_i,G_i} = \mathbf{P}_A \quad \mathbf{P}_{Y_i,U_i} = \mathbf{P}_C \quad \mathbf{P}_{Y_i,D_i} = \mathbf{P}_B\mathbf{R}_{11}^{(i)} \quad \mathbf{P}_{Y_i,L_i} = \mathbf{P}_B\mathbf{R}_{1,a_i} \qquad (5.4.9)$$

In these equations $i,j=0,1$, $i \neq j$, $X_i = G_i$, U_i, I_i, $Y_i = D_i$, L_i, $\mathbf{R}_{a_i,1}$ is the probability of not receiving the ACKi when it was sent, $\mathbf{R}_{11}^{(i)} = \mathbf{R}_{11}+\mathbf{R}_{1,a_j}$ is the probability of not receiving the ACKi when the NAK was sent, $\mathbf{P}_A$ is the probability of correctly decoding the message and its number, and $\mathbf{P}_C$ is the probability of incorrectly decoding the message and correctly decoding its number. We assume that these probabilities do not depend on the message number.

The stationary probabilities of the system states satisfy the following equations:

$$\mathbf{p}(G_i) = \mathbf{c}_j\mathbf{P}_A \quad \mathbf{p}(U_i) = \mathbf{c}_j\mathbf{P}_C \quad \mathbf{p}(I_i) = \mathbf{d}_i\mathbf{P}_F$$
$$\mathbf{p}(S_i) = \mathbf{d}_i\mathbf{P}_B \quad \mathbf{p}(D_i) = \mathbf{c}_i\mathbf{P}_B\mathbf{R}_{11}^{(i)} \quad \mathbf{p}(L_i) = \mathbf{c}_i\mathbf{P}_B\mathbf{R}_{1,a_i}$$

where $i \neq j$ and $i,j=0,1$. One can easily show that the coefficients $\mathbf{c}_i$ and $\mathbf{d}_i$ can be found from the system

$$\mathbf{c}_i = (\mathbf{c}_j+\mathbf{d}_i)\mathbf{P}_F\mathbf{R}_{a_i,a_i} + \mathbf{d}_i\mathbf{P}_B\mathbf{R}_{1,a_i} + \mathbf{c}_i\mathbf{P}_B\mathbf{S}$$
$$\mathbf{d}_i = (\mathbf{c}_j+\mathbf{d}_i)\mathbf{P}_F\mathbf{R}_{a_i,1} + \mathbf{d}_i\mathbf{P}_B\mathbf{R}_{11}^{(i)}$$
$$\mathbf{c}_0 + \mathbf{c}_1 + \mathbf{d}_0 + \mathbf{d}_1 = \boldsymbol{\pi}$$

Solving this system, we obtain

$$\mathbf{c}_0 = \boldsymbol{\pi}(\mathbf{I} + \mathbf{M}_0 + \mathbf{N}_0 + \mathbf{N}_0\mathbf{M}_1)^{-1}$$
$$\mathbf{c}_1 = \mathbf{c}_0\mathbf{N}_0 \quad \mathbf{d}_1 = \mathbf{c}_0\mathbf{M}_0 \quad \mathbf{d}_0 = \mathbf{c}_1\mathbf{M}_1$$
$$\mathbf{M}_j = \mathbf{P}_F\mathbf{R}_{a_i,1}(\mathbf{I} - \mathbf{P}_F\mathbf{R}_{a_i,1} - \mathbf{P}_B\mathbf{R}_{11}^{(i)})^{-1}$$
$$\mathbf{N}_j = [\mathbf{I} + \mathbf{M}_j(\mathbf{P}_F\mathbf{R}_{a_i,a_i} + \mathbf{P}_B\mathbf{R}_{1,a_i})](\mathbf{I} - \mathbf{P}_B\mathbf{S})^{-1}$$

In the set of accepted messages the ARQ performance characteristics can be expressed using stationary probabilities (5.4.9):

$\nu(\mathbf{c}_0+\mathbf{c}_1)\mathbf{P}_A\mathbf{1}$ — The probability that the message is received correctly (the first time)

$\nu(\mathbf{c}_0+\mathbf{c}_1)\mathbf{P}_C\mathbf{1}$ — The probability that the message is received incorrectly (the first time)

$\nu(\mathbf{d}_0+\mathbf{d}_1)\mathbf{P}_F\mathbf{1}$ — The probability that the message is an insertion of the previous message

$p(L)$ — The probability that the message is lost

In these expressions we denoted

$$p(L) = \nu(\mathbf{c}_0\mathbf{P}_B\mathbf{R}_{1,a_0} + \mathbf{c}_1\mathbf{P}_B\mathbf{R}_{1,a_1})\mathbf{1}$$
$$\nu = [1 - (\mathbf{d}_0+\mathbf{d}_1)\mathbf{P}_B\mathbf{1} - (\mathbf{c}_0\mathbf{P}_B\mathbf{R}_{11}^{(0)}+\mathbf{c}_1\mathbf{P}_B\mathbf{R}_{11}^{(1)})\mathbf{1}]^{-1}$$

If the acknowledgements are symmetrical, the probabilities $\mathbf{R}_{a_i,a_i}$, $\mathbf{R}_{a_i,1}$, and $\mathbf{R}_{11}^{(i)}$ do not depend on i and the above solutions can be simplified:

$$\mathbf{c}_0 = \mathbf{c}_1 = 0.5\boldsymbol{\pi}(\mathbf{P}_B\mathbf{R}_{1,a_0} + \mathbf{P}_F\mathbf{R}_{a_0,a_0})(\mathbf{I} - \mathbf{P}_B\mathbf{R}_{11}^{(0)})^{-1}$$
$$\mathbf{d}_0 = \mathbf{d}_1 = 0.5\boldsymbol{\pi}\mathbf{P}_F\mathbf{R}_{a_0,1}(\mathbf{I} - \mathbf{P}_B\mathbf{R}_{11}^{(0)})^{-1}$$

Let us consider now the the possibility of losing synchronization in the direct and return channels. To describe the system's operation we introduce more events:

Λ_d The message is lost in the direct channel

Λ_r The message is lost in the return channel

The corresponding matrix probabilities are denoted $\mathbf{P}_{\Lambda_d}$ and $\mathbf{R}_{\Lambda_r}$.

If the message is lost in the direct channel, it will be retransmitted after a timeout so that the system will return to the state in which it was before transmitting the message. To model this situation, we define for each of the previously defined states X_i ($X = G, U, I, S, D, L$) the direct-channel timeout states X_{i+2} with transition probabilities

$$\mathbf{P}_{X_i,X_{i+2}} = \mathbf{P}_{\Lambda_d} \quad \mathbf{P}_{X_{i+2},X_i} = \mathbf{S}$$

The return-channel message loss can be treated similarly. Suppose that the previous message was in one of the states G_i, U_i, I_i, or S_i. Then the next message without detected errors represents an insertion of the previous one if the acknowledgement has been lost. The probability of this event is

$$\mathbf{P}_{Y_i,I_{i+4}} = \mathbf{R}_{\Lambda_r}\mathbf{P}_F \quad (Y = G, U, I, S)$$

If errors are detected, the next message is discarded, with the probability

$$\mathbf{P}_{Y_i,S_{i+4}} = \mathbf{R}_{\Lambda_r}\mathbf{P}_B$$

If the previous message was in one of the states D_i or L_i, then

$$\mathbf{P}_{Z_i,G_i} = \mathbf{R}_{\Lambda_r}\mathbf{P}_A \quad \mathbf{P}_{Z_i,U_i} = \mathbf{R}_{\Lambda_r}\mathbf{P}_C$$
$$\mathbf{P}_{Z_i,D_i} = \mathbf{R}_{\Lambda_r}\mathbf{P}_B\mathbf{R}_{11}^{(i)} \quad \mathbf{P}_{Z_i,L_i} = \mathbf{R}_{\Lambda_r}\mathbf{P}_B\mathbf{R}_{1,a_i}$$

The transition probabilities from the states G_{i+4}, U_{i+4}, I_{i+4}, S_{i+4}, D_{i+4}, L_{i+4} are the same as from the states G_i, U_i, I_i, S_i, D_i, L_i. The stationary probabilities of the states of the Markov chain can be found similarly to those in previous cases.

5.4.7 Go-Back-N ARQ. The stop-and-wait method is inefficient, because of the idle time that it spends waiting for an acknowledgement. In full-duplex systems it is possible to send messages and receive acknowledgements continuously. The transmitter sends the next message without waiting for the previous message to be acknowledged. As long as all the messages are positively acknowledged, this method is conceptually the same as the stop-and-wait method. When an NAK is received, the transmitter backs up to the negatively acknowledged message and retransmits it and $N-1$ succeeding messages. The receiver blocks for the period of N messages transmission: It discards the negatively acknowledged message and $N-1$ following messages and waits for their retransmission.

Analysis of these systems is complicated by the transmission of the next message before the reception of an acknowledgement for the previous message. Therefore, it is necessary to use multiple Markov chains to describe the system's operation in an SSM channel. The chain order N is defined by the delay between message transmission and decoding of its acknowledgement. Let us consider a piggyback scheme in which messages together with acknowledgements are sent over the channels in both directions. To describe the system functioning, let us introduce the state vector $\mathbf{s} = (m, r, s_1, s_2, \ldots, s_{N-1})$ where

- m is the message sequence number from the beginning of the receiver blocking zone, with $m=0$ if it is outside the blocking zone
- r is the acknowledgement message sequence number from the beginning of the return-channel blocking zone, with $r=0$ if it is outside the blocking zone
- s_i is the logical variable that characterizes the message status: $s_1 = 0$ if the message has not been accepted before, $s_1 = 1$ otherwise; $s_2, \ldots, s_{N-1}$ similarly characterize the succeeding messages

These states represent a Markov function whose matrix transitional probabilities $\mathbf{P}(\boldsymbol{\sigma}; \mathbf{s})$ have the following form:

$$\mathbf{P}(a,b,\mathbf{v};0,0,\boldsymbol{\tau},0) = \mathbf{P}_{F0} \qquad \mathbf{P}(a,b,\mathbf{v};0,1,\boldsymbol{\tau},1) = \mathbf{P}_{F1}$$

$$\mathbf{P}(a,b,\mathbf{v};1,0,\boldsymbol{\tau},0) = \mathbf{P}_{B0} \qquad \mathbf{P}(a,b,\mathbf{v};1,1,\mathbf{w}) = \mathbf{P}_{B1}$$

$$\mathbf{P}(m,b,\mathbf{v};m+1,0,\boldsymbol{\tau},0) = \mathbf{P}_{\Omega 0} \quad \mathbf{P}(m,b,\mathbf{v};m+1,1,\mathbf{w}) = \mathbf{P}_{\Omega 1}$$

$$\mathbf{P}(a,r,\mathbf{v};0,r+1,\boldsymbol{\tau},1) = \mathbf{P}_{F\Omega} \qquad \mathbf{P}(a,r,\mathbf{v};1,r+1,\mathbf{w}) = \mathbf{P}_{B\Omega}$$

$$\mathbf{P}(m,r,\mathbf{v};m+1,r+1,\mathbf{w}) = \mathbf{P}_{\Omega\Omega} \quad \mathbf{P}(m,N-1,\mathbf{v};m+1,N,\mathbf{v}) = \mathbf{P}_{\Omega\Omega}$$

$$\mathbf{P}(a,N-1,\mathbf{v};0,N,\mathbf{v}) = \mathbf{P}_{0\Omega} \qquad \mathbf{P}(a,N-1,\mathbf{v};1,N,\mathbf{v}) = \mathbf{P}_{1\Omega}$$

Here $\mathbf{v} = (s_1, s_2, ..., s_{N-1})$, $\boldsymbol{\tau} = (s_2, s_3, ..., s_{N-1})$, $\mathbf{w} = (s_1, s_2, ..., s_{N-1}, s_1)$, $0 < m \le N-1$, $0 < r < N-1$, $a = 0, N, b = 0, N$. In these equations $\mathbf{P}_{Fv}$ is the matrix probability of not detecting errors in a direct-channel message and receiving an ACK ($v = 0$) or NAK ($v = 1$) over the return channel; $\mathbf{P}_{Bv}$ is the probability of detecting errors in a direct-channel message and receiving the reply v; and $\mathbf{P}_{0\Omega}$ and $\mathbf{P}_{1\Omega}$ are the probabilities of decoding an ACK or NAK from the direct-channel message when the NAK was sent. Also we denoted

$$\mathbf{P}_{\Omega v} = \mathbf{P}_{Fv} + \mathbf{P}_{Bv} \quad \mathbf{P}_{F\Omega} = \mathbf{P}_{F0} + \mathbf{P}_{F1} \quad \mathbf{P}_{B\Omega} = \mathbf{P}_{B0} + \mathbf{P}_{B1} \quad v = 0,\ 1,\ \Omega$$

As in the previous sections, we determine first the stationary probabilities of the system states $\mathbf{p}(\mathbf{s}) = \mathbf{p}(m, r, \mathbf{v})$ that satisfy the equations

$$\mathbf{p}(\mathbf{s}) = \sum_{\boldsymbol{\sigma}} \mathbf{p}(\boldsymbol{\sigma}) \mathbf{P}(\boldsymbol{\sigma};\ \mathbf{s}) \quad \sum_{\mathbf{s}} \mathbf{p}(\mathbf{s}) = \boldsymbol{\pi} \tag{5.4.10}$$

where $\mathbf{P}(\boldsymbol{\sigma};\ \mathbf{s})$ are the above-defined matrix transition probabilities and $\boldsymbol{\pi}$ is the stationary distribution of the SSM model states. It is also the invariant vector of the matrix $\mathbf{P}_{\Omega\Omega}$. The summation is performed over all the states that are reachable from the all-zero state (0,0, ...,0).

If we know the solution of system (5.4.10), then the system's performance characteristics can be expressed by summing the appropriate stationary probabilities:

$\mathbf{u}_0 \mathbf{P}_A \mathbf{1}$	The probability of message correct decoding
$\mathbf{u}_0 \mathbf{P}_C \mathbf{1}$	The probability of message incorrect decoding
$\mathbf{u}_1 \mathbf{P}_F \mathbf{1}$	The probability of message insertion
$\mathbf{w1}$	The probability of message loss where $\mathbf{w} = \mathbf{u}_0 \mathbf{P}_{B0} + \mathbf{v} \mathbf{P}_{\Omega 0}$

In these expressions we denote

$$\mathbf{u}_s = \sum_{r \ne N-1} \sum_{\boldsymbol{\tau}} \mathbf{p}(0, r, s, \boldsymbol{\tau}) + \mathbf{p}(1, r, s, \boldsymbol{\tau}) \quad s = 0, 1$$

$$\mathbf{v} = \sum_{m=1}^{N-1} \sum_{\boldsymbol{\tau}} \mathbf{p}(m, 0, 0, \boldsymbol{\tau}) + \mathbf{p}(m, N, 0, \boldsymbol{\tau}), \quad \boldsymbol{\tau} = (\mathbf{s}_2, ..., \mathbf{s}_{N-1})$$

The system's maximal throughput is greater than that of the stop-and-wait protocol and is equal to

$$\bar{R} = \mathbf{u}_0 \mathbf{P}_F \mathbf{1} k / n$$

To find the probabilities in the set of accepted messages we need to exclude the messages which are retransmitted and perform the appropriate normalization of the above-defined probabilities.

Thus, solving system (5.4.10) is the major difficulty in analyzing the go-back-N scheme. This system has a special structure that simplifies its solution. To simplify demonstration of the method, we assume that the return channel is noiseless. In this case, $\mathbf{P}_{F0} = \mathbf{P}_F$, $\mathbf{P}_{F1} = 0$, $\mathbf{P}_{B1} = \mathbf{P}_B$, $\mathbf{P}_{B0} = 0$, $\mathbf{P}_{\Omega 0} = \mathbf{S}$, $\mathbf{P}_{\Omega 1} = 0$, $r = 0$, $s_i = 0$, and system (5.4.10) takes the form

$$\mathbf{p}(\mathbf{s}_0) = [\mathbf{p}(\mathbf{s}_0) + \mathbf{p}(\mathbf{s}_N)]\mathbf{P}_F$$
$$\mathbf{p}(\mathbf{s}_1) = [\mathbf{p}(\mathbf{s}_0) + \mathbf{p}(\mathbf{s}_N)]\mathbf{P}_B$$
$$\mathbf{p}(\mathbf{s}_{i+1}) = \mathbf{p}(\mathbf{s}_i)\mathbf{S} \quad i = 1,2,..,N$$
$$\sum_{i=0}^{N} \mathbf{p}(\mathbf{s}_i) = \boldsymbol{\pi}$$

where $\mathbf{s}_i = (i, 0,0,...,0)$. Solving this system, we obtain

$$\mathbf{p}(\mathbf{s}_0) = \mathbf{u}_0\mathbf{P}_F \quad \mathbf{p}(\mathbf{s}_i) = \mathbf{u}_0\mathbf{P}_B\mathbf{S}^{i-1} \quad (i = 1,2,...,N)$$

where $\mathbf{u}_0 = \boldsymbol{\pi}(\mathbf{P}_F + \mathbf{P}_B\mathbf{V})^{-1}$

$$\mathbf{V} = \sum_{i=1}^{N} \mathbf{S}^{i-1} = (N-1)\mathbf{1}\boldsymbol{\pi} + (\mathbf{I}-\mathbf{B}^N)(\mathbf{I}-\mathbf{B})^{-1} \quad \text{with} \quad \mathbf{B} = \mathbf{S} - \mathbf{1}\boldsymbol{\pi}$$

Note that there is no need to determine all the elements of the stationary distribution, since the system performance characteristics depend only on sums of the stationary probabilities. In the above-considered particular case, only the sum $\mathbf{u}_0 = \mathbf{p}(\mathbf{s}_0) + \mathbf{p}(\mathbf{s}_1)$ needs to be found to calculate the probabilities of correct and incorrect decoding and of the system throughput. The probability $p_\beta(x)$ that a frame occupies the system for the time necessary to send x frames without retransmission can also be expressed using this sum:

$$p_\beta(lN+1) = \mathbf{u}_0\mathbf{P}_F\mathbf{P}_\beta(lN+1)\mathbf{1} \,/\, \mathbf{u}_0\mathbf{P}_F\mathbf{1}$$

where

$$\mathbf{P}_\beta(lN+1) = (\mathbf{P}_B\mathbf{S}^{N-1})^l\mathbf{P}_F$$

is the corresponding matrix distribution whose generating function is

$$\mathbf{\Phi}_\beta(z) = \sum_{l=0}^{\infty} \mathbf{P}_\beta(lN+1)z^{lN} = z(\mathbf{I}-z^N \mathbf{P}_B \mathbf{S}^{N-1})^{-1}\mathbf{P}_F \tag{5.4.11}$$

The go-back-N protocol is more effective than the stop-and-wait protocol, but its implementation is more complex. It requires a buffer to store at least N messages that have not been acknowledged. If N is large and the channel is noisy, the effectiveness of the scheme may suffer significantly, because it is necessary to retransmit all N messages. In alternative *selective-repeat systems* only the messages with detected errors are retransmitted. These systems use not only a transmitter buffer but also a receiver buffer, since messages may not be accepted in the same order as they were transmitted and therefore may need to be rearranged on the receiver output.

5.4.8 Multiframe Messages. In the systems considered so far, each frame contained only one message. However, in many cases a message may occupy several frames. For example, a message may be a file that is transferred from one computer to another. In these cases long messages are broken up into smaller blocks to fit into transmission frames. The message is transferred correctly if all its blocks are accepted without errors. If at least one of the blocks contains errors, the whole message contains errors.

Suppose that each message is encoded into K blocks which are protected by an (n,k) code. The messages arrive periodically with a transmission period of $\alpha = k+\xi M$ frames. The maximum message transmission delay is equal to the time of transmitting ε frames. Our goal is to calculate the probability of correct and incorrect message reception if the ARQ scheme is used for frame transmission over an SSM channel.

The problem that we are trying to solve is one typical of queueing theory. Messages arrive periodically and the service time is equal to the time that the message occupies the channel, which is equal to the sum of the times taken by the message sub-blocks. The matrix distribution of the service time can be expressed as a convolution of the addends and its generating function equals

$$\mathbf{\Phi}_\tau(z) = [\mathbf{\Phi}_\beta(z)]^K \tag{5.4.12}$$

where $\mathbf{\Phi}_\beta(z)$ is the generating function of the time that a single frame occupies the channel. The message service time distribution can be expressed as

$$\mathbf{P}_\tau(m) = \frac{1}{2\pi j} \oint_\gamma \mathbf{\Phi}_\tau(z) z^{-m-1}\, dz. \tag{5.4.13}$$

Let us now determine the matrix distribution $\mathbf{Q}(j)$ of the message waiting time, which plays a fundamental role in evaluating the system's performance. According to the SSM model, the sequence of waiting times for different messages can be described as a Markov function with the matrix

$$\begin{bmatrix} \mathbf{P}_\tau(\le\alpha) & \mathbf{P}_\tau(\alpha+1) & \cdots & \mathbf{P}_\tau(\varepsilon-1) & \mathbf{P}_\tau(>\varepsilon-1) \\ \mathbf{P}_\tau(\le\alpha-1) & \mathbf{P}_\tau(\alpha) & \cdots & \mathbf{P}_\tau(\varepsilon-2) & \mathbf{P}_\tau(>\varepsilon-2) \\ \cdots & \cdots & \cdots & \cdots & \cdots \\ \mathbf{P}_\tau(1) & \mathbf{P}_\tau(2) & \cdots & \mathbf{P}_\tau(\kappa) & \mathbf{P}_\tau(>\kappa) \\ 0 & \mathbf{P}_\tau(1) & \cdots & \mathbf{P}_\tau(\kappa-1) & \mathbf{P}_\tau(>\kappa-1) \\ \cdots & \cdots & \cdots & \cdots & \cdots \\ 0 & 0 & \cdots & \mathbf{P}_\tau(\alpha-1) & \mathbf{P}_\tau(>\alpha-1) \end{bmatrix} \tag{5.4.14}$$

where $\kappa = \varepsilon-\alpha$,

$$\mathbf{P}_\tau(\le m) = \sum_{j=1}^{m} \mathbf{P}_\tau(j)\mathbf{Q}^{m-j} \tag{5.4.15}$$

$$\mathbf{P}_\tau(>m) = \mathbf{S}^m - \mathbf{P}_\tau(\le m) \tag{5.4.16}$$

The matrix probabilities $\mathbf{Q}(j)$ represent the stationary distribution of the Markov chain with matrix (5.4.14) and can be found from the following system:

$$\begin{aligned} \mathbf{Q}(0) &= \sum_{i=0}^{\alpha-1} \mathbf{Q}(i)\mathbf{P}_\tau(\le\alpha-i) \\ \mathbf{Q}(j) &= \sum_{i=0}^{j+\alpha-1} \mathbf{Q}(i)\mathbf{P}_\tau(\alpha+j-i) \quad 0<j<\kappa \\ \mathbf{Q}(\kappa) &= \sum_{i=0}^{\kappa} \mathbf{Q}(i)\mathbf{P}_\tau(>\varepsilon-i+1) \\ \sum_{j=0}^{\kappa} \mathbf{Q}(j)\mathbf{1} &= 1 \quad \mathbf{Q}(j) = 0 \quad \text{for } j>\kappa \end{aligned} \tag{5.4.17}$$

It is easy to verify that

$$\sum_{j=0}^{\kappa} \mathbf{Q}(j)\mathbf{S}^{\kappa-j} = \boldsymbol{\pi} \tag{5.4.18}$$

We find $\mathbf{Q}(\alpha-1)$ from the first equation

$$\mathbf{Q}(\alpha-1) = [\mathbf{Q}(0) - \sum_{i=0}^{\alpha-2} \mathbf{Q}(i)\mathbf{P}_\tau(\le\alpha-i)]\mathbf{P}_\tau^{-1}(1)$$

and substitute it into the rest of the equations. Then we find $\mathbf{Q}(\alpha)$ from the second equation and replace it in the remaining equations, and so on. Repeating the process, we obtain the system of matrix equations to determine $\mathbf{Q}(0)$, $\mathbf{Q}(1),\ldots,$ $\mathbf{Q}(\alpha-2)$. If α is small, this system can easily be solved. The rest of the matrices $\mathbf{Q}(j)$ are found by directly substituting the solution into the rest of the system.

This system can also be solved using generating functions. We will show that the system can be replaced with an integral equation for the waiting-time distribution generating function

$$\boldsymbol{\rho}(z) = \sum_{j=0}^{\kappa}\mathbf{Q}(j)z^{j}$$

Indeed, we obtain from system (5.4.17)

$$\boldsymbol{\rho}(z) = \sum_{m=0}^{\infty}\mathbf{Q}(m)[\mathbf{P}_{\tau}(\le\alpha-m)+\sum_{j=1}^{\kappa-1}\mathbf{P}_{\tau}(\alpha+j-m)z^{j}+\mathbf{P}_{\tau}(>\varepsilon-m+1)z^{\kappa}] \qquad (5.4.19)$$

Using integral representation (5.4.13) and equations (5.4.15) and (5.4.16) we obtain

$$\mathbf{P}_{\tau}(\le m) = \frac{1}{2\pi \mathrm{j}}\oint_{\gamma}\mathbf{\Phi}_{\tau}(z)(\mathbf{I}-\mathbf{S}z)^{-1}z^{-m-1}\,dz$$

$$\mathbf{P}_{\tau}(>m) = \frac{1}{2\pi \mathrm{j}}\oint_{\gamma}[\mathbf{I}-\mathbf{\Phi}_{\tau}(z)\mathbf{S}z](\mathbf{I}-\mathbf{S}z)^{-1}z^{-m}\,dz$$

Substituting these expressions into equation (5.4.19) we obtain the following integral equation:

$$\boldsymbol{\rho}(z) = z^{\kappa}\boldsymbol{\pi} + \oint_{\gamma}\frac{\boldsymbol{\rho}(w)\mathbf{\Psi}_{\tau}(w)(w^{\kappa}-z^{\kappa})dw}{2\pi \mathrm{j}\,w^{\kappa}(w-z)}(\mathbf{I}-\mathbf{S}z) \qquad (5.4.20)$$

where

$$\mathbf{\Psi}_{\tau}(w) = w^{-\alpha}\mathbf{\Phi}_{\tau}(w)(\mathbf{I}-\mathbf{S}w)^{-1}$$

In deriving this equation we used (5.4.18), which can be rewritten as

$$\frac{1}{2\pi \mathrm{j}}\oint_{\gamma}\boldsymbol{\rho}(w)(\mathbf{I}-\mathbf{S}w)^{-1}w^{-\varepsilon-1}\,dw = \boldsymbol{\pi} \qquad (5.4.21)$$

Integral equation (5.4.20) can be solved by the standard methods of the theory of systems of integral equations. For large κ the solution is close to the case of an infinite delay ($\kappa \to \infty$). Suppose that the message interarrival interval is greater than the average time that the message spends in the system (service time). Then stationary distribution of waiting time exists. Equation (5.4.20) takes the following form when $\kappa \to \infty$:

$$\boldsymbol{\rho}(z) = \boldsymbol{\rho}(z)\boldsymbol{\Psi}_\tau(z) + \oint_\gamma \frac{\boldsymbol{\rho}(w)\boldsymbol{\Psi}_\tau(w)(\mathbf{I}-\mathbf{S}w)^{-1}dw}{2\pi j(w-z)}(\mathbf{I} - \mathbf{S}z)$$

or

$$\boldsymbol{\rho}(z) = \oint_\gamma \frac{\boldsymbol{\rho}(w)\boldsymbol{\Psi}_\tau(w)(\mathbf{I}-\mathbf{S}w)^{-1}dw}{2\pi j(w-z)}(\mathbf{I} - \mathbf{S}z)[\mathbf{I}-\boldsymbol{\Psi}_\tau(z)]^{-1} \tag{5.4.22}$$

It is easy to verify that

$$\xi(z) = \oint_\gamma \frac{\boldsymbol{\rho}(w)\boldsymbol{\Psi}_\tau(w)(\mathbf{I}-\mathbf{S}w)^{-1}dw}{2\pi j(w-z)}$$

is a degree $\alpha+1$ polynomial of z^{-1}

$$\xi(z) = \sum_{m=0}^{\alpha+1} \mathbf{B}_m z^{-m}$$

so that the solution of equation (5.4.22) has the form

$$\boldsymbol{\rho}(z) = \xi(z)(\mathbf{I} - \mathbf{S}z)[\mathbf{I}-\boldsymbol{\Psi}_\tau(z)]^{-1}$$

The coefficients $\mathbf{B}_m$ of the solution may be obtained by using (5.4.21) and the conditions of the function $\boldsymbol{\rho}(z)$ regularity inside the unit circle.

According to (5.4.11) and (5.4.12), $\boldsymbol{\Psi}_\tau(z)$ is a rational function. Therefore, the distribution $\mathbf{Q}(j)$ has the form of (A.2.6), which in the case of simple roots z_m of the equation

$$\det[\mathbf{I} - \Psi_\tau(z)] = 0$$

that lie outside the unit circle has the form

$$\mathbf{Q}(j) = \sum_m \mathbf{A}_m z_m^{-j}$$

Let us now calculate the system performance characteristics under the assumption that probabilities $\mathbf{Q}(j)$ have been found. The probability of receiving a correct message can be expressed as

$$p_c = \sum_{j=0}^{\kappa} \mathbf{Q}(j)\mathbf{Pr}\{A^K;\ \varepsilon-j-1\}\mathbf{1}, \tag{5.4.23}$$

where $\mathbf{P}\{A^K;\ \varepsilon-j-1\}$ is the matrix probability that all K sub-blocks that constitute the message have been received without errors at the time interval of $\varepsilon-j-1$. This matrix represents the probability that the Markov chain for the K-th time reaches one of the states of the set that corresponds to a frame correct reception for the time $\varepsilon-j$ without entering any state other than a state corresponding to detected errors.

The right-hand side of equation (5.4.23) is a convolution, and therefore the probability of receiving a correct message can be expressed using transform methods (see Sec. 2.2):

$$p_c = \frac{1}{2\pi j} \oint_\gamma \boldsymbol{\rho}(z)\mathbf{W}_A(z)\mathbf{1}z^{-\varepsilon+1}\, dz$$

where

$$\mathbf{W}_A(z) = \sum_m \mathbf{Pr}\{A^K;\ m\}z^m$$

is the generating function of the matrix probabilities $\mathbf{P}\{A^K; m\}$. It is not difficult to show, using the convolution theorem, that this generating function has the form

$$\mathbf{W}_A(z) = [\mathbf{\Phi}_A(z)]^K(1-z)^{-1}$$

where $\mathbf{\Phi}_A(z)$ is the generating function of the number of frame retransmissions before its first correct reception:

$$\mathbf{\Phi}_A(z) = (\mathbf{I} - \mathbf{P}_{B1}z)^{-1}\mathbf{P}_A z$$

The probability of incorrect decoding and message erasure (in the case when not all the message sub-blocks were accepted correctly) can be found quite analogously.

5.5 CONCLUSION

In this chapter we evaluated some communication protocols' performance characteristics. Because these protocols are described by finite state sequential machines, the communication system operation can be described using an SSM. The standard method of analyzing a system is to find its state stationary distribution and use this distribution to express the system's performance characteristics.

When one is calculating these characteristics, it is important to define the set of all possible outcomes of transmitting a message. For instance, when we dealt with message delay distribution and channel throughput we considered as a possible outcome the retransmission of a message when the decoder detected errors. We excluded this outcome, though, when we computed the probability of correctly accepting the message.

The purpose of this chapter was to demonstrate matrix probability applications for evaluating some protocol performance, without considering all the details of the protocols. For example, we did not address call processing (link establishment) and transmission termination. We also omitted the details of the existing protocol frame structure. The inclusion of details into analysis does not change the general approach but only makes the analysis more complicated. This statement was illustrated by considering a simple stop-and-wait scheme and then a more complex, modified stop-and-wait protocol.

The message contents may also change our approach to analyzing a system. For example, in some control systems it is important that a command is received at least once, whereas in others the duplicated command may negate the previously received command. Therefore, correct reception of a message should be defined accordingly. We showed that a change in definition requires some changes in the matrix of the system's state transition probabilities without changing the general approach.

APPENDIX 1

1.1 Matrix Processes

The matrix process definition and its application to error source modeling were considered in Sec. 1.3. We provide here proofs of the statements of Sec. 1.3 and also some information that is necessary to understand methods of estimating model parameters. The definitions in the next section play a major role in analyzing matrix processes.

Matrix processes and Markov chains have an interesting common property: There exists a finite set of the so-called basis sequences such that the probability of any sequence can be expressed as a linear combination of the probabilities of these basis sequences.

To prove this statement we consider the sequences of zeroes and ones of the type $s_i t_j$, where s_i is a sequence preceding the sequence t_j. Let $\varnothing$ be an empty sequence that possesses the properties $\varnothing s_i t_j = s_i \varnothing t_j = s_i t_j \varnothing$ and $Pr(\varnothing) = 1$. Denote

$$\mathbf{P}(s_1,...,s_p;\, t_1,...,t_q) = [p(s_i t_j)]_{p,q} \tag{A.1.1}$$

a matrix whose (i,j)-th element is the probability of occurrence of the sequence $s_i t_j$.

Lemma A.1.1: For every matrix process with matrices $\mathbf{M}(e)$ of the order n, there exists an integer $r \le n$ which is equal to the supremum of the ranks of the matrices (A.1.1) (over all possible p and q and sequences $s_i t_j$).

Proof: Indeed, by formula (1.3.5), with $s_i = e_1 e_2 \cdots e_t$, $t_j = e_{t+1} e_{t+2} \cdots e_m$, we have

$$p(s_i t_j) = \mathbf{q}(s_i)\mathbf{r}(t_j) \tag{A.1.2}$$

where $\mathbf{q}(s_i) = \mathbf{yM}(e_1) \cdots \mathbf{M}(e_t)$ and $\mathbf{r}(t_j) = \mathbf{M}(e_{t+1}) \cdots \mathbf{M}(e_m)\mathbf{z}$. If we denote

$$\mathbf{Q}(s_1,...,s_p) = \textit{block col}\{\mathbf{q}(s_i)\} \qquad \mathbf{R}(t_1,...,t_q) = \textit{block row}\{\mathbf{r}(t_j)\}$$

then (A.1.1) can be rewritten as

$$\mathbf{P}(s_1,...,s_p;\, t_1,...,t_q) = \mathbf{Q}(s_1,...,s_p)\mathbf{R}(t_1,...,t_q) \tag{A.1.3}$$

with $\mathbf{q}(\varnothing) = \mathbf{Q}(\varnothing) = \mathbf{y}$, $\mathbf{r}(\varnothing) = \mathbf{R}(\varnothing) = \mathbf{z}$.

According to the Sylvester[21,24] inequalities the rank of this matrix product is not greater than ranks of the matrices $\mathbf{Q}(s_1,...,s_p)$ and $\mathbf{R}(t_1,...,t_q)$. But these ranks are less than the sizes of the matrices, so that the rank of their product $\mathbf{P}(s_1,...,s_p;\, t_1,...,t_q)$ cannot exceed n.

■

The maximal rank of these matrices is called the *rank of the matrix process* and is denoted by r. Sequences for which $\mathbf{P}(s_1,...,s_r;\, t_1,...,t_r) \neq 0$ are called *basis sequences*. They are similar to states of a Markov chain.

Theorem A.1.1: A stationary process is a matrix process if and only if a positive integer r exists such that

$$\det \mathbf{P}(s_1,...,s_r;\, t_1,...,t_r) \neq 0 \tag{A.1.4}$$

and

$$\det \mathbf{P}(s_1,...,s_r,s;\, t_1,...,t_r,t) = 0 \tag{A.1.5}$$

for any sequences s and t.

Proof: The necessity of the conditions follows from the previous lemma.[24] We will prove the conditions' sufficiency if we show that formula (1.3.5) follows from (A.1.4) and (A.1.5). The determinant in the left-hand side of equation (A.1.5) can be expanded by the elements of its last column:

$$p(s\,t) = \sum_{j=1}^{r} a_j(s)p(s_j\,t) \tag{A.1.6}$$

For $s = s_i e$ this equation takes the form

$$p(s_i e\, t) = \sum_{j=1}^{r} a_j(s_i e) p(s_j\, t) \quad i = 1,2,\ldots,r$$

If we denote $\mathbf{A}(e) = [\ a_j(s_i e)\ ]_{r,r}$, then these equations can be presented in the following matrix form

$$\mathbf{P}(s_1,\ldots,s_r;\ e\, t) = \mathbf{A}(e)\mathbf{P}(s_1,\ldots,s_r;\ t) \tag{A.1.7}$$

Denoting $e = e_1$ and $t = e_2 \cdots e_m$, we can rewrite the previous equation as

$$\mathbf{P}(s_1,\ldots,s_r;\ e_1 e_2 \cdots e_m) = \mathbf{A}(e_1)\mathbf{A}(e_2)\mathbf{P}(s_1,\ldots,s_r;\ e_3 \cdots e_m)$$

Repeating this transformation m times, we obtain

$$\mathbf{P}(s_1,\ldots,s_r;\ e_1 e_2 \cdots e_m) = \prod_{i=1}^{m} \mathbf{A}(e_i)\mathbf{P}(s_1,\ldots,s_r;\ \varnothing) \tag{A.1.8}$$

If we choose $s_1 = \varnothing$, then the first element of $\mathbf{P}(s_1,\ldots,s_r;\ e_1 e_2 \cdots e_m)$ becomes $p(\varnothing e_1 e_2 \cdots e_m) = p(e_1 e_2 \cdots e_m)$. This element can be obtained by multiplying $\mathbf{P}(s_1,\ldots,s_r;\ e_1 e_2 \cdots e_m)$ by $\mathbf{y}_l = [\ 1\ 0\ 0 \ldots\ 0]_{1,r}$ from the left. Therefore, denoting $\mathbf{z}_l = \mathbf{P}(s_1,\ldots,s_r;\ \varnothing)$, we obtain from (A.1.8)

$$p(e_1 e_2 \cdots e_m) = \mathbf{y}_l \prod_{i=1}^{m} \mathbf{A}(e_i)\mathbf{z}_l \tag{A.1.9}$$

This equation coincides with (1.3.5), which means that this is a matrix process.

■

If we expand each column of a square matrix $\mathbf{P}(s_1,\ldots,s_r;\ et_1,\ldots,et_r)$ using (A.1.7), we obtain

$$\mathbf{P}(s_1,\ldots,s_r;\ et_1,\ldots,et_r) = \mathbf{A}(e)\mathbf{P}(s_1,\ldots,s_r;\ t_1,\ldots,t_r)$$

Since $\mathbf{P}(s_1,\ldots,s_r;\ t_1,\ldots,t_r)$ is nonsingular, it has an inverse matrix and, therefore,

$$\mathbf{A}(e) = \mathbf{P}(s_1,\ldots,s_r;\ et_1,\ldots,et_r)\mathbf{P}^{-1}(s_1,\ldots,s_r;\ t_1,\ldots,t_r) \tag{A.1.10}$$

This means that the matrices $\mathbf{A}(e)$ for a given process and a fixed basis are determined uniquely. They can be found from the probabilities of the basis sequences $p(s_i t_j)$, $p(s_i e t_j)$, (where $i,j = 1,2,\ldots,r$ and r is the rank of the process).

By expanding the determinant in (A.1.5) by the elements of its last row we arrive at the formula

$$\mathbf{P}(s\,e;\, t_1,...,t_r) = \mathbf{P}(s;\, t_1,...,t_r)\mathbf{B}(e)$$

which is analogous to (A.1.7), and then we obtain equations similar to (A.1.8), (A.1.9), and (A.1.10):

$$\mathbf{P}(e_1 \cdots e_m;\, t_1,...,t_r) = \mathbf{y}_\rho \prod_{i=1}^{m} \mathbf{B}(e_i)$$

$$p(e_1,e_2,...,e_m) = \mathbf{y}_\rho \prod_{i=1}^{m} \mathbf{B}(e_i)\mathbf{z}_\rho \qquad \text{(A.1.11)}$$

$$\mathbf{P}(s_1 e,...,s_r e;\, t_1,...,t_r) = \mathbf{P}(s_1,...,s_r;\, t_1,...,t_r)\mathbf{B}(e)$$

$$\mathbf{B}(e) = \mathbf{P}^{-1}(s_1 e,...,s_r e;\, t_1,...,t_r)\,\mathbf{P}(s_1,...,s_r;\, t_1,...,t_r) \qquad \text{(A.1.12)}$$

where $\mathbf{y}_\rho = \mathbf{P}(\varnothing;\, t_1,...,t_r)$ and $\mathbf{z}_\rho = [\ 1\ 0\ 0\ ...\ 0\ ]'$. Combining these two sets of equations, we can write the generalized formula

$$\mathbf{P}(s_1 e,...,s_r e;\, \varepsilon t_1,...,\varepsilon t_r) = \mathbf{A}(\varepsilon)\mathbf{P}(s_1,...,s_r;\, t_1,...,t_r)\mathbf{B}(e).$$

Formulas (A.1.9) and (A.1.11) show that for a process to be a matrix process, it is necessary and sufficient that the probability of any sequence be a linear combination of the probabilities of its basis sequences. Let us now discuss methods of constructing the process basis.

Theorem A.1.1 requires to test an infinite number of conditions in order to determine the rank of a process. However, it is possible to determine the rank in a finite number of steps. To prove it, we denote $|\,s\,|$ the length of a sequence s; the collection of all sequences having the same length m we call a layer and denote $L_m^{(s)}$. The union of all the layers whose lengths do not exceed m is denoted by

$$W_m^{(s)} = \bigcup_{i=0}^{m} L_i^{(s)}$$

Theorem A.1.2: Let $Z = \{e_j\colon j \in \mathbf{J}, e_j \in (0,1)\}$ be a matrix process, and let the sequences $s_1,s_2,...,s_p$ and $t_1,t_2,...,t_p$ represent the basis for the sets $s \in W_m^{(s)}$ and $t \in W_m^{(t)}$, that is

$$\det \mathbf{P}(s_1,...,s_p;\, t_1,...,t_p) \neq 0 \qquad \text{(A.1.13)}$$

and

$$\det \mathbf{P}(s_1,\ldots,s_p,s;\, t_1,\ldots,t_p,t) = 0 \tag{A.1.14}$$

for any $s \in W_m^{(s)}$ and $t \in W_m^{(t)}$. Then if condition (A.1.14) holds for any sequences $s \in W_{m+1}^{(s)}$ and $t \in W_{m+1}^{(t)}$, the rank of the process is p, and the sequences $s_1,s_2,\ldots,s_p$ and $t_1,t_2,\ldots,t_p$ are basis sequences of the process.

Proof: By (A.1.3) we have

$$\mathbf{P}(s_1,\ldots,s_p,s;t_1,\ldots,t_p,t) = \mathbf{Q}(s_1,\ldots,s_p,s)\mathbf{R}(t_1,\ldots,t_p,t)$$

Since *rank* $\mathbf{P}(s_1,\ldots,s_p,s;t_1,\ldots,t_p,t) = p$, at least one of the matrices $\mathbf{Q}(s_1,\ldots,s_p,s)$ or $\mathbf{R}(t_1,\ldots,t_p,t)$ has rank p for any vectors $s \in W_{m+1}^{(s)}$ or $t \in W_{m+1}^{(t)}$. Let *rank* $\mathbf{R}(t_1,\ldots,t_p,t) = p$ for any $t \in W_{m+1}^{(t)}$. Then the last column of $\mathbf{R}(t_1,\ldots,t_p,t)$ is a linear combination of its other columns:

$$\mathbf{r}(t) = \sum_{i=1}^{p} \alpha_i \mathbf{r}(t_i)$$

An arbitrary sequence τ from $L_{m+2}^{(t)}$ has the form $\tau = e\,t$, where $e \in (0,1)$, $t \in L_{m+1}^{(t)}$. Equation (1.3.5) gives

$$\mathbf{r}(e\,t) = \mathbf{M}(e)\mathbf{r}(t) = \sum_{i=1}^{p} \alpha_i \mathbf{M}(e)\mathbf{r}(t_i) = \sum_{i=1}^{p} \alpha_i \mathbf{r}(et_i)$$

Since $t_i \in W_m^{(t)}$ and $et_i \in W_{m+1}^{(t)}$, vector $\mathbf{r}(et_i)$ can be expressed as a linear combination of the basis vectors

$$\mathbf{r}(et_i) = \sum_{i=1}^{p} \beta_{ij} \mathbf{r}(t_j)$$

Substituting this expression into the previous equation we obtain

$$\mathbf{r}(et) = \sum_{i=1}^{p} \alpha_i \sum_{j=1}^{p} \beta_{ij} \mathbf{r}(t_j) = \sum_{j=1}^{p} \Big(\sum_{i=1}^{p} \alpha_i \beta_{ij}\Big) \mathbf{r}(t_j)$$

This means that the columns of $\mathbf{R}(t_1,t_2,\ldots,t_p,et)$ are linearly dependent so that *rank* $\mathbf{R}(t_1,t_2,\ldots,t_p,et) = p$. According to the Sylvester inequalities,[21,24] the rank of

$$\mathbf{P}(s_1,\ldots,s_p,s\sigma;\, t_1,\ldots,t_p,et) = \mathbf{Q}(s_1,\ldots,s_p,s\sigma)\mathbf{R}(t_1,\ldots,t_p,et)$$

is not greater than p. Thus, $\det \mathbf{P}(s_1,\ldots,s_p,s\sigma;t_1,\ldots,t_p,et) = 0$ for all sequences

$s\sigma \in W_{m+2}^{(s)}$ and $et \in W_{m+2}^{(t)}$.

Using the method of mathematical induction, we prove that this determinant is equal to zero for any sequences s and t, which, according to Theorem A.1.1, proves that the sequences $s_1,...,s_p$ and $t_1,...,t_p$ represent the basis of the process.

■

This theorem gives us a method of building a matrix process basis. We say that a sequence s_1 precedes a sequence s_2 if the length of s_1 is smaller than the length of s_2; if the lengths of the sequences are equal, then s_1 precedes s_2 if the binary number representing s_1 is smaller than the binary number representing s_2. If s_1 precedes s_2, we write $s_1 \propto s_2$. We say that a pair of sequences $(s_1; t_1)$ precedes a pair $(s_2; t_2)$ if $s_1 \propto s_2$ or if $s_1 = s_2$, but $t_1 \propto t_2$.

To build the basis, we start with empty sequences $\sigma_1 = \varnothing$, $\tau_1 = \varnothing$. If the process rank is not less than two, then we can find a pair $(\sigma_2; \tau_2)$ preceding all the pairs that satisfy the condition det $\mathbf{P}(\sigma_1,\sigma_2; \tau_1,\tau_2) \neq 0$. Similarly, we choose a pair $(\sigma_3; \tau_3)$ such that det $\mathbf{P}(\sigma_1,\sigma_2,\sigma_3; \tau_1,\tau_2,\tau_3) \neq 0$, and so on. According to the previous theorem, after a finite number of steps we will construct the basis of the process. This basis is unique by construction; we call it the *process canonic basis*.

Corollary A.1.1: If a sequence $s_1,s_2,...,s_r$ (or $t_1,t_2,...,t_r$) forms a canonic basis of a matrix process and $|s_r| = m$ (or $|t_r| = m$), then this basis includes at least one element of every layer $L_i^{(s)}$ (or $L_i^{(t)}$) for $i<m$.

■

Corollary A.1.2: If the rank of a process is r, then this process is determined uniquely by the multidimensional distributions of sequences $e_1,e_2,...,e_m$ when $m = |e_1 e_2 \cdots e_m| \leq 2r-1$.

Indeed, when building the canonic basis, the most unfavorable case is when, starting with $s_1 = \varnothing$ and $t_1 = \varnothing$, we can include in the basis only one sequence s_i and t_j from every layer in which case max $|s_i e t_j| \leq 2r-1$.

■

Corollary A.1.3: For two matrix processes to be equivalent, it is necessary and sufficient that they have the same rank r and the same multidimensional distribution up to the order $2r-1$.

■

As we saw in Sec. 1.3, it is important to find equivalent matrix processes to simplify a model description. We showed that if conditions (1.3.6) are satisfied, then processes $\{\mathbf{yT},\mathbf{M}(e),\mathbf{z}\}$ and $\{\mathbf{y},\mathbf{M}_1(e),\mathbf{Tz}\}$ are equivalent. We develop here necessary and sufficient conditions for the processes equivalence.

It follows from the results of this Appendix that if the ranks or the canonic bases do not coincide, then the processes are not equivalent. If the ranks and the canonic bases do coincide, it is necessary and sufficient that the matrices $\mathbf{A}(e)$ and $\mathbf{A}_1(e)$ of the canonic representations of these processes coincide which, according to (A.1.3) and (A.1.10), can be written as:

$$\mathbf{QM}(e)\mathbf{R}[\mathbf{QR}]^{-1} = \mathbf{Q}_1\mathbf{M}_1(e)\mathbf{R}_1[\mathbf{Q}_1\mathbf{R}_1]^{-1} \tag{A.1.15}$$

If the matrices $\mathbf{M}(e)$ and $\mathbf{M}_1(e)$ are order r square matrices, then this relation implies the following corollary.

Corollary A.1.4: Suppose that $\mathbf{M}(e)$ and $\mathbf{M}_1(e)$ have orders equal to the ranks of the processes. A necessary and sufficient condition for the matrix processes $\{\mathbf{y},\mathbf{M}(e),\mathbf{z}\}$ and $\{\mathbf{y}_1,\mathbf{M}_1(e),\mathbf{z}_1\}$ to be equivalent is that a matrix $\mathbf{T}$ satisfying the conditions

$$\mathbf{y}_1 = \mathbf{yT} \quad \mathbf{M}_1(e) = \mathbf{T}^{-1}\mathbf{M}(e)\mathbf{T} \quad e = 0,1 \quad \mathbf{z}_1 = \mathbf{T}^{-1}\mathbf{z}$$

exists.

Proof: The sufficiency follows from (1.3.6); let us prove the necessity. It follows from (A.1.15) that

$$\mathbf{M}_1(e) = \mathbf{T}^{-1}\mathbf{M}(e)\mathbf{T}$$

where $\mathbf{T} = \mathbf{Q}^{-1}\mathbf{Q}_1$. Since $\mathbf{q}(\varnothing) = \mathbf{y}$, $\mathbf{q}_1(\varnothing) = \mathbf{y}_1$ are rows of the matrices $\mathbf{Q}$ and $\mathbf{Q}_1$, then

$$\mathbf{y} = \mathbf{y}_l\mathbf{Q} \quad \mathbf{y}_1 = \mathbf{y}_l\mathbf{Q}_1$$

so that

$$\mathbf{y}_1 = \mathbf{yQ}^{-1}\mathbf{Q}_1 = \mathbf{yT}.$$

Similarly, the identities $\mathbf{r}(\varnothing) = \mathbf{z}, \mathbf{r}_1(\varnothing) = \mathbf{z}_1$ imply, because of the relations $\mathbf{z}_l = \mathbf{Qr}(\varnothing)$, $\mathbf{z}_l = \mathbf{Q}_1\mathbf{r}_1(\varnothing)$, that $\mathbf{z}_1 = \mathbf{T}^{-1}\mathbf{z}$. Thus, stationary matrix processes with the matrices of the smallest order (equal to the rank of the process) are equivalent only when their defining matrices are similar.

■

1.1.1 Reduced Canonic Representation. The basis sequences considered in the previous section can also be built using block matrices. To achieve this we associate basis sequences with symbols e, that is, we consider the sequences of the form $s_i^{(e)}et_j^{(e)}$, $s_1(e) = \varnothing$, $t_1^{(e)} = \varnothing$. Let

$$\mathbf{P}_e(s_1^{(e)},s_2^{(e)},\ldots,s_p^{(e)};\, t_1^{(e)},t_2^{(e)},\ldots,t_q^{(e)})$$

denote the matrix whose (i,j)-th element equals $p(s_i^{(e)}et_j^{(e)})$. Then, in accordance with (1.3.8), we can represent it in the form

$$\mathbf{P}_e(s_1^{(e)},...,s_p^{(e)}; t_1^{(e)},...,t_q^{(e)}) = \mathbf{Q}_e(s_1^{(e)},...,s_p^{(e)})\mathbf{R}_e(t_1^{(e)},...,t_q^{(e)}) \qquad \text{(A.1.16)}$$

This equation is analogous to (A.1.3).

Expressions (A.1.16) permit to define ranks r_e corresponding to symbols $e = 0$ or $e = 1$. We can show that the basis sequences exist and that $r_e \le n_e$ (where n_e is the order of the matrix $\mathbf{M}_{ee}$). As in the previous section, we obtain

$$\mathbf{P}_\mu(s_1^{(\mu)},...,s_{r_\mu}^{(\mu)}; e\, t) = \mathbf{A}_\mu(e)\mathbf{P}_e(s_1^{(e)},...,s_{r_e}^{(e)}; t) \qquad \text{(A.1.17)}$$

$$\mathbf{A}_\mu(e) = \mathbf{P}_\mu(s_1^{(\mu)},...,s_{r_\mu}^{(\mu)}; et_1^{(e)},...,et_{r_e}^{(e)})\mathbf{P}^{-1}(s_1^{(e)},...,s_{r_e}^{(e)}; t_1^{(e)},...,t_{r_e}^{(e)}) \qquad \text{(A.1.18)}$$

Note that although these relations are a particular case of those considered in the previous section, the sequences used here in the construction of the basis differ slightly from the sequences used in the previous section. For example, we could have considered previously the sequence $s_1 t_1 = \varnothing$; now the maximal length of a sequence is 1, since we consider sequences of the form $s_i e t_j$, where $e \neq \varnothing$. The matrix $\mathbf{P}(s_1,...,s_p; t_1,...,t_q)$ now has the form

$$\mathbf{P}(s_1^{(0)}0,...,s_{p_0}^{(0)}0,s_1^{(1)}1,...,s_{p_1}^{(1)}1; t_1^{(0)},...,t_{q_0}^{(0)},t_1^{(1)},...,t_{q_1}^{(1)}) =$$

$$\begin{bmatrix} \mathbf{P}_0(s_1^{(0)},...,s_{p_0}^{(0)}; t_1^{(0)},...,t_{q_0}^{(0)}) & 0 \\ 0 & \mathbf{P}_1(s_1^{(1)},...,s_{p_1}^{(1)}; t_1^{(1)},...,t_{q_1}^{(1)}) \end{bmatrix} \qquad \text{(A.1.19)}$$

and

$$\mathbf{P}(s_1^{(0)}0,...,s_{r_0}^{(0)}0,s_1^{(1)}1,...,s_{r_1}^{(1)}1; t) = \begin{bmatrix} \mathbf{P}_0(s_1^{(0)},...,s_{r_0}^{(0)}; t) \\ \mathbf{P}_1(s_1^{(1)},...,s_{r_1}^{(1)}; t) \end{bmatrix} \qquad \text{(A.1.20)}$$

$$\mathbf{y}_l = [\mathbf{y}_{l0} \quad \mathbf{y}_{l1}] \quad \mathbf{z}_l = [\mathbf{z}_{l0} \quad \mathbf{z}_{l1}]$$

$$\mathbf{A}(0) = \begin{bmatrix} \mathbf{A}_{00} & 0 \\ \\ \mathbf{A}_{10} & 0 \end{bmatrix}, \quad \mathbf{A}(1) = \begin{bmatrix} 0 & \mathbf{A}_{01} \\ \\ 0 & \mathbf{A}_{11} \end{bmatrix} \tag{A.1.21}$$

Thus, we see that the use of special basis sequences considered in this section permits to simplify the process description and, in particular, have matrices $\mathbf{A}(e)$, containing zero blocks. We call this representation $\{\mathbf{y}_l, \mathbf{A}(0), \mathbf{A}(1), \mathbf{z}_l\}$ of a matrix process a *reduced canonic representation.*

1.2 Markov Lumpable Chains

The material of this section is an adaptation of the results of Ref. [30] and [40]. It is closely related to that of the previous section and serves as an introduction to the next section.

Let $\mathbf{P} = [\,p_{ij}\,]_{n,n}$ be a transition matrix and $\mathbf{p} = [\,p_i\,]_{1,n}$ be a matrix of initial probabilities of a homogeneous Markov chain s_t. Suppose that $\{A_1, A_2, ..., A_s\}$ is a partition of the set of all states into nonintersecting subsets:

$$\bigcup_{i=1}^{s} A_i = \Omega \quad A_i \cap A_j = \varnothing \quad i \neq j$$

A process y_t is called a *Markov function* or *lumped process*[40] if $y_t = i$ when $s_t \in A_i$.

To simplify notations, let us assume that $A_1 = \{1, 2, ..., m_1\}$, $A_2 = \{m_1+1, ..., m_2\}$, ..., $A_s = \{m_{s-1}+1, ..., m_s\}$. If this assumption is not valid, we can renumber the states to satisfy it (see Appendix 6.1.3). In correspondence with this partition we can write the matrices $\mathbf{P}$ and $\mathbf{p}$ in the block form

$$\mathbf{P} = block\ [\mathbf{P}_{ij}]_{s,s} = \begin{bmatrix} \mathbf{P}_{11} & \mathbf{P}_{12} & \cdots & \mathbf{P}_{1s} \\ \mathbf{P}_{21} & \mathbf{P}_{22} & \cdots & \mathbf{P}_{2s} \\ . & . & \cdots & . \\ \mathbf{P}_{s1} & \mathbf{P}_{s2} & \cdots & \mathbf{P}_{ss} \end{bmatrix}$$

$$\mathbf{p} = block\ row\{\mathbf{p}_i\} = [\,\mathbf{p}_1 \ \ \mathbf{p}_2 \ \ \cdots \ \ \mathbf{p}_s\,]$$

where matrix block $\mathbf{P}_{ij}$ consists of all transition probabilities $p_{\alpha\beta}$, $\alpha \in A_i$, $\beta \in A_j$.

DEFINITION: A Markov chain is called $\mathbf{p}$-lumpable with respect to a partition $\{A_1, A_2, ..., A_s\}$ if there exists a vector $\mathbf{p}$ such that the lumped process y_t is a Markov chain.

■

This means that there are scalars $\hat{p}_u$ and $\hat{p}_{uv}$, $(u,v = 1,2,...,s)$ that, in general, depend on $\mathbf{p}$, such that

$$Pr(y_0, y_1, \ldots y_t) = \hat{p}_{y_0}\hat{p}_{y_0 y_1} \cdots \hat{p}_{y_{t-1}y_t}$$

for any $t \in \mathbf{N}$ and any $y_0, y_1, \ldots, y_t$.

These conditions can be written in matrix form as follows:

$$\mathbf{p}_{y_0}\prod_{i=1}^{t}\mathbf{P}_{y_{i-1}y_i}\mathbf{1} = \hat{p}_{y_0}\prod_{i=1}^{t}\hat{p}_{y_{i-1}y_i} \tag{A.1.22}$$

Theorem A.1.3: If a Markov chain is $\mathbf{p}$-lumpable and $\boldsymbol{\pi}$ is its stationary distribution, then it is $\boldsymbol{\pi}$-lumpable.

Proof: Indeed, the summation of both sides of equation (A.1.22) with respect to y_0 gives

$$\mathbf{p}_{y_1}\prod_{i=2}^{t}\mathbf{P}_{y_{i-1}y_i}\mathbf{1} = \hat{p}_{y_1}\prod_{i=2}^{t}\hat{p}_{y_{i-1}y_i}$$

where

$$\mathbf{p}_1 = \sum_{y_0}\mathbf{p}_{y_0}\mathbf{P}_{y_0 y_1} \qquad \hat{p}_1 = \sum_{y_0}\hat{p}_{y_0}\hat{p}_{y_0 y_1}$$

Therefore, if the chain is $\mathbf{p}$-lumpable, then it is $\mathbf{p}_1$-lumpable, where $\mathbf{p}_1 = \mathbf{pP}$.

Repeating the same transformations, we observe that the chain is $\mathbf{pP}^2$-lumpable, $\mathbf{pP}^3$-lumpable, and consequently for any positive integer m is $\mathbf{pP}^m$-lumpable. Since

$$\boldsymbol{\pi} = \lim_{N\to\infty}\mathbf{p}\sum_{m=0}^{N-1}\mathbf{P}^m / N$$

we conclude that the chain is $\boldsymbol{\pi}$-lumpable. Thus, in order to find out whether a Markov chain is lumpable for any initial distribution it is sufficient to check that the chain is lumpable for the stationary distribution $\boldsymbol{\pi}$.

■

The other necessary conditions of lumpability follow directly from (A.1.22):

$$\hat{p}_i = \mathbf{p}_i \mathbf{1} \tag{A.1.23}$$

$$\hat{p}_{ij} = \mathbf{p}_i \mathbf{P}_{ij} \mathbf{1} / \mathbf{p}_i \mathbf{1}$$

$$\hat{p}_{y_t y_{t+1}} = \mathbf{p}_{y_0} \prod_{i=1}^{t+1} \mathbf{P}_{y_{i-1} y_i} \mathbf{1} / \mathbf{p}_{y_0} \prod_{i=1}^{t} \mathbf{P}_{y_{i-1} y_i} \mathbf{1}$$

Equations (A.1.23) show that if the chain is lumpable, then the initial probabilities of the states of the lumped process are given by

$$\hat{\mathbf{p}}_0 = [\ \mathbf{p}_1 \mathbf{1}\ \ \mathbf{p}_2 \mathbf{1}\ \ \dots\ \ \mathbf{p}_s \mathbf{1}\ ]_{1,s}$$

The matrix of transition probabilities of the lumped process is also uniquely identified as

$$\hat{\mathbf{P}} = [\ \mathbf{q}_i \mathbf{P}_{ij} \mathbf{1} / \mathbf{q}_i \mathbf{1}\ ]_{s,s}$$

where $\mathbf{q}_i$ is the first different from zero matrix in the series

$$\mathbf{p}_i,\ \mathbf{p}_1 \mathbf{P}_{1i},\ \mathbf{p}_2 \mathbf{P}_{2i},\ \dots, \mathbf{p}_s \mathbf{P}_{si}$$

If all these matrices are equal to zero, then we delete the set A_i, because the process lumpability does not depend on this set. From this reasoning the following theorem is derived.

Theorem A.1.4: For a Markov chain to be lumpable with respect to a partition $\{A_1, A_2, \dots, A_s\}$, it is necessary and sufficient that, for any $y_0, y_1, \dots, y_t$ and $t \in \mathbf{N}$,

$$\frac{\mathbf{p}_{y_0} \prod_{i=1}^{t} \mathbf{P}_{y_{i-1} y_i} \mathbf{1}}{\mathbf{p}_{y_0} \prod_{i=1}^{t-1} \mathbf{P}_{y_{i-1} y_i} \mathbf{1}} = \frac{\mathbf{q}_{y_{t-1}} \mathbf{P}_{y_{t-1} y_t} \mathbf{1}}{\mathbf{q}_i \mathbf{1}} = \hat{p}_{y_{t-1} y_t} \tag{A.1.24}$$

in all cases when the denominator is different from zero.

■

This common value (A.1.24) gives the transition probability of the lumped chain, while the initial distribution of the lumped chain is given by (A.1.23).

If we introduce the probability vector

$$\mathbf{g}(y_0,y_1,\dots,y_t) = \mathbf{p}_{y_0}\prod_{i=1}^{t}\mathbf{P}_{y_{i-1}y_i} \Big/ \mathbf{p}_{y_0}\prod_{i=1}^{t}\mathbf{P}_{y_{i-1}y_i}\mathbf{1}$$

then the previous theorem can be formulated in terms of these vectors.

Theorem A.1.5: For a Markov chain to be lumpable with respect to a partition $\{A_1,A_2,\dots,A_s\}$, it is necessary and sufficient that for the above-defined probability vectors $\mathbf{g}(y_0,y_1,\dots,y_{t-1})$ and $t\in\mathbf{N}$ the expression

$$\mathbf{g}(y_0,y_1,\dots,y_{t-1})\mathbf{P}_{y_{t-1}y_t}\mathbf{1} = \hat{p}_{y_{t-1}y_t} \tag{A.1.25}$$

assumes fixed values for any given y_{t-1} and y_t.

■

To test whether the conditions of these theorems are satisfied, one must examine an infinite number of conditions. However, it is sufficient to consider only a finite number of such conditions for the normalized basis of the set

$$\mathbf{Y}_{y_t} = \{\mathbf{p}_{y_t},\ \mathbf{p}_{y_{t-1}}\mathbf{P}_{y_{t-1}y_t},\ \dots,\mathbf{p}_{y_0}\prod_{i=0}^{t}\mathbf{P}_{y_{i-1}y_i},\dots\} \tag{A.1.26}$$

for all possible values of $y_0,y_1,\dots,y_{t-1}$ and fixed y_t. That is, for the vectors

$$\mathbf{g}_{ij} = \mathbf{e}_{ij}/\mathbf{e}_{ij}\mathbf{1} \quad (i = 1,2,\dots,r_j;\ j = 1,2,\dots,s) \tag{A.1.27}$$

where r_j is the rank of the system $\mathbf{Y}_j$, [24] vectors $\mathbf{e}_{1j},\mathbf{e}_{2j},\dots,\mathbf{e}_{r_jj}$ are linearly independent.

Theorem A.1.6: For a Markov chain to be lumpable with respect to a partition $\{A_1,A_2,\dots,A_s\}$ it is necessary and sufficient that the left-hand side of (A.1.25) assumes a constant value for the normalized basis vectors (A.1.27) of the sets $\mathbf{Y}_j$, $(j = 1,2,\dots,s)$.

Proof: The necessity of these conditions is obvious: If condition (A.1.4) is satisfied for all vectors of the set, it is also satisfied for the vectors of their basis.

Let us prove the sufficiency. If we assume that conditions (A.1.25) are satisfied for vectors (A.1.27)

$$\mathbf{g}_{ij}\mathbf{P}_{jm} = \hat{p}_{jm}$$

then we can prove that these conditions are valid for an arbitrary normalized vector $\mathbf{g}_j\in\mathbf{Y}_j$. We decompose this vector with respect to the basis:

$$\mathbf{g}_j = \sum_{i=1}^{r_j} \alpha_i \mathbf{g}_{ij}$$

and since

$$\mathbf{g}_j \mathbf{1} = \mathbf{g}_{ij} \mathbf{1} = 1, \quad \text{then} \quad \sum_{i=1}^{r_j} \alpha_i = 1$$

But, in this case,

$$\mathbf{g}_j \mathbf{P}_{jm} \mathbf{1} = \sum_{i=1}^{r_j} \alpha_i \mathbf{g}_{ij} \mathbf{P}_{jm} \mathbf{1} = \hat{p}_{jm}$$

which proves the theorem.

■

Thus, in order to check the process lumpability we need to build a basis of the set $\mathbf{Y}_j$. To do so it is convenient to consider the layers of the set $\mathbf{Y}_j$:

$L_j^{(0)}: \mathbf{p}_j$

$L_j^{(1)}: \mathbf{p}_1 \mathbf{P}_{1j},\ \mathbf{p}_2 \mathbf{P}_{2j},\ \ldots, \mathbf{p}_s \mathbf{P}_{sj}$

$L_j^{(2)}: \mathbf{p}_1 \mathbf{P}_{11} \mathbf{P}_{1j},\ \mathbf{p}_1 \mathbf{P}_{12} \mathbf{P}_{2j},\ \ldots, \mathbf{p}_s \mathbf{P}_{ss} \mathbf{P}_{sj}$

. . . .

$L_j^{(t)}: \mathbf{p}_{y_0} \prod_{i=1}^{t} \mathbf{P}_{y_{i-1} y_i},$ for all $y_0, y_1, \cdots y_{t-1}$, and $y_t = j$

. . . .

Suppose that we have built the bases of all the sets

$$Y_j^{(t)} = \bigcup_{i=0}^{t} L_j^{(i)}, \qquad \text{(A.1.28)}$$

for $j = 1,2,\ldots,s$, and fixed t. Assume also that by adding the next layer, $L_j^{(t+1)}$, it is possible to keep the basis of $Y_j^{(t+1)}$ unchanged, that is, the ranks of sets $Y_j^{(t)}$ and $Y_j^{(t+1)}$ are the same:

$$rank\ Y_j^{(t)} = rank\ Y_j^{(t+1)} \qquad (j = 1,2,\ldots,s)$$

The following theorem proves that bases of the sets (A.1.28) can serve as bases of the entire sets $\mathbf{Y}_j$.

Theorem A.1.7: If $rank\ Y_j^{(t)} = rank\ Y_j^{(t+1)}$ for all $j = 1,2,...,s$, then

$$rank\ Y_j^{(t)} = rank\ \mathbf{Y}_j$$

Proof: It is sufficient to show that $rank\ Y_j^{(t+1)} = rank\ Y_j^{(t+2)}$, $(j = 1,2,...,s)$. Let $\mathbf{z}_j \in Y_j^{(t+2)}$ be an arbitrary vector from the set $Y_j^{(t+2)}$. By the set $Y_j^{(t+2)}$ composition vector $\mathbf{z}_j$ either belongs to the set $Y_j^{(t+1)}$ or $\mathbf{z}_j = \mathbf{y}_i \mathbf{P}_{ij}$, where $\mathbf{y}_i \in Y_i^{(t+1)}$. But $rank\ Y_i^{(t+1)} = rank\ Y_i^{(t)}$, so that the vector $\mathbf{y}_i$ can be decomposed with respect to the set $Y_i^{(t)}$ basis:

$$\mathbf{y}_i = \sum_{k=1}^{r_i} \beta_{ki} \mathbf{e}_{ki}^{(t)}$$

and, as a result,

$$\mathbf{z}_j = \sum_{k=1}^{r_i} \beta_{ki} \mathbf{e}_{ki}^{(t)} \mathbf{P}_{ij}$$

Therefore, any vector from $Y_j^{(t+2)}$ either coincides with a vector from $Y_j^{(t+1)}$, or is a linear combination of vectors from $Y_j^{(t+1)}$ so that $rank\ Y_j^{(t+1)} = rank\ Y_j^{(t+2)}$.

■

DEFINITION: A Markov chain is called **M**-lumpable with respect to a partition $\{A_1, A_2, ..., A_s\}$ if for any initial distribution $\mathbf{p}$ from the set $\mathbf{M}$ the lumped process is a Markov chain with transition probabilities independent of $\mathbf{p} \in \mathbf{M}$.

■

According to Ref. [40], we say that a Markov chain is *strongly lumpable* if it is Ω-lumpable (where Ω is the set of all probability vectors of n-dimensional space).

Let $\mathbf{h}_1, \mathbf{h}_2, ..., \mathbf{h}_r$ be the basis of $\mathbf{M}$. If we represent these vectors in the matrix-block form $\mathbf{h}_i = [\ \mathbf{h}_{i1}\ \ \mathbf{h}_{i2}\ ...\ \mathbf{h}_{is}\ ]$, corresponding to a partition $\{A_1, A_2, ..., A_s\}$, then the previous results lead to the following theorem.

Theorem A.1.8: A necessary and sufficient condition for a Markov chain to be **M**-lumpable with respect to a partition $\{A_1, A_2, ..., A_s\}$ is that expressions (A.1.25) assume constant value for the normalized vectors $\mathbf{h}_{ij}$:

$$\mathbf{h}_{ij} \mathbf{P}_{im} \mathbf{1} = \hat{p}_{im} \mathbf{h}_{ij} \mathbf{1}$$

and, in addition, that the chain be $\mathbf{h}_i$-lumpable.

■

Theorem A.1.9:[40] A necessary and sufficient condition for a Markov chain to be strongly lumpable with respect to a partition $\{A_1,A_2,...,A_s\}$ is that the sums of elements in the rows of the matrices $\mathbf{P}_{ij}$ be equal; that is,

$$\mathbf{P}_{ij}\mathbf{1} = \hat{p}_{ij}\mathbf{1} \quad (i,j = 1,2,...,s) \tag{A.1.29}$$

Proof: In this case $\mathbf{M} = \Omega$, and the basis can be formed by vectors $\mathbf{h}_1 = [1\ 0\ 0 \cdots 0]$, $\mathbf{h}_2 = [0\ 1\ 0 \cdots 0]$, ..., $\mathbf{h}_s = [0\ 0\ 0 \cdots 1]$. Then it is necessary and sufficient that

$$\mathbf{h}_k\mathbf{P}_{ij}\mathbf{1} = \hat{p}_{ij} \quad i,j,k = 1,2,...,s$$

These expressions are equivalent to (A.1.29). ■

1.3 Semi-Markov Lumpable Chains

The conditions of semi-Markov lumpability of a Markov chain are largely similar to the conditions for Markov lumpability. We assume that sets $A_1,A_2,...,A_s$ are not absorptive. Relationships (1.4.5), which determine the conditions for a semi-Markov process, are similar to (A.1.22). As we mentioned in Sec. 1.4, it is convenient to use the notion of the transition process when dealing with semi-Markov processes. Theorem 1.4.1 can now be rephrased as follows.

Theorem A.1.10: For a Markov chain to be **p**-semi-Markov lumpable according to a partition $\{A_1,A_2,...,A_s\}$, it is necessary and sufficient that for any $y_0,y_1,...,y_t$; $l_0,l_1,...,l_t$; and $t \in \mathbf{N}$ the following conditions be satisfied:

$$f_{y_t y_{t+1}}(l) = \mathbf{p}_{y_0}\prod_{i=0}^{t}\mathbf{Q}_{y_i}(l_i)\mathbf{Q}_{y_i y_{i+1}}\mathbf{1}/\mathbf{p}_{y_0}\prod_{i=0}^{t-1}\mathbf{Q}_{y_i}(l_i)\mathbf{Q}_{y_i y_{i+1}}\mathbf{1} \qquad t = 1,2,3,... \tag{A.1.30}$$

These common expressions represent the transition distributions of the semi-Markov process. ■

Let us consider the sequences of vectors

$$\mathbf{h}_0(y_0) = \mathbf{p}_{y_0}$$

$$\mathbf{h}_t(y_0,y_1,...,y_t;l_0,l_1,...,l_{t-1}) = \mathbf{p}_{y_0}\prod_{i=0}^{t-1}\mathbf{Q}_{y_i}(l_i)\mathbf{Q}_{y_i y_{i+1}} \tag{A.1.31}$$

$$t = 1,2,...$$

and normalized vectors

$$\mathbf{g}_t(y_0,y_1,\ldots,y_t;l_0,l_1,\ldots,l_{t-1}) = \frac{\mathbf{h}_t(y_0,y_1,\ldots,y_t;l_0,l_1,\ldots,l_{t-1})}{\mathbf{h}_t(y_0,y_1,\ldots,y_t;l_0,l_1,\ldots,l_{t-1})\mathbf{1}} \tag{A.1.32}$$

Using these notations, Theorem A.1.10 can be formulated as follows:

Theorem A.1.11: For a Markov chain to be **p**-semi-Markov lumpable according to a partition $\{A_1,A_2,\ldots,A_s\}$, it is necessary and sufficient that for all probability vectors of the form of (A.1.32) the expressions

$$f_{ij}(l) = \mathbf{g}_{t-1}(y_0,y_1,\ldots,y_{t-2},i;l_0,l_1,\ldots,l_{t-2})\mathbf{Q}_i(l)\mathbf{Q}_{ij}\mathbf{1} \tag{A.1.33}$$

keep a constant value for fixed i, j, l $(i,j = 1,2,\ldots,s; l = 1,2,\ldots)$.

■

In other words, the right-hand side of equation (A.1.33) may depend only on i, j, l for all possible values of $y_0,y_1,\ldots,y_{t-2}$; $l_0,l_1,\ldots,l_{t-2}$; and t.

This theorem requires us to test an infinite number of conditions (A.1.33). In some cases it is possible to prove that all these conditions are satisfied. Consider as an illustration the following statement, which gives simple sufficient conditions of semi-Markov lumpability.

Theorem 1.4.2: If the sets of states $\{A_1,A_2,\ldots,A_s\}$ of a Markov chain are not absorbing, and the transition matrices have the form

$$\mathbf{P}_{ij} = \mu_{ij}\mathbf{M}_i\mathbf{N}_j \quad i \neq j \quad (i,j = 1,2,\ldots,s) \tag{A.1.34}$$

where $\mathbf{N}_j$ is a matrix row, $\mathbf{M}_i$ is a matrix column, and μ_{ij} is a number, then this Markov chain is semi-Markov lumpable.

Proof: Since $\mathbf{M}_i\mathbf{N}_j = [m_{ik}n_{jl}]$ and $\mu_{ij}m_{ik}n_{jl} \geq 0$, without loss of generality one can assume that

$$\mathbf{N}_j\mathbf{1} = 1 \quad \mathbf{N}_j \geq 0 \tag{A.1.35}$$

Because $\mathbf{P}$ is a stochastic matrix, $\mathbf{P1} = \mathbf{1}$, which can be written in the following block form

$$\sum_j \mathbf{P}_{ij}\mathbf{1} = \mathbf{1}$$

Then

$$\mathbf{1} - \mathbf{P}_{ii} = \sum_{j \neq i} \mathbf{P}_{ij}\mathbf{1} = \mu_i \mathbf{M}_i$$

where $\mu_i = \sum_j \mu_{ij}$. Thus, since $\mu_i \neq 0$

$$\mathbf{M}_i = (\mathbf{I} - \mathbf{P}_{ii})\mathbf{1}/\mu_i \tag{A.1.36}$$

Substituting this matrix into (A.1.34) we obtain

$$\mathbf{P}_{ii} = \hat{p}_{ij}(\mathbf{I} - \mathbf{P}_{ii})\mathbf{1}\mathbf{N}_j \quad \hat{p}_{ij} = \mu_{ij}/\mu_i \tag{A.1.37}$$

But then

$$\mathbf{Q}_{ij} = (\mathbf{I} - \mathbf{P}_{ii})^{-1}\mathbf{P}_{ij} = \hat{p}_{ij}\mathbf{1}\mathbf{N}_j \tag{A.1.38}$$

To prove the theorem let us show that the chain is **p**-semi-Markov lumpable if the initial distribution has the form

$$\mathbf{p} = [\, \hat{p}_1\mathbf{N}_1 \;\; \hat{p}_2\mathbf{N}_2 \; \dots \; \hat{p}_s\mathbf{N}_s \,]$$

where $\hat{\mathbf{p}} = [\, \hat{p}_1 \;\; \hat{p}_2 \; \dots \; \hat{p}_s \,]$ is any probability vector.

Indeed, vectors (A.1.31) become, after substitutions $\mathbf{p}_{y_0} = \hat{p}_{y_0}\mathbf{N}_{y_0}$ and (A.1.38),

$$\mathbf{h}_0(y_0) = \hat{p}_{y_0}\mathbf{N}_{y_0}$$

$$\mathbf{h}_t(y_0, y_1, \dots, y_t; l_0, l_1, \dots, l_{t-1}) = \hat{p}_{y_0}\mathbf{N}_{y_0}\prod_{i=0}^{t-1}\mathbf{Q}_{y_i}(l_i)\hat{p}_{y_i y_{i+1}}\mathbf{1}\mathbf{N}_{y_{i+1}} \qquad t = 1, 2, \dots .$$

Taking into account that $\mathbf{N}_{y_i}\mathbf{Q}_{y_i}(l_i)\mathbf{1}$ is a number, we obtain

$$\mathbf{g}_t(y_0, y_1, \dots, y_t; l_0, l_1, \dots, l_{t-1}) = \mathbf{N}_{y_t}$$

and therefore

$$\mathbf{g}_{t-1}(y_0, y_1, \dots, y_{t-2}, i; l_0, l_1, \dots, l_{t-2})\mathbf{Q}_i(l)\mathbf{Q}_{ij}\mathbf{1} = \mathbf{N}_i\mathbf{Q}_i(l)\mathbf{Q}_{ij}\mathbf{1}$$

depends only on i, j, l. The conditions of Theorem A.1.11 are satisfied, which proves the theorem.

■

As for Markov lumpable chains, instead of testing the conditions of lumpability for an infinite number of vectors, it is sufficient to check them for the bases of vector sets (A.1.26).

Theorem A.1.12: For a Markov chain to be **p**-semi-Markov lumpable according to a partition $\{A_1, A_2, ..., A_s\}$, it is necessary and sufficient that equations

$$f_{ij}(l) = \mathbf{e}_{ki}\mathbf{Q}_i(l)\mathbf{Q}_{ij}\mathbf{1}/\mathbf{e}_{ki}\mathbf{1} \qquad \text{(A.1.39)}$$

be satisfied for the bases $\{\mathbf{e}_{ki}\}$ of sets $\mathbf{Y}_i$ and for fixed i,j,l, $(i,j = 1,2,...,s;$ $l = 1,2,...)$.

The proof of this theorem is similar to that of Theorem A.1.6. Again, as we have seen, the bases of the sets $\mathbf{Y}_i$ can be built in a finite number of steps by building bases of the expanding unions of layers (A.1.28).

■

Even though we have replaced the infinite number of vectors to be tested by a finite number of vectors (the bases), the number of test conditions (A.1.39) is infinite, since l is an arbitrary positive integer. Let us show that it is sufficient to check only a finite number of conditions (A.1.39).

Conditions (A.1.39) can be written in the form

$$f_{ij}(l) = \mathbf{y}\mathbf{P}_{ii}^{l-1}\mathbf{P}_{ij}\mathbf{1}/\mathbf{y}\mathbf{1}$$

and therefore, as we have seen in Sec. 1.3.4, the transition distribution $f_{ij}(l)$ satisfies a linear difference equation with constant coefficients:

$$f_{ij}(l) = \sum_{m=1}^{k} a_{ijm} f_{ij}(l-m)$$

The order k of this equation is not greater than the size of the matrix $\mathbf{P}_{ii}$. It is well known that this equation has a unique solution that satisfies any fixed initial conditions

$$f_{ij}(1) = y_1,\ f_{ij}(2) = y_2, ...,\ f_{ij}(k) = y_k$$

Thus, if equation (A.1.39) is satisfied for all the basis vectors and all l not greater than the order of the matrix $\mathbf{P}_{ii}$, then this equation is valid for all positive l.

DEFINITION: Let **M** be a set of probability vectors. We shall say that a Markov chain is **M**-semi-Markov lumpable according to a partition $\{A_1, A_2, ..., A_s\}$ if it is

$\mathbf{p}$-semi-Markov lumpable for any vector $\mathbf{p}$ that belongs to the set $\mathbf{M}$ and if transition distributions do not depend on $\mathbf{p} \in \mathbf{M}$.

■

Theorem A.1.13: If a regular Markov chain is $\mathbf{M}$-semi-Markov lumpable, then the stationary distribution of the transition process belongs to $\mathbf{M}$.

Indeed, if a Markov chain is $\mathbf{M}$-semi-Markov lumpable, there exists a vector $\mathbf{p}$ such that

$$\mathbf{p}_{y_0} \prod_{i=0}^{t} \mathbf{Q}_{y_i}(l_i) \mathbf{Q}_{y_i y_{i+1}} \mathbf{1} = \hat{p}_{y_0} \prod_{i=0}^{t} f_{y_i y_{i+1}}(l_i)$$

If we sum these identities with respect to $y_0, y_1, \ldots, y_{k-1}$ and $l_0, l_1, \ldots, l_{k-1}$, we will have

$$\mathbf{p}_{y_k}^{(k)} \prod_{i=k}^{t} \mathbf{Q}_{y_i}(l_i) \mathbf{Q}_{y_i y_{i+1}} \mathbf{1} = \hat{p}_{y_k}^{(k)} \prod_{i=k}^{t} f_{y_i y_{i+1}}(l_i)$$

where $\mathbf{p}_{y_k}^{(k)}$ is the block of the matrix $\mathbf{p}\mathbf{Q}^k$, while $\hat{p}_{y_k}^{(k)}$ is an element of the matrix $\hat{\mathbf{p}}\hat{\mathbf{P}}^k$. But, due to matrix multiplication linearity, the previous equality is valid when we replace the block $\mathbf{p}_{y_k}^{(k)}$ with the corresponding block of the matrix

$$\mathbf{p} \sum_{k=0}^{N-1} \mathbf{Q}^k / N$$

For $t = m + N$ and $N \to \infty$, we have

$$\mathbf{q}_{z_0} \prod_{i=0}^{m} \mathbf{Q}_{z_i}(\lambda_i) \mathbf{Q}_{z_i z_{i+1}} \mathbf{1} = \hat{q}_{z_0} \prod_{i=0}^{m} f_{z_i z_{i+1}}(\lambda_i)$$

where

$$\mathbf{q} = \lim_{N \to \infty} \mathbf{p} \sum_{k=0}^{N-1} \mathbf{Q}^k / N.$$

The previous equation shows that the process is $\mathbf{q}$-semi-Markov lumpable so that $\mathbf{q} \in \mathbf{M}$.

■

The invariant vector $\mathbf{q} = [\ \mathbf{q}_1 \ \ \mathbf{q}_2 \ \ldots \ \mathbf{q}_s\ ]$ of the matrix $\mathbf{Q}$ can be found as the solution of the system

$$\mathbf{q}(\mathbf{Q}-\mathbf{I})=0 \qquad \mathbf{q1}=1$$

This system has a unique solution, because

$$rank\ (\mathbf{Q}-\mathbf{I})=rank\ (\mathbf{P}-\mathbf{I})=n-1$$

as a result of the relationship

$$(\mathbf{Q}-\mathbf{I})=(\mathbf{I}-\mathbf{R})^{-1}(\mathbf{P}\text{–}\mathbf{I}) \qquad \text{where } \mathbf{R}=block\ diag\{\mathbf{P}_{ii}\}$$

It is easy to verify that the unique solution has the form

$$\mathbf{q}_i=\boldsymbol{\pi}_i(\mathbf{1}-\mathbf{P}_{ii})/\sum_{k=1}^{s}\boldsymbol{\pi}_k(\mathbf{1}-\mathbf{P}_{kk})\mathbf{1}$$

where $\boldsymbol{\pi}$ is the stationary distribution of the Markov chain with the transition matrix **P**.

APPENDIX 2

2.1 Asymptotic Expansion of Matrix Probabilities

If elements of the z-transform

$$\mathbf{\Phi}(z) = \sum_{m=0}^{\infty} \mathbf{W}_m z^{-m} \tag{A.2.1}$$

are rational fractions

$$\mathbf{\Phi}(z) = [\, p_{ij}(z) \,/\, q_{ij}(z) \,]_{k,k}$$

where $p_{ij}(z)$ and $q_{ij}(z)$ are polynomials, then if we denote $q(z)$ a common denominator for all the fractions we obtain

$$\mathbf{\Phi}(z) = \mathbf{V}(z) \,/\, q(z) \tag{A.2.2}$$

where $\mathbf{V}(z)$ is a polynomial with matrix coefficients.

If series (A.2.1) converges outside the disk $|\,z\,| < r$, then it is possible to find $\mathbf{W}_m$ by using the inverse z-transform:

$$\mathbf{W}_m = \frac{1}{2\pi \mathrm{j}} \oint_{\gamma} \mathbf{\Phi}(z) z^{m-1} dz \qquad \gamma = \{z : |\,z\,| = \rho\} \quad \rho > r. \tag{A.2.3}$$

It is possible to calculate this integral, using the residue theorem. If we denote by $z_1, z_2, \ldots, z_n$ all the roots of the polynomial $q(z)$, then

$$q(z) = (z-z_1)^{m_1}(z-z_2)^{m_2} \cdots (z-z_n)^{m_n}$$

From the residue theorem we have

$$\mathbf{W}_m = \sum_{i=1}^{n} res\{\mathbf{\Phi}(z)z^{m-1};z_i\} \tag{A.2.4}$$

Since all the singularities of the function $\mathbf{\Phi}(z)$ are poles, then

$$res\{\mathbf{\Phi}(z)z^{m-1};z_i\} = \frac{1}{(m_i-1)!}\frac{d^{m_i-1}}{dz^{m_i-1}}[\mathbf{\Phi}_i(z)z^{m-1}]\,\Big|_{z=z_i}$$

where $\mathbf{\Phi}_i(z) = (z - z_i)^{m_i}\mathbf{\Phi}(z)$.

After differentiation using the Leibnitz formula, we obtain

$$res\{\mathbf{\Phi}(z)z^{m-1};z_i\} = \sum_{\nu=0}^{m_i-1} \mathbf{A}_{i\nu}\binom{m-1}{\nu}z_i^{m-\nu-1} \tag{A.2.5}$$

where

$$\mathbf{A}_{i\nu} = \frac{1}{(m_i-\nu-1)!}\mathbf{\Phi}_i^{(m_i-\nu-1)}(z_i)$$

Formula (A.2.2) takes the form

$$\mathbf{W}_m = \sum_{i=1}^{n}\sum_{\nu=0}^{m_i-1} \mathbf{A}_{i\nu}\binom{m-1}{\nu}z_i^{m-\nu-1} \tag{A.2.6}$$

In the particular case when all the roots of the polynomial $q(z)$ are simple, the previous formula becomes

$$\mathbf{W}_m = \sum_{i=1}^{n} \mathbf{A}_{i0}z_i^{m-1} \tag{A.2.7}$$

where

$$\mathbf{A}_{i0} = \mathbf{\Phi}_i(z_i) = \mathbf{V}(z_i) \,/\, q'(z_i) \tag{A.2.8}$$

Note that the component corresponding to the residue at $z = 0$ is included only in a finite number of the first terms. Therefore, to find asymptotic formulas for matrices $\mathbf{W}_m$ it is possible to drop the components corresponding to the residues at $z = 0$.

Theorem A.2.1: Let the generating function of the sequence $\mathbf{W}_m$ have the form of (A.2.1) and $z_1, z_2, \ldots, z_a$ be the roots of the polynomial $q(z)$ with the maximal absolute value that have the highest multiplicity factor b, then $\mathbf{W}_m$ has the following asymptotic representation

$$\mathbf{W}_m \sim \frac{m^{b-1}}{(b-1)!} \sum_{i=1}^{a} \mathbf{A}_{i\,b-1} z_i^{m-b} \tag{A.2.9}$$

Proof: Denote $\lambda = \max_i |z_i|$. Since

$$\lim_{m \to \infty} \binom{m-1}{\nu} z_j^m \lambda^{-m} = 0$$

for $|z_j| < \lambda$, we can retain in the asymptotic formula for (A.2.6) only the components corresponding to roots z_i with modules equal to λ. Furthermore, since for $|z_i| = |z_j|$ and $\alpha > \nu$,

$$\lim_{m \to \infty} \frac{\binom{m-1}{\nu} z_i^m}{\binom{m-1}{\alpha} z_j^m} = 0$$

we can ignore in the remaining sum all the components for which binomial coefficients $\binom{m-1}{\nu} z_i^m$ have ν that is less than the maximum (which is $b-1$). Thus, we obtain the asymptotic representation

$$\mathbf{W}_m \sim \binom{m-1}{b-1} \sum_{i=1}^{a} \mathbf{A}_{i\,b-1}\, z_i^{m-b} \tag{A.2.10}$$

And, finally, since

$$\lim_{m \to \infty} \binom{m-1}{b-1} (b-1)!/m^{b-1} = 1$$

we obtain (A.2.9).

■

Note that formula (A.2.9) gives the main term of the asymptotic expansion. It is not difficult to obtain the whole asymptotic expansion using (A.2.6). To do this it is necessary to write formula (A.2.6) in the form of an asymptotic series for the significance of addends:

$$\mathbf{W}_m \sim \mathbf{W}_m^{(0)} + \mathbf{W}_m^{(1)} + \cdots \tag{A.2.11}$$

where $\mathbf{W}_m^{(0)}$ is defined by formula (A.2.10); $\mathbf{W}_m^{(1)}$ is an expression of the similar form

$$\mathbf{W}_m^{(1)} = \binom{m-1}{b-2} \sum_{i=1}^{a} \mathbf{A}_{i\,b-2} z_i^{m-b+1} \tag{A.2.12}$$

and so on. After we exhaust all components corresponding to the roots with the largest module, we consider the roots with the largest module among the remaining roots, and so on. In this way we obtain asymptotic expansion with respect to sequences of the form $(m-1)...(m-v)q^m$. It is possible to obtain an expansion with respect to $m^v q^m$ if we use Sterling's numbers. [118]

When we use any approximation formulas, it is important to estimate the error, which is equal to the absolute value of the difference between the exact value and its approximation. As we have seen, the main term of an asymptotic expansion corresponds to the sum of the residues of the z-transform at the poles with the maximal absolute value

$$\mathbf{W}_m \sim \sum_{j=1}^{a} res\{\mathbf{\Phi}(z) z^{m-1}, z_j\} \qquad |z_j| = \max_i |z_i| = \lambda$$

The error we calculated by replacing $\mathbf{W}_m$ by the right-hand side of this formula can be written in the integral form

$$\Delta_m = \mathbf{W}_m - \sum_{j=1}^{a} res\{\mathbf{\Phi}(z) z^{m-1}, z_j)\} = \frac{1}{2\pi \mathrm{j}} \oint_{|z|=r} \mathbf{\Phi}(z) z^{m-1} dz$$

where $\lambda > r > \lambda_1$, $\lambda_1 = \max_{i>a} |z_i|$. For elements of Δ_m we obtain

$$\Delta_{mij} = \frac{1}{2\pi \mathrm{j}} \oint_{|z|=r} \phi_{ij}(z) z^{m-1} dz$$

If we denote

$$M_{ij}(r) = \max_{|z|=r} \phi_{ij}(z)$$

then by replacing the integral on the right-hand side of the last formula with its upper bound, we obtain

$$|\Delta_{mij}| \leq M_{ij}(r) r^m$$

Note that to find $M_{ij}(r)$ it is convenient to use the following lemma: If $a_m \geq 0$ then

$$\max_{|z|=r} \left| \sum_m a_m z^m \right| = \sum_m a_m r^m$$

If we consider residues at the poles inside the circle of radius λ_1, then we can replace r by $r_1 < \lambda_1$ in the error upper bound.

2.2 Chernoff Bounds.

The upper bound of probabilities

$$\mathbf{P}(\xi \geq k) = \sum_{x=k}^{\infty} \mathbf{P}_{\xi}(x) \tag{A.2.13}$$

can be found using the z-transform of the distribution $\mathbf{P}_{\xi}(x)$:

$$\mathbf{\Phi}(z) = \sum_{x=0}^{\infty} \mathbf{P}_{\xi}(x) z^{-x} \tag{A.2.14}$$

For real z meeting the conditions $R < z \leq 1$, where R is the radius of convergence of the series in (A.2.14), we have

$$\mathbf{\Phi}(z) \geq \sum_{x=k}^{\infty} \mathbf{P}_{\xi}(x) z^{-x} \geq z^{-k} \sum_{x=k}^{\infty} \mathbf{P}_{\xi}(x) = z^{-k} \mathbf{P}(\xi \geq k)$$

This equation gives us the bound

$$\mathbf{P}(\xi \geq k) \leq z^k \mathbf{\Phi}(z) \tag{A.2.15}$$

from which we obtain the following inequalities for the elements of $\mathbf{P}(\xi \geq k)$:

$$p_{ij}(\xi \geq k) \leq z^k \phi_{ij}(z) \tag{A.2.16}$$

If, for $z = z_{ij} \in (R, 1]$ the function $z^k \phi_{ij}(z)$ reaches the minimum, then

$$p_{ij}(\xi \geq k) \leq z_{ij}^k \phi_{ij}(z_{ij}) \tag{A.2.17}$$

gives us the desired bound. If we denote z_0 such that

$$| z_0 | = \min_{ij} | z_{ij} |$$

then the last formula can be written in the matrix form

$$\mathbf{P}(\xi \geq k) \leq z_0^k \mathbf{F}_k(z_0) \tag{A.2.18}$$

where $\mathbf{F}_k(z_0) = [\ (z_{ij} / z_0)^k \phi_{ij}(z_{ij}) \ ]$. Bounds (A.2.16) through (A.2.18) can be generalized to the case of multidimensional distributions.

2.3 Block Graphs

A finite graph $G = (X,U)$ [50,120] is called a *block graph* (or *general flow graph*),[50] if it can be partitioned into nonintersecting subgraphs $G_1 = (X_1,U_1), G_2 = (X_2,U_2), \ldots, G_s = (X_s,U_s)$

$$G = \bigcup_{m=1}^{s} G_m \quad G_i \cap G_j = \varnothing \quad i \neq j, \quad (i,j = 1,2,\ldots,s) \tag{A.2.19}$$

The partitioning of a signal graph G into subgraphs corresponds to the partition of the matrix of transmission into sub-blocks

$$\mathbf{T} = block[\mathbf{T}_{ij}]_{s,s} = \begin{bmatrix} \mathbf{T}_{11} & \mathbf{T}_{12} & \cdots & \mathbf{T}_{1s} \\ \mathbf{T}_{21} & \mathbf{T}_{22} & \cdots & \mathbf{T}_{2s} \\ \cdots & \cdots & \cdots & \cdots \\ \mathbf{T}_{s1} & \mathbf{T}_{s2} & \cdots & \mathbf{T}_{ss} \end{bmatrix} \tag{A.2.20}$$

Here the diagonal blocks $\mathbf{T}_{11}, \mathbf{T}_{22}, \ldots, \mathbf{T}_{ss}$ are matrices of internal transmissions of subgraphs $\mathbf{G}_{11}, \mathbf{G}_{22}, \ldots, \mathbf{G}_{ss}$, whereas nondiagonal blocks correspond to transmissions between different subgraphs.

We will depict block graphs in the same way as normal graphs by writing under an arrow the corresponding matrix branch transmission. Matrices of internal transmissions between nodes of subgraphs are represented by loops (Fig. A.1).

Consider now some transformations of signal-flow graphs. The basic transformation of the block graph is the *absorption*[50] of a subgraph, which is frequently used throughout this book when dealing with Markov functions.

Suppose that we are given a graph G with a matrix $\mathbf{T} = [\ \mathbf{T}_{ij} \ ]$ and we want to find the matrix of $G^{(k)} = G \setminus G_k$ after absorption of the subgraph G_k. Then by the definition of a matrix of a signal-flow graph [50] we have

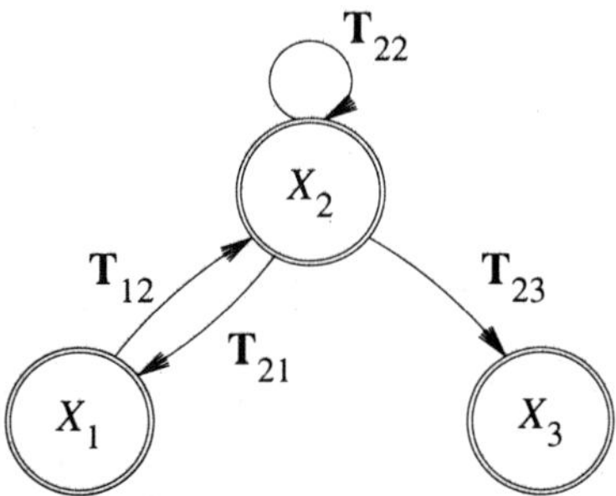

Figure A.1. Block graph.

$$\mathbf{X}_i = \sum_{j=1}^{s} \mathbf{T}_{ij}\mathbf{X}_j \qquad (i=1,2,...,s) \tag{A.2.21}$$

For $i = k$ we have

$$(\mathbf{I} - \mathbf{T}_{kk})\mathbf{X}_k = \sum_{j \neq k} \mathbf{T}_{kj}\mathbf{X}_j$$

If we assume that the matrix $\mathbf{I} - \mathbf{T}_{kk}$ is non-singular, then

$$\mathbf{X}_k = \sum_{j \neq k} (\mathbf{I} - \mathbf{T}_{kk})^{-1}\mathbf{T}_{kj}\mathbf{X}_j$$

Substituting the value of $\mathbf{X}_k$ in the remaining equations of system (A.2.21), we obtain

$$\mathbf{X}_i = \sum_{j \neq k} [\mathbf{T}_{ij} + \mathbf{T}_{ik}(\mathbf{I} - \mathbf{T}_{kk})^{-1}\mathbf{T}_{kj}]\mathbf{X}_j. \tag{A.2.22}$$

This leads to the conclusion that if $\det(\mathbf{I} - \mathbf{T}_{kk}) \neq 0$, the matrix branch transmission of the graph $G^{(k)}$ can be found by the formula

$$\mathbf{T}_{ij}^{(k)} = \mathbf{T}_{ij} + \mathbf{T}_{ik}(\mathbf{I} - \mathbf{T}_{kk})^{-1}\mathbf{T}_{kj} \tag{A.2.23}$$

Note: The particular cases of transformation (A.2.23) are addition and multiplication of matrices and finding the inverse matrix.

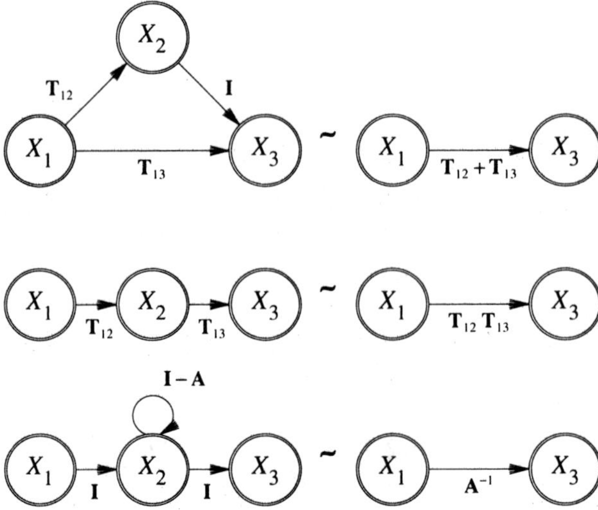

Figure A.2. Elementary matrix operations.

These operations are illustrated in Fig. A.2. Therefore, we can perform elementary operations of matrix algebra, using signal-flow graphs.

APPENDIX 3

3.1 Matrix Process Parameter Estimation

In this appendix we build a matrix process model using its basis sequences introduced in Appendix 1. The method of model parameter estimation consists in finding a set of basis sequences and then estimating the process matrices.

Matrix process rank can be estimated by building different sequences of the form $s_i t_j$ according to the rule of ordering introduced in Appendix 1 (see Theorem A.1.2). If we succeed in constructing a basis from vectors of the sets $W_m^{(s)}$ and $W_m^{(t)}$, and the introduction of vectors of the next layer does not change this basis, then the number of vectors constituting the basis determines the process rank. Therefore, to estimate the rank in accordance with canonical representation (A.1.21), we build sequences of matrices

$$\hat{\mathbf{P}}_e^{(1)} = \hat{p}(e) \quad \hat{\mathbf{P}}_e^{(2)} = \begin{bmatrix} \hat{p}(e) & \hat{p}(e0) \\ \hat{p}(0e) & \hat{p}(0e0) \end{bmatrix}$$

$$\hat{\mathbf{P}}_e^{(3)} = \begin{bmatrix} \hat{p}(e) & \hat{p}(e1) \\ \hat{p}(1e) & \hat{p}(1e1) \end{bmatrix}, \ldots, \quad \hat{\mathbf{P}}_e^{(k)} = \left[\hat{p}(s_i^{(e)} e t_j^{(e)})\right], \ldots$$

where $\hat{p}(s_i^{(e)} e t_j^{(e)})$ is the relative frequency of the sequence $s_i^{(e)} e t_j^{(e)}$. According to Theorem A.1.2, the basis building is completed when

$$\det \left[p(s_i^{(e)} e t_j^{(e)})\right]_{r,r} \neq 0 \qquad \det \left[p(s_i^{(e)} e t_j^{(e)})\right]_{r+1,r+1} = 0$$

for any $s_{r+1}^{(e)}$ and $t_{r+1}^{(e)}$. Therefore, to estimate the process rank we should be able to test a hypothesis that a random matrix determinant is equal to zero.

To test this hypothesis we need to find the distribution of determinants of above-defined random matrices. Let $\nu(s_i^{(e)}et_j^{(e)})$ be a frequency of the sequence $s_i^{(e)}et_j^{(e)}$ in the experimental data. We will prove that this frequency is asymptotically normal if the SSM satisfies conditions of regularity. Indeed, consider all sequences of the same length l as $s_i^{(e)}et_j^{(e)}$. They can be described as a Markov function whose transition matrix probabilities are given by

$$\mathbf{P}(e_1e_2\cdots e_l|g_0g_1\cdots g_{l-1}) = \begin{cases}\mathbf{P}(e_l) & \text{if } e_1=g_1,\ldots,e_{l-1}=g_{l-1}\\ 0 & \text{otherwise}\end{cases}$$

If we assume that this chain is regular (see Appendix 6), then [48,55] the frequencies of visiting any state of the chain are asymptotically normal random variables. Since the sequence $s_i^{(e)}et_j^{(e)}$ is one of the states of this chain, $\nu(s_i^{(e)}et_j^{(e)})$ is asymptotically normal.

Therefore, to test the hypothesis that the rank of the process is equal to r, we can use the distribution of random determinants whose elements have normal distributions.[82—85] Denote the determinant distribution density as $f_k(x)$. The hypothesis is rejected if

$$|\det \hat{\mathbf{P}}_e^{(k)}| > x_\alpha$$

where x_α is a root of the equation

$$\int_{-x_\alpha}^{x_\alpha} f_k(x)dx = 1-\alpha$$

After estimating ranks r_0 and r_1, it is necessary to estimate matrices (A.1.20) and (A.1.21), which determine the matrix process. We estimate them by the corresponding relative frequencies

$$\mathbf{A}_\mu(e) = \hat{\mathbf{P}}_\mu(s_1^{(\mu)},\ldots,s_{r_\mu}^{(\mu)};\, et_1^{(e)},\ldots,et_{r_e}^{(e)})\hat{\mathbf{P}}^{-1}(s_1^{(e)},\ldots,s_{r_e}^{(e)};\, t_1^{(e)},\ldots,t_{r_e}^{(e)}) \qquad \text{(A.3.1)}$$

where

$$\hat{\mathbf{P}}_\mu(s_1^{(\mu)},\ldots,s_{r_\mu}^{(\mu)};\, et_1^{(e)},\ldots,et_{r_e}^{(e)}) = \left[\hat{p}(s_i^{(\mu)}\mu et_j^{(e)})\right]_{r_\mu,r_e}$$

$$\hat{\mathbf{P}}^{-1}(s_1^{(e)},\ldots,s_{r_e}^{(e)};\, t_1^{(e)},\ldots,t_{r_e}^{(e)}) = \left[\hat{p}(s_i^{(e)}et_j^{(e)})\right]_{r_e,r_e}$$

and

$$s_1^{(e)},...,s_{r_e}^{(e)};\ t_1^{(e)},...,t_{r_e}^{(e)} \quad s_1^{(e)} = \varnothing \quad t_1^{(e)} = \varnothing$$

are corresponding left and right basis sequences. Matrix $\mathbf{Z}_{le}$ is estimated by the relation

$$\hat{\mathbf{Z}}_{le} = col\ \{\, p(s_i^{(e)}) \}.$$

This method of parameter estimation, arising from results in Sec. 1.3 and Appendix 1, is in essence a variant of the method of moments. The most difficult part of this method is estimating ranks r_0 and r_1.

APPENDIX 4

4.1 Sums with Binomial Coefficients

It is convenient to use contour integrals to find sums that contain binomial coefficients. From the binomial formula

$$(1+z)^n = \sum_{m=0}^{n} \binom{n}{m} z^m \tag{A.4.1}$$

we obtain

$$\binom{n}{m} = \oint_{\gamma} (1+z)^n z^{-m-1} dz \tag{A.4.2}$$

where γ is a contour on the complex plane that contains $z = 0$. By applying analytical expansion we may use this equation as a binomial coefficient definition if we assume that the contour γ does not contain $z = -1$. For example, it follows from (A.4.2) that $\binom{n}{m} = 0$ when $m > n$ or $m < 0$. If $n < 0$, equation (A.4.2) gives

$$\binom{n}{m} = (-1)^m \binom{m-n-1}{m}$$

which is identical to the definition of Ref. [39].

The utility of the equation for summation is in replacement of binomial coefficients with the powers of $(1+z)$ and z under the sign of the integral. Usually it is easier to find a sum of a power series than the original sum.

4.2 Maximum-Distance-Separable Code Weight Generating Function

As an example, let us find the generating function of the number A_w of code words of the maximum-distance-separable (MDS) codes whose Hamming weight is equal to w. It is well known [75,106] that the Hamming weight of the MDS $(n,n-r)$ code over GF(q) can be expressed as

$$A_w = \binom{n}{w} \sum_{i=0}^{w-r-1} (-1)^i \binom{w}{i} (q^{w-r-i}-1) =$$
$$\binom{n}{w}(q-1) \sum_{i=0}^{w-r-1} (-1)^i \binom{w-1}{i} q^{w-r-i-1} \quad \text{for } w > r \tag{A.4.3}$$

$A_0 = 1$ and $A_w = 0$ for $w = 1,2,...,r$. Our goal is to find the generating function

$$A(u) = \sum_{w=0}^{n} A_w u^w \tag{A.4.4}$$

It follows from equation (A.4.2) that

$$(-1)^i \binom{w}{i} = \oint_{\gamma} (1-z)^w z^{-i-1} dz \tag{A.4.5}$$

If we replace the coefficient $(-1)^i \binom{w}{i}$ in equation (A.4.3) with contour integral (A.4.5) and sum the geometric progression under the sign of the integral, we obtain, after some elementary transformations,

$$A_w = \frac{(q-1)}{2\pi j} \binom{n}{w} \oint_{\gamma} \frac{(1-z)^{w-1}}{z^{w-r}(1-qz)} dz. \tag{A.4.6}$$

It follows from Cauchy's theorem that this integral is equal to zero for $w \leq r$. After substituting (A.4.6) into (A.4.4) and summing under the sign of the integral, using binomial formula (A.4.1), we obtain

$$A(u) = 1 + \frac{(q-1)}{2\pi j} \oint_{\gamma} \frac{[u+(1-u)z]^n dz}{(1-qz)(1-z)z^{n-r}}$$

This equation represents the generating function in the integral form. We can evaluate this integral by the residue theorem and obtain the following relation:

$$A(u) = 1 + \sum_{i=r+1}^{n} \binom{n}{i} u^i (1-u)^{n-i} (q^{i-r} - 1) \tag{A.4.7}$$

Now that we know the expression of the generating function, we can easily verify its correctness by a direct expansion into power series.

Usually $n-r$ is much greater than r, and we can simplify the expression by replacing the summation from $r+1$ to n by the summation from 0 to r if we make use of the identity

$$\sum_{i=r+1}^{n} \binom{n}{i} a^i b^{n-i} = (a+b)^n - \sum_{i=0}^{r} \binom{n}{i} a^i b^{n-i}$$

The generating function $A(u)$ then takes the form

$$A(u) = q^{-r}(1-u+uq)^n + \sum_{i=0}^{r-1} \binom{n}{i} u^i (1-u)^{n-i} (1-q^{i-r}) \tag{A.4.8}$$

4.3 Union Bounds on Viterbi Algorithm Performance

In this appendix we use transform methods to derive a closed-form expression for the error-event probability and performance union bound of the Viterbi algorithm.

Consider a discrete-time system [121]

$$\begin{aligned} \mathbf{y}_k &= \mathbf{f}(\mathbf{w}_k) \\ \mathbf{s}_{k+1} &= \mathbf{g}(\mathbf{w}_k) \\ \mathbf{w}_k &= (\mathbf{x}_k, \mathbf{s}_k) \qquad -\infty < k < \infty \end{aligned} \tag{A.4.9}$$

where $\mathbf{x}_k$ is a source symbol, $\mathbf{y}_k$ is the transmitted channel symbol, and $\mathbf{s}_k$ is the corresponding system (transmitter) state. System (4.2.24) is a particular case of this system.

Symbols $\mathbf{y}_k$ are transmitted over a noisy SSM channel, which outputs symbols

$$\mathbf{z}_k = \mathbf{h}(\mathbf{y}_k, \mathbf{n}_k) \tag{A.4.10}$$

where $\mathbf{n}_k$ is the noise.

The receiver outputs symbols $\hat{\mathbf{x}}_k$ if $\hat{\mathbf{w}}_k = (\hat{\mathbf{s}}_k, \hat{\mathbf{x}}_k)$ minimizes the sum

$$M(\mathbf{z}, \mathbf{w}) = \sum_k m(\mathbf{z}_k, \mathbf{W}_k)$$

where $\mathbf{W}_k = (\mathbf{w}_k, \mathbf{w}_{k+1})$ (see Sec. 4.1.6.1). If we interpret $\mathbf{w}_k$ as a node on a trellis diagram (see Fig. 4.5), then $m(\mathbf{z}_k, \mathbf{W}_k)$ can be interpreted as a metric of the branch that connects nodes $\mathbf{w}_k$ and $\mathbf{w}_{k+1}$. Now the problem of minimizing $M(\mathbf{z}, \mathbf{w})$ can be formulated as the problem of finding the shortest path on the trellis diagram. The problem can be solved by the Viterbi algorithm described in Sec. 4.2.6.2:

> Let $M(\mathbf{w}_t)$ be the length of the shortest path (the survivor) that ends at node $\mathbf{w}_t$ at moment t. Then the survivor that ends at node $\mathbf{w}_{t+1}$ is obtained from one of the survivors at moment t and its length is given by
>
> $$M(\mathbf{w}_{t+1}) = \min_{\mathbf{w}_t}\{M(\mathbf{w}_t) + m(\mathbf{z}_t, \mathbf{W}_t)\}$$

In other words, to find the shortest path at level $t+1$ we need only consider the shortest paths at level t. On the trellis diagram we select a survivor at node $\mathbf{w}_{t+1}$ If several paths have the minimal length, one of them is selected randomly. That is why the shortest path is called a survivor. This algorithm is illustrated in Example 4.2.7.

Depending on the application, we may want to evaluate Viterbi's receiver performance by using some distortion measure $\bar{d} = \mathbf{E}\{d(\mathbf{W}_k, \hat{\mathbf{W}}_k)\}$, where $d(\mathbf{W}_k, \hat{\mathbf{W}}_k)$ is the distortion characteristic of the symbol $\mathbf{W}_k$, which the receiver identifies as $\hat{\mathbf{W}}_k$.

In order to find the distortion measure union bound, let us consider an error event [95] of length L which is a pair of the correct path $\sigma_L = \{\mathbf{w}_k\}$ and a path $\hat{\sigma}_L = \{\hat{\mathbf{w}}_k\}$, erroneously selected by the Viterbi algorithm, such that $\mathbf{s}_k \neq \hat{\mathbf{s}}_k$ for $k = 1,2,...,L-1$ and $\mathbf{s}_k = \hat{\mathbf{s}}_k$ otherwise.

The distortion measure is upper bounded by the union bound: [121]

$$\bar{d} \leq \mathop{\mathbf{E}}_{\mathbf{x},\, \mathbf{s}_0} \sum_{L=1}^{\infty} \sum_{\hat{\sigma}_L} P_L(\hat{\sigma}_L \,|\, \sigma_L) d_L(\sigma_L, \hat{\sigma}_L)$$

where the average is taken over all source sequences and system stationary states, $P_L(\hat{\sigma}_L \,|\, \sigma_L)$ is the conditional probability that the incorrect path $\hat{\sigma}_L$ has a smaller metric than the correct path σ_L, and one-half of the probability of metrics' equality if a tie is resolved randomly:

$$P_L(\hat{\sigma}_L \,|\, \sigma_L) = Pr\{\sum_{k=0}^{L-1} \Delta m(\mathbf{z}_k, \mathbf{W}_k, \hat{\mathbf{W}}_k) > 0\} + 0.5\, Pr\{\sum_{k=0}^{L-1} \Delta m(\mathbf{z}_k, \mathbf{W}_k, \hat{\mathbf{W}}_k) = 0\} \tag{A.4.11}$$

where $\Delta m(\mathbf{z}_k, \mathbf{W}_k, \hat{\mathbf{W}}_k) = m(\mathbf{z}_k, \mathbf{W}_k) - m(\mathbf{z}_k, \hat{\mathbf{W}}_k)$ and $d_L(\sigma_L, \hat{\sigma}_L)$ is the total distortion along the incorrect path:

$$d_L(\sigma_L, \hat{\sigma}_L) = \sum_{k=0}^{L-1} d(\mathbf{W}_k, \hat{\mathbf{W}}_k)$$

For the SSM channel model the error-event probability can be found using matrix probabilities. Define a generating function of a variable $\Delta m(\mathbf{z}_k, \mathbf{W}_k, \hat{\mathbf{W}}_k)$:

$$\mathbf{D}(v;\mathbf{W}_k,\hat{\mathbf{W}}_k) = \sum_{\mathbf{z}_k}\mathbf{P}(\mathbf{z}_k\,|\,\mathbf{W}_k)v^{\Delta m(\mathbf{z}_k,\mathbf{W}_k,\hat{\mathbf{W}}_k)} \tag{A.4.12}$$

According to the convolution theorem, the generating function of the sum of variables $\Delta m(\mathbf{z}_k,\mathbf{W}_k,\hat{\mathbf{W}}_k)$ is equal to the product of generating functions (A.4.12), and the matrix probability $\mathbf{P}_L(\hat{\sigma}_L\,|\,\sigma_L)$ is equal to the sum of the coefficients of the positive power terms of the product and one-half of the zero power term coefficient. Using contour integrals, we can express the sum as

$$\mathbf{P}_L(\hat{\sigma}_L\,|\,\sigma_L) = \frac{1}{2\pi j}\oint_{|v|=\rho}\left(\frac{1}{v-1}-\frac{1}{2v}\right)\prod_{k=0}^{L-1}\mathbf{D}(v;\mathbf{W}_k,\hat{\mathbf{W}}_k)dv \tag{A.4.13}$$

where $\rho > 1$. Using the formula of the derivative of the product we obtain

$$d_L(\sigma_L,\hat{\sigma}_L) = \frac{d}{dv}\prod_{k=0}^{L-1}u^{d(\mathbf{W}_k,\hat{\mathbf{W}}_k)}\,\Big|_{u=1} \tag{A.4.14}$$

and therefore the average distortion is bounded by

$$\bar{d} \le \frac{1}{2\pi j}\frac{d}{du}\left\{\oint_C\left(\frac{1}{v-1}-\frac{1}{2v}\right)\mathbf{pG}(u,v)\mathbf{1}dv\right\}\Big|_{u=1} \tag{A.4.15}$$

where

$$\mathbf{G}(u,v) = \underset{\mathbf{X},\,\mathbf{s}_0}{\mathbf{E}}\sum_{L=1}^{\infty}\sum_{\hat{\sigma}_L}\prod_{k=0}^{L-1}\mathbf{D}(v;\mathbf{W}_k,\hat{\mathbf{W}}_k)u^{d(\mathbf{W}_k,\hat{\mathbf{W}}_k)} \tag{A.4.16}$$

and $C \in \{v : R_G \cap |\,v\,| > 1\}$ and R_G is the region of convergence of (A.4.16).

The generating function $\mathbf{G}(u,v)$ can be found from the system state $(\mathbf{W}_k,\hat{\mathbf{W}}_k)$ transition graph with branch weights

$$\mathbf{D}(v;\mathbf{W}_k,\hat{\mathbf{W}}_k)u^{d(\mathbf{W}_k,\hat{\mathbf{W}}_k)} \tag{A.4.17}$$

Indeed, for any path on the graph the product

$$\prod_{k=0}^{L-1}\mathbf{D}(v;\mathbf{W}_k,\hat{\mathbf{W}}_k)u^{d(\mathbf{W}_k,\hat{\mathbf{W}}_k)}$$

represents the transition weight (which is similar to matrix probabilities and generating functions considered in Sec. 2.1.1, 2.2.1, and 2.4.4). A length L error event may be interpreted as a path that starts at a node with $\mathbf{s}_0 = \hat{\mathbf{s}}_0$ and returns to some node with $\mathbf{s}_L = \hat{\mathbf{s}}_L$ for the first time. The generating function

$$\sum_{L=1}^{\infty} \sum_{\hat{\sigma}_L} \prod_{k=0}^{L-1} \mathbf{D}(v;\mathbf{W}_k,\hat{\mathbf{W}}_k) u^{d(\mathbf{W}_k,\hat{\mathbf{W}}_k)}$$

represents the transition weight for all error events that start at $\mathbf{s}_0 = \hat{\mathbf{s}}_0$ and end at $\mathbf{s}_L = \hat{\mathbf{s}}_L$ for $L = 1,2,\dots$.

To find this generating function let us denote $\mathbf{0}$ the set of nodes with $\mathbf{s} = \hat{\mathbf{s}}$. As in the derivations of the Markov function interval distributions in Sec. 2.4.3, we obtain the weight generating function for all the paths that start at $\mathbf{0}$ and return to $\mathbf{0}$ for the first time:

$$\mathbf{T}(u,v) = \mathbf{R}_{\mathbf{0},\mathbf{0}} + \mathbf{R}_{\mathbf{0},\mathbf{\Theta}}[\mathbf{I} - \mathbf{R}_{\mathbf{\Theta},\mathbf{\Theta}}]^{-1}\mathbf{R}_{\mathbf{\Theta},\mathbf{0}}$$

where $\mathbf{R}$ is the matrix whose elements are equal to the graph branch weights (A.4.17) that correspond to erroneous transitions and $\mathbf{\Theta}$ is the complement of the set $\mathbf{0}$. In the absence of parallel transitions, $\mathbf{R}_{\mathbf{0},\mathbf{0}} = 0$. This equation coincides with equation (A.2.38) and therefore the generating function $\mathbf{T}(z)$ can be obtained by absorbing the subgraph $\mathbf{\Theta}$ from the system graph.

Sometimes it is convenient to duplicate the set $\mathbf{0}$ and create a new graph in which all transitions from $\mathbf{\Theta}$ to $\mathbf{0}$ are diverted to the duplicate set $\mathbf{0}'$. In this case $\mathbf{T}(z)$ can be found as a transfer function from $\mathbf{0}$ to $\mathbf{0}'$ (rather than as a transfer function from $\mathbf{0}$ to itself, as in the original graph).

Generating function (A.4.16) can be found from the generating function $\mathbf{T}(z)$ by averaging. As illustrated in Sec. 4.4.1, the system's symmetry simplifies the construction of the generating function.

APPENDIX 5

5.1 Matrices

This appendix provides a review of matrix properties relevant to the material discussed in this book.

5.1.1 Basic Definitions. A matrix is a rectangular table of numbers:

$$\mathbf{A} = [a_{ij}]_{m,n} = \begin{bmatrix} a_{11} & a_{12} & \cdots & a_{1n} \\ a_{21} & a_{12} & \cdots & a_{2n} \\ \cdots & \cdots & \cdots & \cdots \\ a_{m1} & a_{m2} & \cdots & a_{mn} \end{bmatrix}$$

Some particular cases of matrices and elementary operations are listed on pages xii and xiv and are illustrated by the following simple examples.

EXAMPLE A.5.1: Suppose that we have two matrices

$$\mathbf{A} = \begin{bmatrix} 1 & 2 \\ 3 & 4 \end{bmatrix} \quad \mathbf{B} = \begin{bmatrix} 5 & 6 \\ 7 & 8 \end{bmatrix}$$

then (see definitions on pages xii and xiv)

$$\mathbf{A} + \mathbf{B} = \begin{bmatrix} 6 & 8 \\ 10 & 12 \end{bmatrix} \quad \mathbf{A\,B} = \begin{bmatrix} 1{\cdot}5{+}2{\cdot}7 & 1{\cdot}6{+}2{\cdot}8 \\ 3{\cdot}5{+}4{\cdot}7 & 3{\cdot}6{+}4{\cdot}8 \end{bmatrix} = \begin{bmatrix} 19 & 22 \\ 43 & 50 \end{bmatrix}$$

$$3\,\mathbf{A} = \begin{bmatrix} 3 & 6 \\ 9 & 12 \end{bmatrix} \quad \mathbf{A}' = \begin{bmatrix} 1 & 3 \\ 2 & 4 \end{bmatrix} \quad \mathbf{B\,1} = \begin{bmatrix} 11 \\ 15 \end{bmatrix} \quad \mathbf{B\,I} = \mathbf{I\,B} = \mathbf{B}$$

■

The *determinant* of a square $(n \times n)$ matrix can be defined recursively as

$$|\,\mathbf{A}\,| = \det \mathbf{A} = \sum_{k=1}^{n} a_{ik} A_{ik} \tag{A.5.1}$$

where A_{ik} is an *adjunct* or *cofactor* of an element a_{ik}:

$$A_{ik} = (-1)^{i+k} M_{ik}$$

M_{ik} is the determinant of the matrix obtained by exclusion of i-th row and k-th column from **A**.

EXAMPLE A.5.2: Let us calculate a determinant using an expansion related to its last row.

$$\begin{vmatrix} 1 & 2 & 3 \\ 4 & 5 & 6 \\ 7 & 8 & 9 \end{vmatrix} = 7\cdot \begin{vmatrix} 2 & 3 \\ 5 & 6 \end{vmatrix} - 8\cdot \begin{vmatrix} 1 & 3 \\ 4 & 6 \end{vmatrix} + 9\cdot \begin{vmatrix} 1 & 2 \\ 4 & 5 \end{vmatrix} = 7\cdot(-3) - 8\cdot(-6) + 9\cdot(-3) = 0$$

■

A matrix whose determinant equals 0 is called *singular*. Every nonsingular matrix **A** has a unique *inverse matrix* $\mathbf{A}^{-1}$, which is defined by the relation $\mathbf{A}\,\mathbf{A}^{-1} = \mathbf{A}^{-1}\,\mathbf{A} = \mathbf{I}$. The inverse matrix can be found by the frequently used equation

$$\mathbf{A}^{-1} = \mathbf{A}^{*} \,/\, \det \mathbf{A} \tag{A.5.2}$$

where $\mathbf{A}^{*} = [A_{ij}]'_{n,n}$ is the so-called *adjoint matrix* (a transposed matrix of adjuncts of the elements of **A**). One can prove the correctness of (A.5.2) using formula (A.5.1).

EXAMPLE A.5.3: Let us find the inverse matrix for

$$\mathbf{A} = \begin{bmatrix} 3 & 2 \\ 1 & 1 \end{bmatrix}$$

Since $\det \mathbf{A} = 1 \neq 0$ the inverse matrix exists. We have

$$\mathbf{A}^* = \begin{bmatrix} 1 & -2 \\ -1 & 3 \end{bmatrix}$$

In this case $\mathbf{A}^{-1} = \mathbf{A}^*$, because $\det \mathbf{A} = 1$. We can test the correctness of this result using the inverse matrix definition

$$\mathbf{A}\,\mathbf{A}^{-1} = \begin{bmatrix} 3 & 2 \\ 1 & 1 \end{bmatrix} \begin{bmatrix} 1 & -2 \\ -1 & 3 \end{bmatrix} = \begin{bmatrix} 1 & 0 \\ 0 & 1 \end{bmatrix} = \mathbf{I}$$ ■

5.1.2 Systems of Linear Equations. A system of linear equations has the form

$$\begin{aligned} a_{11}x_1 + a_{12}x_2 + \cdots + a_{1n}x_n &= b_1 \\ a_{21}x_1 + a_{22}x_2 + \cdots + a_{2n}x_n &= b_2 \\ \cdots\cdots\cdots\cdots\cdots\cdots &\quad \cdots \\ a_{m1}x_1 + a_{m2}x_2 + \cdots + a_{mn}x_n &= b_m \end{aligned}$$

Using matrix notations, this system can be expressed as

$$\mathbf{A}\,\mathbf{x} = \mathbf{b}$$

Where $\mathbf{A} = [a_{ij}]_{m,n}$, $\mathbf{x} = [x_1 \;\; x_2 \;\; \ldots \;\; x_n]'$, and $\mathbf{b} = [b_1 \;\; b_2 \;\; \ldots \;\; b_m]'$. (This system can also be expressed as $\mathbf{x}'\,\mathbf{A}' = \mathbf{b}'$.)

Each solution of this system is a matrix column $\mathbf{x}$ that satisfies it (turns it to numerical identity). A system of linear equations may have a unique solution, an infinite number of solutions, or none. To analyze solutions of a linear system we need to introduce the notion of matrix rank.

An order-k minor of the $n \times m$ matrix $\mathbf{A}$ is a determinant of a $k \times k$ matrix

$$\begin{vmatrix} a_{i_1 j_1} & a_{i_1 j_2} & \cdots & a_{i_1 j_k} \\ a_{i_2 j_1} & a_{i_2 j_2} & \cdots & a_{i_2 j_k} \\ \cdots & \cdots & \cdots & \cdots \\ a_{i_k j_1} & a_{i_k j_2} & \cdots & a_{i_k j_k} \end{vmatrix}$$

where $i_1, i_2, \ldots, i_k$ and $j_1, j_2, \ldots, j_k$ represent the selected rows and columns of $\mathbf{A}$, respectively. If all the minors of orders greater than r are equal to 0, but there is an order-r minor different from 0, then r is called matrix $\mathbf{A}$ *rank*. That is, the rank is the highest order of all the matrix minors that are different from zero.

EXAMPLE A.5.4: Let us calculate the rank of the matrix

$$\mathbf{A} = \begin{bmatrix} 1 & 2 & 3 \\ 4 & 5 & 6 \\ 7 & 8 & 9 \end{bmatrix}$$

As we saw in Example A.5.2, $\det \mathbf{A} = 0$. Therefore, *rank* $\mathbf{A} < 3$. Since the determinant, which is composed out of the first and the second rows and columns

$$\begin{vmatrix} 1 & 2 \\ 4 & 5 \end{vmatrix} = -3$$

is not equal to 0, *rank* $\mathbf{A} = 2$.

■

It is difficult to calculate matrix rank using its definition, because the number of the minors to be tested might be very large. We can calculate it much faster using the so-called matrix *elementary operations*: multiplying an arbitrary row (column) of the matrix by a number and adding the result to its other row (column). It is easy to prove that these operations do not change matrix rank. (They also do not change the determinant of a square matrix.) To find the rank we try to make as many zero rows (or columns) as possible by applying the elementary operations. If we were able to transform an $m \times n$ matrix $\mathbf{A}$ to matrix $\mathbf{B}$ that has $m-r$ zero rows and also has an order-r minor different from 0, then *rank* $\mathbf{A} =$ *rank* $\mathbf{B} = r$. We say that $\mathbf{A}$ has *full rank* if *rank* $\mathbf{A} = \min(m,n)$.

EXAMPLE A.5.5: Let us calculate the rank of the matrix of Example A.5.4 using elementary operations. After multiplying the first column of $\mathbf{A}$ by (–2) and adding it to the second column; and then multiplying the first column by (–3) and adding it to the third column, we obtain

$$\mathbf{A} = \begin{bmatrix} 1 & 2 & 3 \\ 4 & 5 & 6 \\ 7 & 8 & 9 \end{bmatrix} \sim \begin{bmatrix} 1 & 0 & 0 \\ 4 & -3 & -6 \\ 7 & -6 & -12 \end{bmatrix} \sim \mathbf{B} = \begin{bmatrix} 1 & 0 & 0 \\ 4 & -3 & 0 \\ 7 & -6 & 0 \end{bmatrix}$$

We obtained matrix $\mathbf{B}$ after multiplying the second column of the previously transformed matrix by (–3) and adding the result to the third column.

We observe that $\mathbf{B}$ has two nonzero columns and it has the minor composed out of its first two rows and columns, which is equal to –3. Thus *rank* $\mathbf{A} =$ *rank* $\mathbf{B} = 2$.

■

EXAMPLE A.5.6: Vandermonde's determinant, which we often use in this book, has the form

$$\det \mathbf{V} = \begin{vmatrix} 1 & 1 & \cdots & 1 \\ \lambda_1 & \lambda_2 & \cdots & \lambda_n \\ \cdots & \cdots & \cdots & \cdots \\ \lambda_1^{n-1} & \lambda_2^{n-1} & \cdots & \lambda_n^{n-1} \end{vmatrix}$$

Let us calculate third-order Vandermonde's determinant using elementary operations:

$$\begin{vmatrix} 1 & 1 & 1 \\ \lambda_1 & \lambda_2 & \lambda_3 \\ \lambda_1^2 & \lambda_2^2 & \lambda_3^2 \end{vmatrix} = \begin{vmatrix} 1 & 0 & 0 \\ \lambda_1 & \lambda_2-\lambda_1 & \lambda_3-\lambda_1 \\ \lambda_1^2 & \lambda_2^2-\lambda_1^2 & \lambda_3^2-\lambda_1^2 \end{vmatrix} = \begin{vmatrix} \lambda_2-\lambda_1 & \lambda_3-\lambda_1 \\ \lambda_2^2-\lambda_1^2 & \lambda_3^2-\lambda_1^2 \end{vmatrix} = (\lambda_2-\lambda_1)(\lambda_3-\lambda_1)(\lambda_3-\lambda_2)$$

In the general case this formula can be written as

$$\det \mathbf{V} = \prod_{i \neq j} (\lambda_j - \lambda_i)$$

This means that $\det \mathbf{V} \neq 0$ if and only if all λ_i are different.

■

Let us denote $\mathbf{B} = [\mathbf{A}\ \mathbf{b}]$ the matrix consisting of $\mathbf{A}$ with matrix column $\mathbf{b}$ attached at the end. System (A.5.3) has a solution if and only if $rank\ \mathbf{A} = rank\ \mathbf{B}$.[24] If $r = rank\ \mathbf{A} = rank\ \mathbf{B}$, then this system has a unique solution if and only if $r = n$. It has an infinite number of solutions if $r < n$.

In particular, if $\mathbf{A}$ is a square matrix ($m = n$), then $rank\ A = n$ if only if $\det \mathbf{A} \neq 0$, which means that the system has a unique solution if and only if its matrix is nonsingular. This unique solution can be obtained by multiplying both sides of (A.5.3) by $\mathbf{A}^{-1}$: $\mathbf{A}^{-1}\,\mathbf{A}\,\mathbf{x} = \mathbf{A}^{-1}\,\mathbf{b}$, which, because of $\mathbf{A}^{-1}\,\mathbf{A} = \mathbf{I}$, yields

$$\mathbf{x} = \mathbf{A}^{-1}\,\mathbf{b}$$

The system is called *homogeneous* when $\mathbf{b} = 0$. Any homogeneous system has a trivial solution $\mathbf{x} = 0$. If $r = rank\ A = n$, this trivial solution is the only solution to the system. Otherwise the system has $n - r$ linearly independent solutions $\mathbf{x}_1$, $\mathbf{x}_2$, ..., $\mathbf{x}_r$, which are called *basis solutions*. Any solution can be expressed as a linear combination of these basis solutions:

$$\mathbf{x} = \alpha_1\,\mathbf{x}_1 + \alpha_2\,\mathbf{x}_2 + \cdots + \alpha_r\,\mathbf{x}_r$$

where $\alpha_1, \alpha_2, \ldots, \alpha_r$ are real numbers.

If $\mathbf{A}$ is a square nonsingular matrix ($\det \mathbf{A} \neq 0$), the homogeneous system $\mathbf{A}\,\mathbf{x} = 0$ has only the trivial solution $\mathbf{x} = 0$. Therefore, this system has a nonzero solution if and only if its determinant equals 0.

EXAMPLE A.5.7: Let us find a solution to the following system of $n = 2$ homogeneous equations

$$0.75x_1 - 0.5x_2 = 0$$
$$-0.75x_1 + 0.5x_2 = 0$$

In this case

$$\det \mathbf{A} = \begin{vmatrix} 0.75 & -0.5 \\ 0.75 & 0.5 \end{vmatrix} = 0$$

so that $r = rank\ \mathbf{A} = 1$. This system has $n - r = 1$ linearly independent solutions. We see that the second equation is proportional to the first equation. Therefore, we can find all the solutions from the first equation, which can be rewritten as $x_2 = 1.5\,x_1$. We may assign any value to x_1 and determine x_2 from this equation. For $x_1 = 0.4$ we have $x_2 = 0.6$ so that the basis solution is $\mathbf{x} = [0.4 \quad 0.6]$. Every other solution can be expressed as $\alpha\,\mathbf{x}$.

■

5.1.3 Calculating Powers of a Matrix. Matrix powers are defined recursively: $\mathbf{A}^2 = \mathbf{A}\,\mathbf{A}$, $\mathbf{A}^3 = \mathbf{A}^2\,\mathbf{A}$, ..., $\mathbf{A}^n = \mathbf{A}^{n-1}\,\mathbf{A}$. Since matrix multiplication is a complex operation, it is difficult to find matrix powers multiplying them directly. Even for a simple matrix

$$\mathbf{A} = \begin{bmatrix} 0 & 1 \\ 1 & 1 \end{bmatrix}$$

it is difficult to compute $\mathbf{A}^{100}$ by direct multiplication.

Powers of matrices play an important role in analyzing Markov chains (see Appendix 6) and error source models. Therefore, we need to develop methods of finding matrix powers.

To find $\mathbf{A}^m$ we use its z-transform

$$\mathbf{S}(z) = \sum_{m=0}^{\infty} \mathbf{A}^m z^{-m-1}$$

which converges for large z. This sum can be found in a manner similar to the usual geometric progression. After multiplying both sides of this equation by

$\mathbf{A}z^{-1}$ and subtracting it from $\mathbf{S}(z)$, all power terms, except for $\mathbf{I}z^{-1}$, cancel out and we obtain

$$\mathbf{S}(z)(\mathbf{I} - \mathbf{A}z^{-1}) = \mathbf{I}z^{-1}$$

This equation can be rewritten as $\mathbf{S}(z) = (\mathbf{I}z - \mathbf{A})^{-1}$. Thus, we obtain the following formula for the sum of the matrix-geometric series

$$\sum_{m=0}^{\infty} \mathbf{A}^m z^{-m-1} = (\mathbf{I}z - \mathbf{A})^{-1} \tag{A.5.3}$$

Matrix $\mathbf{A}^m$ is a coefficient of z^{-m-1} in the expansion of this z-transform. To find it we present the inverse matrix as a ratio of its adjoint matrix $\mathbf{B}(z)$ and its determinant $\Delta(z)$:

$$(\mathbf{I}z - \mathbf{A})^{-1} = \mathbf{B}(z) / \Delta(z)$$

The denominator of this formula $\Delta(z) = \det(\mathbf{I}z - \mathbf{A})$ is called a *characteristic polynomial* of $\mathbf{A}$, and its roots $\lambda_1, \lambda_2, ..., \lambda_r$ are called matrix *characteristic numbers*:

$$\Delta(z) = (z-\lambda_1)^{m_1}(z-\lambda_2)^{m_2} \cdots (z-\lambda_r)^{m_r}$$

Since $\mathbf{B}(z) / \Delta(z)$ is a rational function we can find its series expansion by first decomposing it into partial fractions:

$$(\mathbf{I}z - \mathbf{A})^{-1} = \mathbf{B}(z) / \Delta(z) = \sum_{j=1}^{r} \sum_{i=1}^{m_j} \mathbf{D}_{ij} / (z-\lambda_j)^i \tag{A.5.4}$$

Matrices $\mathbf{D}_{ij}$ can be determined by multiplying both sides of (A.5.4) by $\Delta(z)$ and comparing the coefficients of the same powers of z or by using formula (A.2.4), which gives

$$\mathbf{D}_{ij} = \frac{1}{(m_j-i)!} \frac{d^{m_j-i}}{dz^{m_j-i}} \left[\frac{\mathbf{B}(z)(z-\lambda_j)^{m_j}}{\Delta(z)} \right]_{z=\lambda_j}$$

To find power series expansion of (A.5.4) we can apply a negative binomial formula

$$(z-\lambda)^{-k} = \sum_{m=k-1}^{\infty} \binom{m}{k-1} \lambda^{m-k+1} z^{-m-1}$$

which can be obtained by either expanding $(1-\lambda x)^{-k}$ into power series using Taylor's formula, or by differentiating a geometric progression sum

$$(z-\lambda)^{-1} = \sum_{m=0}^{\infty} \lambda^m z^{-m-1}$$

Expanding (A.5.4) we find that the coefficient of z^{-m-1} is equal to

$$\mathbf{A}^m = \sum_{j=1}^{r} \sum_{i=1}^{m_j} \mathbf{D}_{ij} \binom{m}{i-1} \lambda_j^{m-i+1} \tag{A.5.5}$$

Here $\binom{m}{i-1}=1$ when $i=1$ and $\binom{m}{i-1}=0$ when $i>m+1$ (see Appendix 4.1).

EXAMPLE A.5.8: Let us find $\mathbf{A}^m$ for

$$\mathbf{A} = \begin{bmatrix} 0.25 & 0.75 \\ 0.5 & 0.5 \end{bmatrix}$$

We can find $\mathbf{A}^m$ by formula (A.5.5) using decomposition of $(\mathbf{I}z - \mathbf{A})^{-1}$ into partial fractions.

To find the inverse matrix for

$$\mathbf{I}z - \mathbf{A} = \begin{bmatrix} 1 & 0 \\ 0 & 1 \end{bmatrix} z - \begin{bmatrix} 0.25 & 0.75 \\ 0.5 & 0.5 \end{bmatrix} = \begin{bmatrix} z-0.25 & -0.75 \\ -0.5 & z-0.5 \end{bmatrix}$$

we calculate its determinant

$$\Delta(z) = \det(\mathbf{I}z - \mathbf{A}) = \det \begin{bmatrix} z-0.25 & -0.75 \\ -0.5 & z-0.5 \end{bmatrix} = z^2-0.75z-0.25 = (z-1)(z+0.25)$$

and the adjoint matrix

$$\mathbf{B}(z) = \begin{bmatrix} z-0.5 & 0.75 \\ 0.5 & z-0.25 \end{bmatrix} = \begin{bmatrix} 1 & 0 \\ 0 & 1 \end{bmatrix} z + \begin{bmatrix} -0.5 & 0.75 \\ 0.5 & -0.25 \end{bmatrix}$$

(Readers can find computationally efficient methods of simultaneously evaluating

$\mathbf{B}(z)$ and $\Delta(z)$ in Ref. [24].) We decompose the inverse matrix into partial fractions

$$(\mathbf{I}z - \mathbf{A})^{-1} = \mathbf{B}(z)/\Delta(z) = \mathbf{D}_{11} / (z-1) + \mathbf{D}_{12} / (z+0.25) \tag{A.5.6}$$

Matrices $\mathbf{D}_{11}$ and $\mathbf{D}_{12}$ can be determined by multiplying both sides of (A.5.6) by $\Delta(z)=(z-1)(z+0.25)$

$$\mathbf{B}(z) = \mathbf{D}_{11}(z+0.25) + \mathbf{D}_{12}(z-1)$$

and comparing the coefficients of the same powers of z:

$$\mathbf{D}_{11}+\mathbf{D}_{12} = \begin{bmatrix} 1 & 0 \\ 0 & 1 \end{bmatrix} \quad 0.25\,\mathbf{D}_{11} - \mathbf{D}_{12} = \begin{bmatrix} -0.5 & 0.75 \\ 0.5 & -0.25 \end{bmatrix}$$

Solving this system we obtain

$$\mathbf{D}_{11} = \begin{bmatrix} 0.4 & 0.6 \\ 0.4 & 0.6 \end{bmatrix} \quad \mathbf{D}_{12} = \begin{bmatrix} 0.6 & -0.6 \\ -0.4 & 0.4 \end{bmatrix}$$

[This solution can also be found by using formula (A.2.13) with $q(z)=\Delta(z)$.]

Thus, for this example, we have according to (A.5.5)

$$\mathbf{A}^m = \mathbf{D}_{11} + \mathbf{D}_{12}(-0.25)^m \tag{A.5.7}$$

which after substitutions of $\mathbf{D}_{11}$ and $\mathbf{D}_{12}$ takes the form

$$\mathbf{A}^m = \begin{bmatrix} 0.4+0.6(-0.25)^m & 0.6-0.6(-0.25)^m \\ 0.4-0.4(-0.25)^m & 0.6+0.4(-0.25)^m \end{bmatrix}$$

■

5.1.4 Eigenvectors and Eigenvalues. An alternative approach to finding $\mathbf{A}^m$ is based on the notion of its eigenvectors and eigenvalues.

We say that matrix column $\mathbf{x} \neq 0$ is a *right eigenvector* of $\mathbf{A}$ if there is a number λ called an *eigenvalue*, such that $\mathbf{A}\mathbf{x}=\lambda\mathbf{x}$. Similarly, matrix row $\mathbf{y}\neq 0$ is called a *left eigenvector* if $\mathbf{y}\mathbf{A}=\mu\mathbf{y}$.

It follows from these definitions that $(\mathbf{I}\lambda - \mathbf{A})\mathbf{x}=0$ and $\mathbf{y}(\mathbf{I}\lambda - \mathbf{A})=0$. This means that the eigenvectors satisfy linear homogeneous equations. As we know, these equations have nonzero solutions if and only if their determinants are equal to zero:

$$\Delta(\lambda) = \det(\mathbf{I}\lambda - \mathbf{A}) = 0 \quad \Delta(\mu) = \det(\mathbf{I}\mu - \mathbf{A}) = 0$$

Therefore, both eigenvalues are characteristic numbers of **A** and there is no need to make a distinction between them.

EXAMPLE A.5.9: Let us find eigenvectors and eigenvalues for the matrix of Example A.5.8. Since $\Delta(z)=(z-1)(z+0.25)$ the eigenvalues are $\lambda_1=1$ and $\lambda_2=-0.25$. The left eigenvectors corresponding to $\lambda_1=1$ can be found from the system $\mathbf{y}_1(\mathbf{I}\lambda_1 - \mathbf{A}) = 0$, which takes the form

$$[y_{11} \quad y_{12}]\begin{bmatrix}0.75 & -0.75\\ -0.5 & 0.5\end{bmatrix} = 0 \quad \text{or} \quad \begin{matrix}0.75y_{11} - 0.5y_{12} = 0\\ -0.75y_{11} + 0.5y_{12} = 0\end{matrix}$$

This system has an infinite number of solutions, which can be expressed as $\mathbf{y}_1 = \nu_1\,[0.4 \quad 0.6]$ where ν_1 is an arbitrary real number (see Example A.5.7). Thus all the eigenvectors corresponding to λ_1 are proportional (colinear). The rest of the eigenvectors can be found similarly. The left eigenvectors corresponding to λ_2 have the form $\mathbf{y}_2=\nu_2\,[1 \quad -1]$. The right eigenvectors are $\mathbf{x}_1=\nu_3\,[1 \quad 1]'$ and $\mathbf{x}_2=\nu_4\,[0.6 \quad -0.4]'$ (these vectors are matrix columns, we express them as transposed matrix rows to save space).

One can observe that the left eigenvectors are proportional to the rows of the matrices $\mathbf{D}_{11}$ and $\mathbf{D}_{12}$ of Example A.5.8, while the right eigenvectors are proportional to their columns. These properties allow us to construct $\mathbf{D}_{11}$ and $\mathbf{D}_{12}$ using eigenvectors.

■

We say that matrix **A** has a *simple structure* if equation (A.5.4) has the following form

$$(\mathbf{I}z - \mathbf{A})^{-1} = \mathbf{D}_{11}/(z-\lambda_1) + \mathbf{D}_{12}/(z-\lambda_2) + \cdots + \mathbf{D}_{1r}/(z-\lambda_r) \qquad \text{(A.5.8)}$$

In this case (A.5.5) becomes

$$\mathbf{A}^m = \mathbf{D}_{11}\lambda_1^m + \mathbf{D}_{12}\lambda_2^m + \cdots + \mathbf{D}_{1r}\lambda_r^m \qquad \text{(A.5.9)}$$

and the coefficients $\mathbf{D}_{1j}$ can be expressed by formula (A.2.8):

$$\mathbf{D}_{1j} = \mathbf{B}(\lambda_j)/\Delta'(\lambda_j)$$

It follows from the results of the previous section that if all the roots of the characteristic polynomial $\Delta(z)=(\mathbf{I}z - \mathbf{A})$ are different, then **A** has a simple structure. Matrix **A** of Example A.5.8 has a simple structure.

Theorem A.5.1: If matrix **A** has a simple structure, then rows and columns of the matrices $\mathbf{D}_{1j}$ defined by equation (A.5.8) are its eigenvectors.

Proof: To simplify notations let us assume that the eigenvalues are sorted according to their absolute values:

$$|\lambda_1| > |\lambda_2| > \cdots > |\lambda_r|$$

Multiplying both sides of (A.5.9) by λ_1^{-m} we obtain

$$\lambda_1^{-m}\mathbf{A}^m = \mathbf{D}_{11} + \mathbf{D}_{12}(\lambda_2 / \lambda_1)^m + \cdots + \mathbf{D}_{1r}(\lambda_r / \lambda_1)^m \qquad \text{(A.5.10)}$$

Since $|\lambda_j / \lambda_1| < 1$

$$\lim_{m\to\infty} \lambda_1^{-m}\mathbf{A}^m = \mathbf{D}_{11} \qquad \text{(A.5.11)}$$

Passing to the limit in the identity $\lambda_1^{-m}\mathbf{A}^m\,\mathbf{A} = \lambda_1^{-m}\mathbf{A}^{m+1}$, we obtain

$$\mathbf{D}_{11}\,\mathbf{A} = \lambda_1\,\mathbf{D}_{11} \qquad \text{(A.5.12)}$$

This means that each row of $\mathbf{D}_{11}$ is a left eigenvector corresponding to λ_1. Using the identity $\lambda_1^m\mathbf{A}\,\mathbf{A}^m = \lambda_1^m\mathbf{A}^{m+1}$, we can prove that each column of $\mathbf{D}_{11}$ is a right eigenvector of **A**: $\mathbf{A}\,\mathbf{D}_{11} = \lambda_1\,\mathbf{D}_{11}$. Thus, we have proven the theorem for the matrix $\mathbf{D}_{11}$ which corresponds to the largest eigenvalue λ_1. To prove it for the rest of the matrices we need to study some properties of $\mathbf{D}_{11}$.

Property I: $\mathbf{D}_{11}\mathbf{A}^m = \lambda_1^m\mathbf{D}_{11}$ and $\mathbf{A}^m\mathbf{D}_{11} = \lambda_1^m\mathbf{D}_{11}$.
We can prove these equations by applying (A.5.12) to $\mathbf{D}_{11}\mathbf{A}^m$:

$$\mathbf{D}_{11}\mathbf{A}^m = \mathbf{D}_{11}\mathbf{A}\,\mathbf{A}^{m-1} = \lambda_1\mathbf{D}_{11}\mathbf{A}^{m-1} = \lambda_1^2\mathbf{D}_{11}\mathbf{A}^{m-2} = \cdots = \lambda_1^m\mathbf{D}_{11}$$

Property II: $\mathbf{D}_{11}^2 = \mathbf{D}_{11}$.
This property can be proven by passing to the limit when $m \to \infty$ in the identity $\mathbf{D}_{11}\lambda_1^{-m}\mathbf{A}^m = \mathbf{D}_{11}$ and using (A.5.11).

Property III: $\mathbf{D}_{11}\mathbf{D}_{1j} = 0$ and $\mathbf{D}_{1j}\mathbf{D}_{11} = 0$ for $j = 2, 3, \ldots, r$.
Multiplying both sides of (A.5.9) by $\mathbf{D}_{11}$ and using Property I, we obtain

$$\mathbf{D}_{11}\mathbf{D}_{12}\lambda_2^m + \cdots + \mathbf{D}_{11}\mathbf{D}_{1r}\lambda_r^m = 0 \quad \text{for} \quad m = 1, 2, \ldots \qquad \text{(A.5.13)}$$

This is possible only when $\mathbf{D}_{11}\mathbf{D}_{12} = 0$, ..., $\mathbf{D}_{11}\mathbf{D}_{1r} = 0$, because the determinants of the first $r-1$ systems are Vandermonde's determinants, which are not singular (see

Example A.5.6), so that these systems have only trivial solutions. We can prove similarly that $\mathbf{D}_{12}\mathbf{D}_{11}=0, \ldots, \mathbf{D}_{1r}\mathbf{D}_{11}=0$.

To complete the proof of the theorem consider the matrix $\mathbf{A}_1 = \mathbf{A} - \mathbf{D}_{11}\lambda_1$, which, according to (A.5.9) for $m = 1$ can be written as

$$\mathbf{A}_1 = \mathbf{A} - \mathbf{D}_{11}\lambda_1 = \mathbf{D}_{12}\lambda_2 + \cdots + \mathbf{D}_{1r}\lambda_r \tag{A.5.14}$$

It follows from the above properties of $\mathbf{D}_{11}$ that $\mathbf{A}^m = \mathbf{A}_1^m + \lambda_1^m \mathbf{D}_{11}$. Substituting it into (A.5.9) we obtain the equation

$$\mathbf{A}_1^m = \mathbf{D}_{12}\lambda_2^m + \cdots + \mathbf{D}_{1r}\lambda_r^m \tag{A.5.15}$$

which is similar to (A.5.9). Therefore, applying the previous results, we can prove that the rows and columns of $\mathbf{D}_{12}$ are the eigenvectors of $\mathbf{A}_1$ corresponding to λ_2:

$$\mathbf{D}_{12}\,\mathbf{A}_1 = \lambda_2\mathbf{D}_{12} \qquad \mathbf{A}_1\,\mathbf{D}_{12} = \lambda_2\mathbf{D}_{12}$$

But they are also the eigenvectors of $\mathbf{A}$, which is seen from the following transformations

$$\mathbf{D}_{12}\,\mathbf{A} = \mathbf{D}_{12}\,\mathbf{A}_1 + \lambda_1\mathbf{D}_{12}\,\mathbf{D}_{11} = \lambda_2\,\mathbf{D}_{12}$$

Thus, we have proven the theorem for $\mathbf{D}_{11}$ and $\mathbf{D}_{12}$. We can prove it for $\mathbf{D}_{13}$ by considering $\mathbf{A}_2 = \mathbf{A}_1 - \lambda_2\mathbf{D}_{12}$, and so on. Finally, we will prove it for all $\mathbf{D}_{1j}$.

■

In conclusion, we would like to point out that all $\mathbf{D}_{1j}$ possess properties similar to I, II, and III. This can be written as

$$\mathbf{D}_{1j}\,\mathbf{D}_{1j} = \mathbf{D}_{1j} \qquad \mathbf{D}_{1i}\,\mathbf{D}_{1j} = 0 \quad \text{for} \quad i \neq j \ \text{and} \ i,j = 1,2,\ldots,r \tag{A.5.16}$$

In the particular case when all the roots of the characteristic polynomial are different, all the eigenvectors corresponding to the same eigenvalue are proportional (colinear). Therefore, matrices $\mathbf{D}_{1j}$ can be expressed as

$$\mathbf{D}_{1j} = \mathbf{x}_j\,\mathbf{y}_j \tag{A.5.17}$$

where $\mathbf{x}_j$ is the right eigenvector and $\mathbf{y}_j$ is the left eigenvector corresponding to λ_j. It follows from (A.5.16) that

$$\mathbf{y}_j\,\mathbf{x}_j = 1 \quad \text{and} \quad \mathbf{y}_i\,\mathbf{x}_j = 0 \quad \text{for } i \neq j \tag{A.5.18}$$

EXAMPLE A.5.10: It is easy to verify that the matrices $\mathbf{D}_{11}$ and $\mathbf{D}_{12}$ of Example A.5.9 can be expressed as

$$\mathbf{D}_{11} = \begin{bmatrix} 0.4 & 0.6 \\ 0.4 & 0.6 \end{bmatrix} = \begin{bmatrix} 1 \\ 1 \end{bmatrix} [0.4 \quad 0.6] \quad \mathbf{D}_{12} = \begin{bmatrix} 0.6 & -0.6 \\ -0.4 & 0.4 \end{bmatrix} = \begin{bmatrix} 0.6 \\ -0.4 \end{bmatrix} [1 \quad -1]$$

These equations together with the results of Example A.5.7 illustrate (A.5.16). ∎

5.1.5 Similar Matrices. Matrices **A** and **B** are called *similar* if there is a matrix **T** satisfying the following equation

$$\mathbf{T}\,\mathbf{A}\,\mathbf{T}^{-1} = \mathbf{B} \tag{A.5.19}$$

Similar matrices play an important in simplifying error source models. The following theorem is often used in this book.

Theorem A.5.2: If matrix **A** has a simple structure, then it is similar to a diagonal matrix.

Proof: To simplify the proof, let us assume that all the eigenvalues are different. In this case matrices $\mathbf{D}_{1j}$ can be expressed by (A.5.17).

Consider the matrix **T** whose rows are the eigenvectors of **A**:

$$\mathbf{T} = block\ col\{\mathbf{y}_i\} = \begin{bmatrix} \mathbf{y}_1 \\ \mathbf{y}_2 \\ \cdots \\ \mathbf{y}_n \end{bmatrix} \tag{A.5.20}$$

Because of (A.5.18), its inverse matrix is given by

$$\mathbf{T}^{-1} = block\ row\{\mathbf{x}_i\} = [\mathbf{x}_1 \quad \mathbf{x}_2 \ \dots \ \mathbf{x}_n]$$

To complete the proof of the theorem let us calculate $\mathbf{T}\,\mathbf{A}\,\mathbf{T}^{-1}$. Since the columns of $\mathbf{T}^{-1}$ are the matrix **A** eigenvectors, we have

$$\mathbf{A}\,\mathbf{T}^{-1} = [\mathbf{A}\,\mathbf{x}_1 \quad \mathbf{A}\,\mathbf{x}_2 \ \dots \ \mathbf{A}\,\mathbf{x}_n] = [\lambda_1\,\mathbf{x}_1 \quad \lambda_2\,\mathbf{x}_2 \ \dots \ \lambda_r\,\mathbf{x}_r]$$

Multiplying this equation from the left by **T** and using (A.5.20), we obtain

$$
\mathbf{T}\,\mathbf{A}\,\mathbf{T}^{-1} = \begin{bmatrix} \lambda_1\,\mathbf{y}_1\,\mathbf{x}_1 & \lambda_2\,\mathbf{y}_1\,\mathbf{x}_2 & \cdots & \lambda_n\,\mathbf{y}_1\,\mathbf{x}_n \\ \lambda_1\,\mathbf{y}_2\,\mathbf{x}_1 & \lambda_2\,\mathbf{y}_2\,\mathbf{x}_2 & \cdots & \lambda_n\,\mathbf{y}_2\,\mathbf{x}_n \\ \cdots & \cdots & \cdots & \cdots \\ \lambda_1\,\mathbf{y}_n\,\mathbf{x}_1 & \lambda_2\,\mathbf{y}_n\,\mathbf{x}_2 & \cdots & \lambda_n\,\mathbf{y}_n\,\mathbf{x}_n \end{bmatrix}
$$

According to (A.5.18) this equation can be rewritten as

$$
\mathbf{T}\,\mathbf{A}\,\mathbf{T}^{-1} = \begin{bmatrix} \lambda_1 & 0 & \cdots & 0 \\ 0 & \lambda_2 & \cdots & 0 \\ \cdots & \cdots & \cdots & \cdots \\ 0 & 0 & \cdots & \lambda_n \end{bmatrix} \qquad \text{(A.5.21)}
$$

This formula is called matrix **A** *spectral representation.*

■

EXAMPLE A.5.11: Let us show that the matrix **A** of Example A.5.8 is similar to a diagonal matrix. Using the results of Example A.5.9 and the previous theorem, we have

$$
\mathbf{T} = \begin{bmatrix} 0.4 & 0.6 \\ 1 & -1 \end{bmatrix} \qquad \mathbf{T}^{-1} = \begin{bmatrix} 1 & 0.6 \\ 1 & -0.4 \end{bmatrix}
$$

By direct multiplication we obtain

$$
\mathbf{TAT}^{-1} = \begin{bmatrix} 1 & 0 \\ 0 & -0.25 \end{bmatrix}
$$

■

We would like to point out that if **A** does not have simple structure it is similar to the so-called Jordan normal matrix, which is given by

$$
\mathbf{J} = \begin{bmatrix} \mathbf{J}_1 & 0 & \cdots & 0 \\ 0 & \mathbf{J}_2 & \cdots & 0 \\ \cdot\cdot & \cdot\cdot & \cdots & \cdot\cdot \\ 0 & 0 & \cdots & \mathbf{J}_s \end{bmatrix} \quad \text{where} \quad \mathbf{J}_i = \begin{bmatrix} \lambda_i & 1 & \cdots & 0 & 0 \\ 0 & \lambda_i & \cdots & 0 & 0 \\ \cdot\cdot & \cdot\cdot & \cdots & \cdot\cdot & \cdot\cdot \\ 0 & 0 & \cdots & \lambda_i & 1 \\ 0 & 0 & \cdots & 0 & \lambda_i \end{bmatrix} \qquad \text{(A.5.22)}
$$

Let us illustrate the application of similar matrices to finding A^m. Equation (A.5.19) can be rewritten as $\mathbf{A} = \mathbf{T}^{-1}\,\mathbf{B}\,\mathbf{T}$. Therefore

$$\mathbf{A}^2 = \mathbf{T}^{-1}\mathbf{B}\mathbf{T}\,\mathbf{T}^{-1}\mathbf{B}\mathbf{T} = \mathbf{T}^{-1}\mathbf{B}^2\mathbf{T}$$

Repeating these transformations we prove that $\mathbf{A}^m = \mathbf{T}^{-1}\mathbf{B}^m\mathbf{T}$. Thus, if we know how to find $\mathbf{B}^m$, we can obtain $\mathbf{A}^m$ using the previous equation. In particular, if $\mathbf{B}$ is a diagonal matrix, we have

$$\mathbf{A}^m = \mathbf{T}^{-1}\begin{bmatrix} \lambda_1^m & 0 & \cdots & 0 \\ 0 & \lambda_2^m & \cdots & 0 \\ \cdots & \cdots & \cdots & \cdots \\ 0 & 0 & \cdots & \lambda_n^m \end{bmatrix}\mathbf{T} \tag{A.5.23}$$

EXAMPLE A.5.12: Let us find $\mathbf{A}^m$ for the matrix $\mathbf{A}$ of Example A.5.8. According to the results of Example A.5.11 and equation (A.5.21), we have

$$\mathbf{A}^m = \mathbf{T}^{-1}\begin{bmatrix} 1 & 0 \\ 0 & (-0.25)^m \end{bmatrix}\mathbf{T} = \begin{bmatrix} 0.4+0.6\,(-0.25)^m & 0.6-0.6\,(-0.25)^m \\ 0.4-0.4\,(-0.25)^m & 0.6+0.4\,(-0.25)^m \end{bmatrix}$$

■

5.1.6 Matrix Series Summation. A matrix

$$\mathbf{S} = \sum_{m_1,\ldots,m_k} \mathbf{A}_{m_1,\ldots,m_k} \tag{A.5.24}$$

is called a matrix series sum if its elements are the sums of the corresponding elements of the matrices $\mathbf{A}_{m_1,\ldots,m_k}$:

$$s_{ij} = \sum_{m_1,\ldots,m_k} a_{m_1,\ldots,m_k}^{(ij)} \tag{A.5.25}$$

For series (A.5.24) to converge, it is necessary and sufficient that all scalar series of type (A.5.25) be convergent. It is convenient to use integral form (2.2.14) to replace a summation of complex expressions by the summation of a geometric progression under the integral sign.

For example, it follows from the matrix-geometric series (A.5.3) that

$$\mathbf{A}^m = \frac{1}{2\pi \mathrm{j}} \oint_{\gamma} (\mathbf{I}z - \mathbf{A})^{-1} z^m dz \tag{A.5.26}$$

where $\gamma = \{z : |z| = \rho\}$, $\rho > \max|\lambda_i|$, and λ_i are the matrix $\mathbf{A}$ eigenvalues. This formula is convenient to use for replacing $\mathbf{A}^m$ in sums. It can also be used to find an arbitrary function of the matrix.

An arbitrary function of a matrix is defined [24] as a sum of the series

$$f(\mathbf{A}) = \sum_{m=0}^{\infty} a_m \mathbf{A}^m$$

if eigenvalues of **A** lie inside the disk of convergence of the series

$$f(z) = \sum_{m=0}^{\infty} a_m z^m$$

Then, from formula (A.5.26), we obtain

$$f(\mathbf{A}) = \sum_{m=0}^{\infty} a_m \frac{1}{2\pi \mathrm{j}} \oint_{\gamma} (\mathbf{I}z - \mathbf{A})^{-1} z^m dz = \frac{1}{2\pi \mathrm{j}} \oint_{\gamma} (\mathbf{I}z - \mathbf{A})^{-1} f(z) dz \quad \text{(A.5.27)}$$

Using formula (A.5.4), we obtain

$$f(\mathbf{A}) = \frac{1}{2\pi \mathrm{j}} \sum_{j=1}^{r} \sum_{i=1}^{m_j} \mathbf{D}_{ij} \oint_{\gamma} \frac{f(z)}{(z-\lambda_j)^i} dz$$

which, after applying Cauchy's theorem, gives

$$f(\mathbf{A}) = \sum_{j=1}^{r} \sum_{i=1}^{m_j} \mathbf{D}_{ij} f^{(i-1)}(\lambda_j) \,/\, (i-1)! \quad \text{(A.5.28)}$$

This formula can be used for any function of matrix calculation. Formula (A.5.5) is a particular case of this formula, when $f(z) = z^m$.

EXAMPLE A.5.13: As an example, let us calculate various functions of

$$\mathbf{A} = \begin{bmatrix} 1 & 1 \\ 0 & 1 \end{bmatrix}$$

For any function of **A** we need to find matrices $\mathbf{D}_{ij}$ using decomposition (A.5.4). We have

$$\mathbf{I}z - \mathbf{A} = \begin{bmatrix} 1 & 0 \\ 0 & 1 \end{bmatrix} z - \begin{bmatrix} 1 & 1 \\ 0 & 1 \end{bmatrix} = \begin{bmatrix} z-1 & -1 \\ 0 & z-1 \end{bmatrix}$$

$$\Delta(z) = (z-1)^2 \qquad \mathbf{B}(z) = \begin{bmatrix} z-1 & 1 \\ 0 & z-1 \end{bmatrix}$$

so that

$$(\mathbf{I}z - \mathbf{A})^{-1} = \mathbf{B}(z) / \Delta(z) = \begin{bmatrix} (z-1)^{-1} & (z-1)^{-2} \\ 0 & (z-1)^{-1} \end{bmatrix}$$

The partial fraction decomposition has the form

$$(\mathbf{I}z - \mathbf{A})^{-1} = \mathbf{D}_{11} / (z-1) + \mathbf{D}_{21} / (z-1)^2$$

where

$$\mathbf{D}_{11} = \begin{bmatrix} 1 & 0 \\ 0 & 1 \end{bmatrix} \qquad \mathbf{D}_{21} = \begin{bmatrix} 0 & 1 \\ 0 & 0 \end{bmatrix}$$

and therefore any function of this matrix is given by

$$f(\mathbf{A}) = \mathbf{D}_{11} f(1) + \mathbf{D}_{21} f'(1)$$

For calculating matrix powers we use $f(z)=z^m$, which gives

$$\mathbf{A}^m = \mathbf{D}_{11} + \mathbf{D}_{21} m = \begin{bmatrix} 1 & m \\ 0 & 1 \end{bmatrix}$$

For calculating $e^{\mathbf{A}t}$ we use $f(z) = e^{zt}$:

$$e^{\mathbf{A}t} = \mathbf{D}_{11} e^t + \mathbf{D}_{21} t e^t = \begin{bmatrix} e^t & te^t \\ 0 & e^t \end{bmatrix}$$

■

In conclusion, let us prove the Cayley-Hamilton theorem, which we use in Sec. 1.3.3. This theorem states that every square matrix is a root of its characteristic polynomial: $\Delta(\mathbf{A})=0$.

Indeed, our function is $f(z)=\Delta(z)=\det(\mathbf{I}z-\mathbf{A})$ so that equation (A.5.27) becomes

$$\Delta(\mathbf{A}) = \frac{1}{2\pi j}\oint_{\gamma}(\mathbf{I}z-\mathbf{A})^{-1}\,\Delta(z)dz = \frac{1}{2\pi j}\oint_{\gamma}\mathbf{B}(z)dz = 0 \qquad \text{(A.5.29)}$$

This integral is equal to 0, because the adjoint matrix $\mathbf{B}(z)$ is a polynomial and does not have poles inside γ.

APPENDIX 6

6.1 Markov Chains and Graphs

This appendix provides a brief description of the most important properties of Markov chains, which are used in this book.

6.1.1 Transition Probabilities. A sequence of random variables $y_0, y_1, y_2, \ldots$ is called a *finite Markov chain* if

$$Pr(y_t = i_t \mid y_0 = i_0, y_1 = i_1, \ldots, y_{t-1} = i_{t-1}) = Pr(y_t = i_t \mid y_{t-1} = i_{t-1})$$

for any $t \in \mathbf{N} = \{1,2,\ldots\}$ and $i_0, i_1, \ldots, i_t$ from a finite set Ω (usually, $i_k \in \Omega = \{1,2,\ldots,n\}$). In other words, the probability that a Markov chain is in state i_t at the moment t depends only on its state at the moment $t-1$.

If the transition probability

$$p_{ij} = Pr(y_t = j \mid y_{t-1} = i)$$

does not depend on t, the chain is called *homogeneous*. Any joint distribution of a homogeneous Markov chain is completely determined by the matrix of its transition probabilities

$$\mathbf{P} = [p_{ij}]_{n,n}$$

and the initial distribution

$$\mathbf{p} = [p_1 \ p_2 \ \cdots \ p_n]$$

Indeed, the chain starts from state i_0 with the probability p_{i_0}. Then it transfers to state i_1 with the probability $p_{i_0 i_1}$. Next it transfers to state i_2 with the probability $p_{i_1 i_2}$, and so on. Therefore, the joint probability of the sequence of states can be expressed as

$$Pr(y_0 = i_0, y_1 = i_1, ..., y_t = i_t) = p_{i_0} p_{i_0 i_1} \cdots p_{i_{t-1} i_t} = p_{i_0} \prod_{k=1}^{t} p_{i_{k-1} i_k}$$

Clearly, the elements of the transition matrix are non-negative and the sum of the elements of each row equals 1:

$$\sum_{j=1}^{n} p_{ij} = 1 \quad i=1,2,...,n$$

This can be rewritten as $\mathbf{P}\,\mathbf{1} = \mathbf{1}$, where $\mathbf{1}$ is the matrix column of ones. Therefore, $\mathbf{1}$ is an eigenvector of $\mathbf{P}$ with the eigenvalue $\lambda = 1$. This vector is sometimes called a *right invariant vector* of the matrix $\mathbf{P}$, since the multiplication $\mathbf{P}\,\mathbf{1}$ does not change $\mathbf{1}$.

We shall use the notation $p_{ij}^{(m)}$ for the m-step transition probability from state i to state j. This probability can be found recursively by the total probability formula

$$p_{ij}^{(m)} = \sum_{k=1}^{n} p_{ik} p_{kj}^{(m-1)} \tag{A.6.1}$$

which can be written in the matrix form as $\mathbf{P}^{(m)} = \mathbf{P}\mathbf{P}^{(m-1)}$. Therefore, $\mathbf{P}^{(2)} = \mathbf{P}\mathbf{P} = \mathbf{P}^2$, $\mathbf{P}^{(3)} = \mathbf{P}\mathbf{P}^{(2)} = \mathbf{P}^3$, and so on; finally, $\mathbf{P}^{(m)} = [\, p_{ij}^{(m)} \,]_{n,n} = \mathbf{P}^m$. In other words, m-step transition probabilities are the elements of the m-th power of $\mathbf{P}$.

We can find $\mathbf{P}^m$ using the method described in Appendix 5.1.3. In particular, equation (A.5.5) gives

$$\mathbf{P}^m = \sum_{j=1}^{r} \sum_{i=1}^{m_j} \mathbf{D}_{ij} \binom{m}{i-1} \lambda_j^{m-i+1} \tag{A.6.2}$$

EXAMPLE A.6.1: Suppose that we have a Markov chain with two states and the transition matrix

$$\mathbf{P} = \begin{bmatrix} 0.25 & 0.75 \\ 0.5 & 0.5 \end{bmatrix}$$

Let us find the m-step transition probabilities.

The m-step transition matrix is equal to $\mathbf{P}^m$. Fortunately, we have found this matrix in Example A.5.8. It is given by equation (A.5.7). Thus,

$$\begin{aligned} p_{11}^{(m)} &= 0.4+0.6\,(-0.25)^m & p_{12}^{(m)} &= 0.6-0.6\,(-0.25)^m \\ p_{21}^{(m)} &= 0.4-0.4\,(-0.25)^m & p_{22}^{(m)} &= 0.6+0.4\,(-0.25)^m \end{aligned} \tag{A.6.3}$$

■

The probability distribution of states after m steps is given by

$$p_i^{(m)} = \sum_{k=1}^{n} p_k p_{ki}^{(m-1)}$$

This can be written using matrix notations as

$$\mathbf{p}^{(m)} = \mathbf{p}\,\mathbf{P}^{(m)} = \mathbf{p}\mathbf{P}^m \tag{A.6.4}$$

EXAMPLE A.6.2: For the Markov chain considered in Example A.6.1 with the initial distribution $\mathbf{p}=[\,0.2 \quad 0.8\,]$ let us find the probability distribution of its states after m steps.

Using equations (A.5.5) and (A.6.4), we obtain the required result:

$$\mathbf{p}^{(m)} = \mathbf{p}\mathbf{D}_{11} + \mathbf{p}\mathbf{D}_{12}\,(-0.25)^m = [\,0.4\text{–}0.2(-0.25)^m \quad 0.6\text{+}0.2(-0.25)^m\,]$$

■

6.1.2 Stationary Markov Chains. Matrix $\mathbf{P}$ is called a *regular matrix* if its characteristic polynomial $\Delta(z)=\det(\mathbf{I}z-\mathbf{P})$ has only one root $z=1$, whose absolute value equals 1. In other words, $\mathbf{P}$ is regular if $\Delta(1)=0$, $\Delta'(1)\neq 0$, and the absolute values of the remaining roots of the equation $\Delta(z)=0$ are not equal to 1. If a transition matrix of a Markov chain is regular, then the chain also is called regular.

Theorem A.6.1: If a Markov chain is regular, then its m-step transition probability has a limit, which does not depend on the chain's initial state:

$$\lim_{m\to\infty} p_{ij}^{(m)} = \pi_j \tag{A.6.5}$$

These limits are called *equilibrium* or *stationary* probabilities of states of the Markov chain.

Proof: Since $z=1$ is a simple root of $\Delta(z)$,

$$\Delta(z) = (z-1)(z-\lambda_2)^{m_2}(z-\lambda_3)^{m_3} \cdots (z-\lambda_r)^{m_r}$$

and formula (A.5.4) becomes

$$(\mathbf{I}z - \mathbf{P})^{-1} = \mathbf{B}(z) / \Delta(z) = \mathbf{D}_{11} / (z-1) + \sum_{j=2}^{r} \sum_{i=1}^{m_j} \mathbf{D}_{ij} / (z-\lambda_j)^i \qquad \text{(A.6.6)}$$

Expanding the right-hand side of this equation, we obtain from (A.5.5)

$$\mathbf{P}^m = \mathbf{D}_{11} + \sum_{j=2}^{r} \sum_{i=1}^{m_j} \mathbf{D}_{ij} \binom{m}{i-1} \lambda_j^{m-i+1} \qquad \text{(A.6.7)}$$

If we assume that $|\lambda_j| > 1$, then $|\lambda_j^m| \to \infty$ when $m \to \infty$, which is impossible since all the elements of $\mathbf{P}^m$, being probabilities, are bounded: $0 \le p_{ij}^{(m)} \le 1$. Thus, $|\lambda_j| < 1$ for $j=2,3,...,r$, because $z=1$ is the only root whose absolute value is equal to 1. But in this case

$$\lim_{m \to \infty} \binom{m}{i-1} \lambda_j^{m-i+1} = 0 \qquad \text{(A.6.8)}$$

and we obtain

$$\lim_{m \to \infty} \mathbf{P}^m = \mathbf{D}_{11}$$

This proves that each m-step transition probability has a limit when $m \to \infty$.

Passing to the limit in the identity $\mathbf{P}^m \mathbf{P} = \mathbf{P}^{m+1}$, we obtain $\mathbf{D}_{11} \mathbf{P} = \mathbf{D}_{11}$. This means that each row of $\mathbf{D}_{11}$ is an eigenvector corresponding to the eigenvalue $\lambda = 1$, so that all the rows of $\mathbf{D}_{11}$ are proportional. Passing to the limit as $m \to \infty$ in the identity $\mathbf{P}^m \mathbf{1} = \mathbf{1}$, we obtain $\mathbf{D}_{11} \mathbf{1} = \mathbf{1}$, which means that all the rows of $\mathbf{D}_{11}$ are identical and equal to the stationary probability vector $\boldsymbol{\pi} = [\pi_1, \pi_2, ..., \pi_n]$. Therefore, the limits in (A.6.5) do not depend on the initial state (the number of a row of $\mathbf{D}_{11}$). It is easy to check that [see also (A.5.17)]

$$\mathbf{D}_{11} = \mathbf{1}\boldsymbol{\pi}$$

■

EXAMPLE A.6.3: For the Markov chain of Example A.6.1 we had $\Delta(z) = (z-1)(z+0.25)$. This polynomial has a simple root $z=1$, therefore the chain is regular. Passing to the limit as $m \to \infty$ in equations (A.6.3) we obtain

$$p_{11}^{(\infty)} = 0.4 \qquad p_{12}^{(\infty)} = 0.6$$
$$p_{21}^{(\infty)} = 0.4 \qquad p_{22}^{(\infty)} = 0.6 \tag{A.6.9}$$

So that the stationary distribution has the form

$$\boldsymbol{\pi} = [\,\pi_1 \quad \pi_2\,] = [\,0.4 \quad 0.6\,]$$

It follows from (A.5.7) that (see also Example A.5.10)

$$\lim_{m\to\infty} \mathbf{P}^m = \mathbf{D}_{11} = \begin{bmatrix} 0.4 & 0.6 \\ 0.4 & 0.6 \end{bmatrix} = \begin{bmatrix} 1 \\ 1 \end{bmatrix} [\,0.4 \quad 0.6\,] = \mathbf{1}\,\boldsymbol{\pi}$$

■

The stationary distribution can be found as a solution of the system:

$$\boldsymbol{\pi}\mathbf{P} = \boldsymbol{\pi} \qquad \boldsymbol{\pi}\mathbf{1} = 1 \tag{A.6.10}$$

since $\boldsymbol{\pi}$ is the left eigenvector of $\mathbf{P}$ corresponding to $\lambda=1$ (see Appendix 5.1.4). The second equation in (A.6.10) simply means that the total sum of the stationary probabilities is equal to 1:

$$\sum_{j=1}^{n} \pi_j = \boldsymbol{\pi}\mathbf{1} = 1$$

In its explicit form system (A.6.10) can be written as

$$\pi_j = \sum_{i=1}^{n} \pi_i p_{ij} \qquad \sum_{i=1}^{n} \pi_i = 1 \quad j = 1,2,...,n$$

It follows from Theorem A.6.1 that this system has a unique solution if the Markov chain is regular.

If the matrix of the initial probabilities $\mathbf{p}$ coincides with the matrix of the stationary probabilities $\boldsymbol{\pi}$, the chain is called *stationary*.

EXAMPLE A.6.4: System (A.6.10) for the Markov chain of Example A.6.1 has the following form

$$[\,\pi_1 \quad \pi_2\,] \begin{bmatrix} 0.25 & 0.75 \\ 0.5 & 0.5 \end{bmatrix} = [\,\pi_1 \quad \pi_2\,] \qquad [\,\pi_1 \quad \pi_2\,] \begin{bmatrix} 1 \\ 1 \end{bmatrix} = 1$$

In its explicit form this system is

$$\begin{aligned} 0.25\,\pi_1 + 0.5\,\pi_2 &= \pi_1 \\ 0.75\,\pi_1 + 0.5\,\pi_2 &= \pi_2 \\ \pi_1 + \pi_2 &= 1 \end{aligned}$$

The unique solution of this system is

$$\boldsymbol{\pi} = [\,\pi_1 \quad \pi_2\,] = [\,0.4 \quad 0.6\,]$$ ■

6.1.3 Partitioning States of a Markov Chain. It is convenient to partition the set of all states of a Markov chain into subsets of states $\{A_1, A_2, ..., A_s\}$, such that

$$\Omega = \bigcup_{i=1}^{s} A_i \qquad A_i \cap A_j = \varnothing \quad \text{for } i \neq j$$

As will be seen shortly, it is convenient to describe the process of transition between the subsets if the numbers of the states belonging to the same subset are consecutive. For example, $A_1 = (1,2,3)$, $A_2 = (4,5)$, $A_3 = (6,7,8,9)$. If a partition does not satisfy this condition, we can always renumber states that do not satisfy it. Exchanging numbers of two states requires permuting the corresponding rows and columns of the transition matrix. This operation can also be performed by the transformation **UPU** in which **U** is obtained from the unit matrix by permuting its i-th and j-th rows.

After reordering all the states according to the selected partition, the transition matrix can be rewritten in the following block form

$$\mathbf{P} = \begin{bmatrix} \mathbf{P}_{11} & \mathbf{P}_{12} & \cdots & \mathbf{P}_{1s} \\ \mathbf{P}_{21} & \mathbf{P}_{22} & \cdots & \mathbf{P}_{2s} \\ \cdots & \cdots & \cdots & \cdots \\ \mathbf{P}_{s1} & \mathbf{P}_{s2} & \cdots & \mathbf{P}_{ss} \end{bmatrix}$$

EXAMPLE A.6.5: Consider a Markov chain with three states an the transition probability matrix

$$\mathbf{P} = \begin{bmatrix} 0.5 & 0 & 0.5 \\ 0.3 & 0.3 & 0.4 \\ 0.2 & 0 & 0.8 \end{bmatrix}$$

Suppose that we want to rewrite this matrix in the block form corresponding to the partition $\{A_1, A_2\}$ in which A_1 is composed of the first and the last states of the chain and A_2 is the second state. In this case we need to permute the second and third states of the chain. This can be achieved by permuting the second and third columns and rows of the transition matrix. This operation can also be performed by the transformation **UPU** where **U** is obtained from the unit matrix $\mathbf{I}=diag\ \{1, 1, 1\}$ by permuting its second and third rows:

$$\mathbf{UPU} = \begin{bmatrix} 0.5 & 0.5 & 0 \\ 0.2 & 0.8 & 0 \\ 0.3 & 0.4 & 0.3 \end{bmatrix} = \begin{bmatrix} \mathbf{P}_{11} & \mathbf{P}_{12} \\ \mathbf{P}_{21} & \mathbf{P}_{22} \end{bmatrix} \tag{A.6.11}$$

where

$$\mathbf{P}_{11} = = \begin{bmatrix} 0.5 & 0.5 \\ 0.2 & 0.8 \end{bmatrix} \quad \mathbf{P}_{12} = \begin{bmatrix} 0 \\ 0 \end{bmatrix}$$

$$\mathbf{P}_{21} = [\,0.3 \quad 0.4\,] \quad \mathbf{P}_{22} = 0.3$$

■

As an illustration of usage of state partitioning, let us consider the problem of exclusion of unobserved states of a Markov chain.[40] Suppose that states of a Markov chain with the transition matrix **P** and initial distribution **p** are partitioned into two subsets A_1 and A_2. We can observe only the states that belong to A_1, and we want to describe the process of their transition.

Clearly, the observed process is a Markov chain. Therefore, to describe it we need to find its transition matrix $\mathbf{Q}=[q_{ij}]_{s,s}$ and initial distribution **q**. Let x and y be two states that belong to A_1. The transition probability of the visible process can be expressed as

$$q_{xy} = p_{xy} + \sum_{j\in A_2} p_{xj}p_{jy} + \sum_{j,k\in A_2} p_{xj}p_{jk}p_{ky} + \cdots$$

In this equation p_{xy} corresponds to the transfer $x \to y$ in one step in the original chain, the first sum over A_2 corresponds to the first transition $x \to y$ in two steps (from the visible state x to an invisible state j and back to the visible state y), the second sum corresponds to the first transition $x \to y$ in three steps , and so on. Obviously, these sums can be expressed as matrix products, so that

$$\mathbf{Q} = \mathbf{P}_{11} + \mathbf{P}_{12}\mathbf{P}_{21} + \mathbf{P}_{12}\mathbf{P}_{22}\mathbf{P}_{21} + \mathbf{P}_{12}\mathbf{P}_{22}^2\mathbf{P}_{21} + \cdots$$

Using (A.5.3) we can perform summation in the right-hand side of the previous equation

$$\mathbf{Q} = \mathbf{P}_{11} + \mathbf{P}_{12}(\mathbf{I} - \mathbf{P}_{22})^{-1}\mathbf{P}_{21}$$

The initial probability vector $\mathbf{q}$ of the observed process can be obtained similarly to its transition matrix:

$$\mathbf{q} = \mathbf{p}_1 + \mathbf{p}_2(\mathbf{I} - \mathbf{P}_{22})^{-1}\mathbf{P}_{21}$$

where $\mathbf{p} = [\mathbf{p}_1 \;\; \mathbf{p}_2]$ is the initial vector of the original process.

6.1.4 Signal Flow Graphs. It is convenient to represent state transitions using signal flow graphs or state-transition diagrams. States are represented by the graph vertices and state transitions are represented by the edges.

EXAMPLE A.6.6: The signal flow graph of the Markov chain of Example A.6.1 is shown in Fig. A.3.

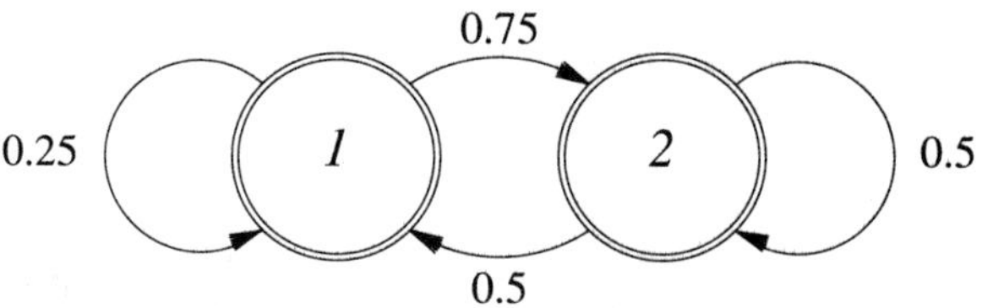

Figure A.3. State diagram of a Markov chain.

In this figure, numbers inside the circles represent states, while numbers near arcs represent the corresponding transition probabilities.

If states of a Markov chain are partitioned into subsets of states $\{A_1, A_2, \ldots, A_s\}$, then the corresponding vertices can be partitioned accordingly to form a block graph whose vertices represent the subsets and the edges represent all the transitions between the subsets.

EXAMPLE A.6.7: The block graph of the Markov chain of Example A.6.4 is shown in Fig. A.4. In this figure, symbols inside the circles represent subsets A_1 and A_2 while the symbols near the arcs represent the corresponding sub-blocks of the transition matrix.

It is convenient to use graphs for testing the conditions of a Markov chain regularity.

Matrix $\mathbf{P}$ is called *reducible* if by permutation of its rows and corresponding columns it can be transformed into

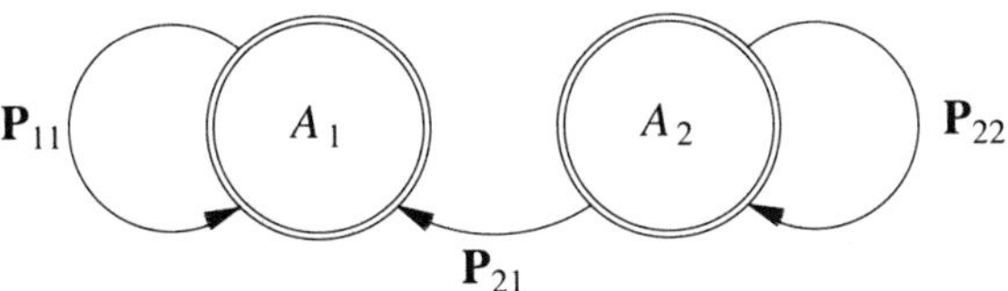

Figure A.4. Block graph of a Markov chain.

$$\begin{bmatrix} \mathbf{P}_{11} & 0 \\ \mathbf{P}_{21} & \mathbf{P}_{22} \end{bmatrix}$$

where $\mathbf{P}_{11}$ and $\mathbf{P}_{22}$ are square matrices. Otherwise, $\mathbf{P}$ is *irreducible*. The transition matrix of Example A.6.1 is irreducible, while the transition matrix of Example A.6.4 is reducible. Any Markov chain with a reducible transition matrix can be represented with the graph shown in Fig. A.4. There is no path leading from the states of the subset A_1 to the subset A_2, so that a Markov chain with a reducible matrix once entered the subset A_1 never leaves it. If $\mathbf{P}$ is irreducible, then on the corresponding graph there is a path leading from any vertex to any other vertex.

Matrix $\mathbf{P}$ is called *cyclic* if by permutation of its rows and corresponding columns it can be transformed into

$$\begin{bmatrix} 0 & \mathbf{P}_{12} & 0 & \cdots & 0 \\ 0 & 0 & \mathbf{P}_{23} & \cdots & 0 \\ \cdots & \cdots & \cdots & \cdots & \cdots \\ 0 & 0 & 0 & \cdots & \mathbf{P}_{s-1,s} \\ \mathbf{P}_{s1} & 0 & 0 & \cdots & 0 \end{bmatrix}$$

where the diagonal blocks are square matrices consisting of zeroes. Otherwise, $\mathbf{P}$ is *acyclic*. If a transition matrix $\mathbf{P}$ is cyclic then the Markov chain is called *periodic*, otherwise it is *aperiodic*. A block graph of a periodic Markov chain is depicted in Fig. A.5.

If a transition matrix $\mathbf{P}$ is irreducible and acyclic, then it is regular and all the stationary probabilities are positive. This result follows from the classical Perron-Frobenius theorem.[24]

EXAMPLE A.6.8: Transition matrix (A.6.11) of Example A.6.5 is reducible

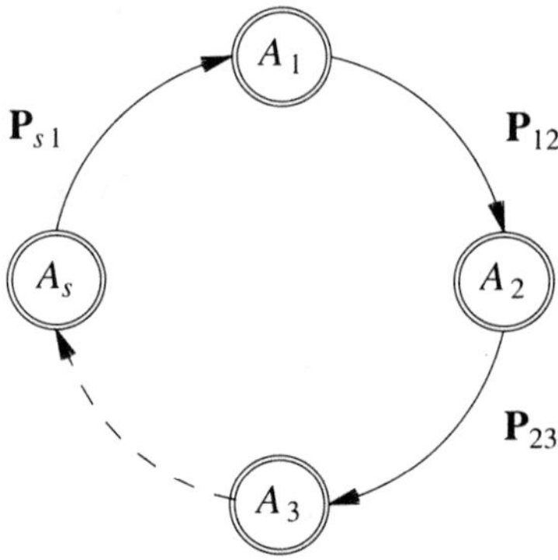

Figure A.5. Block graph of a periodic Markov chain.

$$\mathbf{P} = \begin{bmatrix} 0.5 & 0.5 & 0 \\ 0.2 & 0.8 & 0 \\ 0.3 & 0.4 & 0.3 \end{bmatrix}$$

Its characteristic polynomial $\Delta(z) = (z-1)(z-0.3)^2$ does not have roots, except for $z = 1$ with absolute value 1. Therefore, this chain is regular. But not all the stationary probabilities are positive: $\boldsymbol{\pi} = [\,2/7 \quad 5/7 \quad 0\,]$.

■

REFERENCES

[1] J. W. Carlyle, "Reduced forms for stochastic sequential machines," *J. Math. Anal. and Appl.*, vol. 7, no. 2, 1963, pp. 167—175.

[2] E. N. Gilbert, "Capacity of a burst-noise channel," *Bell Syst. Tech. J.*, vol. 39, Sept. 1960, pp. 1253—1266.

[3] A. G. Usol'tsev and W. Ya. Turin, "Investigation of the error distribution laws in tone telegraph channels with frequency modulation," *Telecommunications,* no. 7, 1963.

[4] E. O. Elliott, "Estimates of error rates for codes on burst-noise channels," *Bell Syst. Tech. J.*, vol. 42, Sept. 1963, pp. 1977—1997.

[5] B. D. Fritchman, "A binary channel characterization using partitioned Markov chains," *IEEE Trans. Inform. Theory,* vol. IT-13, Apr. 1967, pp. 221—227.

[6] O. V. Popov and W. Ya. Turin, "The probability distribution law of different numbers of errors in a combination," *Telecommunications and Radioengineering,* no. 5, 1967.

[7] A. A. Amosov and V. V. Kolpakov, "On decomposition of a binary communication channel into binomial components" (in Russian), *Third Conf. on Theor. Inform. Transm. and Coding,* Tashkent, 1967.

[8] R. H. McCullough, "The binary regenerative channel," *Bell Syst. Tech. J.*, vol. 47, Oct. 1968, pp. 1713—1735.

[9] E. L. Cohen and S. Berkovits, "Exponential distributions in Markov chain models for communication channels," *Inform. Control,* vol. 13, 1968, pp. 134—139.

[10] K. Müller, "Simulation Buschelatiger Storimpulse," *Nachrichtechn. Z.*, vol 21, no. 11, 1968, pp. 688—692.

[11] S. Tsai, "Markov characterization of the HF channel," *IEEE Trans. Commun. Technol.*, vol. COM-17, Feb. 1969, pp. 24—32.

[12] J. Swoboda, "Ein Statistischen Modell für die Fehler bei Binarer Datenübertragung auf Fernsprechknälen," *Arch. Elektr. Ubertag.*, no. 6, 1969.

[13] E. L. Bloch, O. V. Popov, W. Ya. Turin, *Models of Error Sources in Channels for Digital Information Transmission* (in Russian), Sviaz Publishers, Moscow, 1971.

[14] P. J. Trafton, H. A. Blank, and N. F. McAllister, "Data transmission network computer-to-computer study," *Proc. ACM/IEEE 2nd Symp. on Problems in the Optimization of Data Communication Systems* (Palo Alto, CA), Oct. 1971, pp. 183—191.

[15] R. T. Chien, A. H. Haddad, B. Goldberg, and E. Moyes, "An analytic error model for real channels," in *Conf. Rec., IEEE Int. Conf. Communications (ICC)*, June 1972, pp. 15-7—15-12.

[16] E. L. Bloch, O. V. Popov, and W. Ya. Turin, "Error number distribution for the stationary symmetric channel with memory," *Second International Symposium on Information Theory,* Academiai Kiado, Budapest, 1972.

[17] W. Ya. Turin, "Probability distribution laws for the number of errors in several combinations," *Telecommunications and Radioengineering*, vol. 27, no. 12, 1973.

[18] ———, "Estimation of the probability of the delay exceeding in decision feedback systems," *Problems of Information Transmission,* vol. 10, no. 3, 1974.

[19] L. N. Kanal and A. R. K. Sastry, "Models for channels with memory and their applications to error control," *Proc. of the IEEE,* vol. 66, no. 7, July 1978.

[20] R. G. Bucharaev, *Probabilistic Automata* (in Russian), Kazan University Press, Kasan, 1970.

[21] M. Marcus and H. Minc, *Survey of Matrix Theory and Matrix Inequalities*, Allyn & Bacon, Needham Heights, MA, 1964.

[22] A. Graham, *Kronecker Products and Matrix Calculus with Applications*, Ellis Horwood Ltd., 1981.

[23] I. I. Gichman and A. V. Scorochod, *The Theory of Stochastic Processes* (in Russian), vol. 1, Science Publishers, Moscow, 1971.

[24] F. R. Gantmacher, *The Theory of Matrices*, Chelsea Publ. Co., New York, 1959.

[25] J. J. Spilker, Jr., *Digital Communications by Satellite,* Prentice-Hall, Englewood Cliffs, NJ, 1977.

[26] M. Schwartz, W. R. Bennett, and S. Stein, *Communication Systems and Techniques,* McGraw-Hill, New York, 1966.

[27] M. L. Steinberger, P. Balaban, K. S. Shanmugan, "On the effect of uplink noise on a nonlinear digital satellite channel," in *Conf. Rec., 1981 Int. Conf. Commun.*, paper 20.2, Denver, June 1981.

[28] R. V. Ericson, "Functions of Markov chains," *Ann. Math. Stat.*, vol. 41, no. 3, 1970, pp. 843—850.

[29] D. Blackwell and L. Koopmans, "On the identifiability problem for functions of finite Markov chains," *Ann. Math. Stat.*, vol. 28, no. 4, 1957, pp.

1011—1015.

[30] C. J. Burke and M. A. Rosenblatt, "Markovian function of a Markov chain," *Ann. Math. Stat.*, vol. 29, no. 4, 1958, pp. 1112—1122.

[31] M. Rosenblatt, "Functions of a Markov process that are Markovian," *J. Math. Mech.*, vol. 8, no. 4, 1959, pp. 585—596.

[32] S. W. Dharmadhikary, "Functions of finite Markov chains," *Ann. Math. Stat.*, vol. 34, no. 3, 1963, pp. 1022—1032.

[33] ———, "Sufficient conditions for a stationary process to be a function of a finite Markov chain," *Ann. Math. Stat.*, vol. 34, no.3, 1963, pp. 1033—1041.

[34] ———, "Characterization of class of functions of finite Markov chains," *Ann. Math. Stat.*, vol. 35, no. 2, 1965, pp. 524—528.

[35] A. Heller, "On stochastic processes derived from Markov chains," *Ann. Math. Stat.*, vol. 36, no. 4, 1966, pp. 1286—1291.

[36] F. W. Leysieffer, "Functions of finite Markov chains," *Ann. Math. Stat.*, vol. 38, no. 1, 1967, pp. 206—212.

[37] R. A. Howard, *Dynamic Probabilistic Systems*, vol. II: *Semi-Markov and Decision Processes*. John Wiley & Sons, New York, 1971.

[38] E. Cilnar, *Introduction to Stochastic Processes,* Prentice Hall, Englewood Cliffs, NJ, 1975.

[39] W. Feller, *An Introduction to Probability Theory and Its Applications*, vol. 1, John Wiley & Sons, 1962.

[40] J. G. Kemeny and J. L. Snell, *Finite Markov Chains,* Van Nostrand, Princeton, NJ, 1960.

[41] M. F. Neuts, *Matrix-Geometric Solutions in Stochastic Models*, Johns Hopkins University Press, Baltimore, 1981.

[42] E. O. Elliott, "A model for the switched telephone network for data communications," *Bell Syst. Tech. J.*, vol. 44, Jan. 1965, pp. 89—119.

[43] A. A. Alexander, R. M. Gryb, and D. W. Nast, "Capabilities of the telephone network for data transmission," *Bell Syst. Tech. J.*, vol. 39, no. 3, May 1963, pp. 471—476.

[44] W. R. Bennett and F. E. Froelich, "Some results on the effectiveness of error control procedures in digital data transmission," *IRE Trans. on Comm. Syst.*, vol. CS-9, no. 1, 1961, pp. 58—65.

[45] O. V. Popov and W. Ya. Turin, "On the nature of errors in binary communication over standard telephone channels" (in Russian), *Second All-Union Conf. on Coding Theory and Its Applications*, sec. 3, part II, 1965.

[46] B. A. Fuchs and B. V. Shabat, *Functions of a Complex Variable*, vol. 1, Addison-Wesley Publishing Co., Reading, MA, 1964.

[47] L. R. Rabiner and C. M. Rader, *Digital Signal Processing*, IEEE Press, New York, 1972.

[48] M. S. Bartlett, *Introduction to Stochastic Processes with Special Reference to Methods and Applications,* Cambridge Univ. Press, Cambridge, 1978.

[49] K. L. Chung, *Markov Chains with Stationary Transition Probabilities,* Springer-Verlag, Berlin, 1960.

[50] S. J. Mason and H. J. Zimmermann, *Electronic Circuits, Signals, and Systems,* John Wiley & Sons, New York, 1965.

[51] W. Huggins, "Signal Flow Graphs and Random Signals," *Proc. IRE,* vol. 45, 1957, pp. 74—86.

[52] G. A. Medvedev and V. P. Tarasenko, *Probabilistic Methods in Extremal Systems Investigation* (in Russian), Science Publishers, Moscow, 1967.

[53] S. Kullback, *Information Theory and Statistics*, John Wiley & Sons, New York, 1959.

[54] H. Cramer, *Mathematical Methods of Statistics*, Princeton Univ. Press, Princeton, NJ, 1946

[55] M. S. Bartlett, "The frequency goodness of fit test for probability chains," *Proc. Cambridge Philos. Soc.,* vol. 47, 1951, pp. 86—95.

[56] P. G. Hoel, "A test for Markoff chains," *Biometrica*, vol. 41, 1954, pp. 430—433.

[57] T. W Anderson and L. A. Goodman, "Statistical inference about Markov chains," *Ann. Math. Statist.*, vol. 28, 1957, pp. 89—110.

[58] P. Billingsley, "Statistical Methods in Markov Chains," *Ann. Math. Statist.*, vol. 32, 1961, pp. 12—40.

[59] C. Chatfield, "Statistical inference regarding Markov chain models," *Appl. Statist.*, vol. 22, 1973, pp. 7—20.

[60] P. Billingsley, *Statistical Inference for Markov Processes,* University of Chicago Press, Chicago, 1961.

[61] C. R. Rao, *Linear Statistical Inference and Its Applications,* John Wiley & Sons, New York, 1965.

[62] J. L. Walsh, *Interpolation and Approximation by Rational Functions in the Complex Domain*, Amer. Math. Soc., vol. 20, 1969.

[63] H. Padé, "Sur la Representation Aprochee d'une Fonction par des Fractions Rationelles," *Ann. Ecole Norm.,* supp., vol. 3, no. 9, 1892, pp. 1—93 (supplement).

[64] W. Leighton and W. T. Scott, "A general continued fraction expansion," *Bull. Am. Math. Soc.,* vol. 45, 1935, pp. 596—605.

[65] E. Frank, "Corresponding type continued fractions," *Amer. Jour. of Math.*, vol. 68, 1946, pp. 89—108.

[66] W. H. Mills, "Continued fractions and linear recurrences," *Math. of Comp.*, vol. 29, no. 129, 1975, pp. 173—180.

[67] R. de Prony, "Essai experimentale et analytique," *J. Ecole Polytechnique* 1795, pp. 24—76.

[68] H. B. Mann and A. Wald, "On the statistical treatment of linear stochastic difference equations," *Econometrica*, vol. 11, no. 3, 1943, pp. 173—220.

[69] T. Kailath, "A view of three decades of linear filtering theory," *IEEE Trans. Inform. Theory,* vol. IT-20, no. 2, 1974, pp. 146—181.

[70] J. Makhoul, "Linear prediction, pp. a tutorial review," *Proc. IEEE,* vol. 63, no. 4, April 1975, pp. 561—580.

[71] G. Szegö, "Orthogonal Polynomials," Colloquium Publications, no. 23, *Amer. Math. Society,* 1939.

[72] N. Levinson, "The Wiener RMS (Root Mean Square) error criterion in filter design and prediction," *J. Math. Phys.,* vol. 25, no. 4, 1947, pp. 261—278. Also Appendix B in N. Wiener, *Extrapolation, Interpolation and Smoothing of Stationary Time Series,* M.I.T. Press, Cambridge, MA, 1964.

[73] R. E. Blahut, *Fast Algorithms for Digital Signal Processing,* Addison-Wesley, Reading, MA, 1984.

[74] Y. Sugiyama, "An algorithm for solving discrete-time Wiener-Hopf equations based upon Euclid's algorithm," *IEEE Trans. Inform. Theory,* vol. IT-32, 1986, pp. 394—409.

[75] E. R. Berlekamp, *Algebraic Coding Theory,* McGraw-Hill, New York, 1968.

[76] J. L. Massey, "Shift register synthesis and BCH decoding," *IEEE Trans. Inform. Theory,* vol. IT-15, 1969, pp. 122—127.

[77] Y. Sugiyama, M. Kasahara, S. Hirsawa, and T. Namekawa, "A method for solving key equations for decoding Goppa codes," *Inform. Control*, vol. 27, 1975, pp. 87—99.

[78] R. E. Blahut, *Theory and Practice of Error Control Codes,* Addison-Wesley, Reading, MA, 1983.

[79] J. Durbin, "The fitting of time-series models," *Rev. Inst. Int. Stat.,* vol. 28, no. 3, 1960, pp. 233—243.

[80] R. P. Brent, F. G. Gustavson, and D. Y. Y. Yun, "Fast solution of Toeplitz systems of equations and computation of Pade approximants," *J. Algorithms*, vol. 1, 1980, pp. 259—295.

[81] J. K. Wolf, "Decoding of Bose-Chaudhuri-Hocquenghem codes and Prony's method of curve fitting," *IEEE Trans. Inform. Theory,* vol. IT-13, Oct. 1967, pp. 608.

[82] R. Fortret, "Random determinants," *J. Res. Nat. Bureau Standards*, vol. 47, 1951, pp. 465—470.

[83] H. Nuquist, S. Rice, J. Riordan, "The distribution of random determinants," *Quart. Appl. Math.*, vol. 12, no. 2, 1954.

[84] J. Komlós, "On the determinant of random matrices," *Studia Sci. Math. Hungary*, vol. 3, no. 4, 1968.

[85] V. L. Girco, *Random Matrices* (in Russian), Vischa Shcola Publishers, Kiev, 1975.

[86] Yu. A. Shreider, *Method of Statistical Testing*, Elsevier Publishing Co., Amsterdam, 1964.

[87] I. V. Dunin-Barkovsky and K. V. Smirnov, *Probability Theory and Mathematical Statistics* (in Russian), Gostechizdat, Moscow, 1955.

[88] W. W. Peterson and E. J. Weldon, Jr., *Error-Correcting Codes*, M.I.T. Press, Cambridge, MA, 1961.

[89] A. S. Tanenbaum, *Computer Networks*, Prentice-Hall, Englewood Cliffs, NJ, 1981.

[90] R. G. Gallager, *Information Theory and Reliable Communication,* John Wiley & Sons, New York, 1968.

[91] Y. Sugiyama, M. Kasahara, S. Hirasawa, and T. Namekawa, "A method of solving key equation for decoding Goppa codes," *Inform. Control,* vol. 27, Jan. 1975, pp. 173—180.

[92] J. Justesen, "On the complexity of decoding Reed-Solomon codes," *IEEE Trans. Inform. Theory,* vol. IT-22, Mar. 1976, pp. 237—238.

[93] D. V. Sarwate, "On the complexity of decoding Goppa codes," *IEEE Trans. Inform. Theory,* vol. IT-22, July 1976, pp. 515—516.

[94] A. Gill, *Linear Sequential Circuits,* McGraw-Hill Book Co., New York, 1967.

[95] A. J. Viterbi, "Error bounds for convolutional codes and asymptotically optimum decoding algorithm," *IEEE Trans. Inform. Theory,* vol. IT-12, Apr. 1967, pp. 260—269.

[96] G. D. Forney, Jr.,"The Viterbi Algorithm,"*Proc. IEEE*, vol. 61, no. 3, Mar. 1973, pp. 268—278.

[97] J. L. Massey, *Threshold Decoding,* M.I.T. Press, Cambridge, MA, 1963.

[98] W. W. Wu, "New convolutional codes - Part I," *IEEE Trans. Commun.,* vol. COM-23, Sept. 1975, pp. 442—456.

[99] ———, "New convolutional codes - Part II," *IEEE Trans. Commun.*, vol. COM-24, Jan. 1976, pp. 19—32.

[100] ———, "New convolutional codes - Part III," *IEEE Trans. Commun.*, vol. COM-24, Sept. 1976, pp. 946—955.

[101] F. J. MacWilliams, "A Theorem on the distribution of weights in a systematic code," *Bell System Tech. J.*, vol. 42, 1963, pp. 79—84.

[102] S. C. Chang and J. K. Wolf, "A simple derivation of the MacWilliams' identity for linear codes," *IEEE Trans. Inform. Theory*, vol. IT-26, no. 4, July 1980, pp. 476—477.

[103] W. Feller, *An Introduction to Probability Theory and Its Applications*, vol. 2, John Wiley & Sons, New York, 1971.

[104] E. L. Bloch and V. V. Zyablov, *Generalized Concatenated Codes* (in Russian), Sviaz Publishers, Moscow, 1976.

[105] F. Oberhettinger and L. Badii, *Tables of Laplace Transform*, Springer-Verlag, New York, 1973, pp. p. 16, eq. 2.34.

[106] G. D. Forney, *Concatenated Codes*, Research Monograph 37, MIT Press, Cambridge, MA, 1966.

[107] W. K. Pratt, *Digital Image Processing*, John Wiley & Sons, New York, 1978.

[108] W. Turin, "Union Bounds on Viterbi Algorithm Performance," *AT&T Tech. J.*, vol. 64, no. 10, 1985, pp. 2375—2385.

[109] K. A. Post, "Explicit evaluation of Viterbi's union bounds on convolutional code performance for the binary symmetric channel," *IEEE Trans. Inform. Theory*, vol. IT-23, May 1977, pp. 403—404.

[110] T. N. Morrissey, Jr., "Analysis of decoders for convolutional codes by stochastic sequential machine methods," *IEEE Trans. Inform. Theory*, vol IT-16, July 1970, pp. 460—469.

[111] P. M. Dollard, "A new algorithm for decoding convolutional self-orthogonal codes," *ICC '83 Conf. Rec.*, 1983, pp. 318—322.

[112] E. R. Berlekamp, *Interleaved Coding for Bursty Channels*, Final Project Report, NSF Program SBIR-1982, Phase 1.

[113] J. P. Gray, "Line control procedures," *Proc. IEEE*, vol. 60, Nov. 1972, pp. 1301—1312.

[114] IBM Corp. *General Information—Binary Synchronous Communications*, File TP-09, Order GA 27-3004-2, 1970.

[115] J. J. Metzner and K. C. Morgan, "Coded binary decision-feedback communication systems," *IRE Trans. Commun. Syst.*, vol. CS-8, June 1960, pp. 101—113.

[116] R. J. Benice and A. H. Frey, Jr., "An analysis of retransmission systems," *IEEE Trans. Commun. Technol.*, vol. COM-12, Dec. 1964, pp. 135—145.

[117] H. O. Burton and D. D. Sullivan, "Errors and error control," *Proc. IEEE*, vol. 60, Nov. 1972, pp. 1293—1301.

[118] J. Riordan, *An Introduction to Combinatorial Analysis*, John Wiley & Sons, New York, 1958.

[119] M. A. Evgrafov, *Asymptotic Estimates and Entire Functions*, Gordon & Breach Science Publishers, New York, 1961.

[120] C. Berge, *Theory of Graphs and Its Applications*, John Wiley & Sons, New York, 1962.

[121] J. K. Omura, "Performance bounds for Viterbi algorithms," ICC '81 Conference Record, Denver, June 1981, pp. 2.2.1—2.2.5.

[122] M. B. Brilliant, "Observations of errors and error rates on T1 digital repeatered lines," *Bell Syst. Tech. J.*, vol. 57, no. 3, March 1978 , pp. 711—746.

[123] D. M. Lucantoni, *An Algorithmic Analysis of a Communication Model with Retransmissions of Flawed Messages*, Pitman, London, 1983.

[124] H. Heffes and D. M. Lucantoni, "A Markov modulated characterization of packetized voice and data Traffic and Related Statistical Multiplexer Performance," *IEEE J. on Selected Areas in Comm.*, vol. SAC-4, no. 6, Sept. 1986, pp. 856—868.

[125] D. M. Lucantoni, K. S. Meier-Hellstern, and M. F. Neuts, "A single server queue with server vacations and a class of non-renewal arrival processes," to appear in *Adv. in Appl. Probability*, Sep. 1990.

INDEX